Biological and Medical Physics, Biomedical Engineering

The fields of biological and medical physics and biomedical engineering are broad, multidisciplinary and dynamic. They lie at the crossroads of frontier research in physics, biology, chemistry, and medicine. The Biological & Medical Physics/Biomedical Engineering Series is intended to be comprehensive, covering a broad range of topics important to the study of the physical, chemical and biological sciences. Its goal is to provide scientists and engineers with textbooks, monographs, and reference works to address the growing need for information.

Books in the series emphasize established and emergent areas of science including molecular, membrane, and mathematical biophysics; photosynthetic energy harvesting and conversion; information processing; physical principles of genetics; sensory communications; automata networks, neural networks, and cellular automata. Equally important will be coverage of applied aspects of biological and medical physics and biomedical engineering such as molecular electronic components and devices, biosensors, medicine, imaging, physical principles of renewable energy production, advanced prostheses, and environmental control and engineering.

For further volumes:
http://www.springer.com/series/3740

Jiyuan Tu • Kiao Inthavong
Kelvin Kian Loong Wong

Computational Hemodynamics—Theory, Modelling and Applications

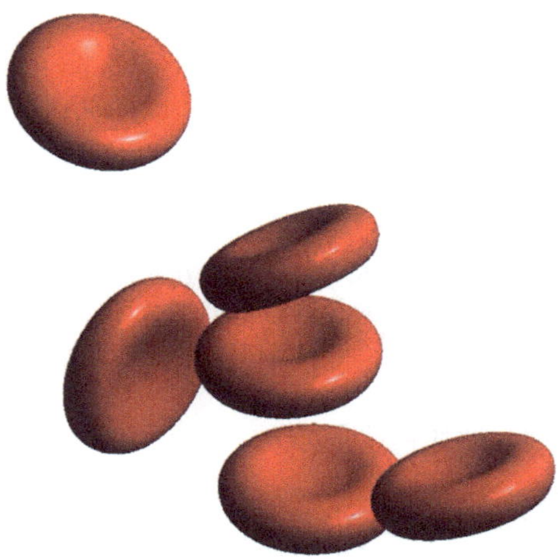

Springer

Jiyuan Tu
School of Aerospace Mechanical and Manufacturing Engineering
RMIT University
Bundoora
Victoria
Australia

Kelvin Kian Loong Wong
School of Aerospace Mechanical and Manufacturing Engineering
RMIT University
Bundoora
Victoria
Australia

Kiao Inthavong
School of Aerospace Mechanical and Manufacturing Engineering
RMIT University
Bundoora
Victoria
Australia

ISSN 1618-7210 ISSN 2197-5647 (electronic)
Biological and Medical Physics, Biomedical Engineering
ISBN 978-94-017-9593-7 ISBN 978-94-017-9594-4 (eBook)
DOI 10.1007/978-94-017-9594-4

Library of Congress Control Number: 2014954751

Springer Dordrecht Heidelberg New York London

Printed on acid-free paper

Springer is part of Springer Science+Business Media (www.springer.com)

Preface

Traditionally, cardiovascular studies have been limited to medical textbooks and courses, taught by medical experts. Now, computational modeling and simulation has infiltrated the medical field which has advanced to a point where it can be categorized into its own sub fields. The use of Computational Fluid Dynamics (CFD) for modeling cardiovascular flow leads to a new field of study known as computational biomedical engineering. In particular when we model blood as a biofluid and compute its flow, we establish a branch termed as *Computational Haemodynamics* (CHD). This new field of engineering constitutes important knowledge in the study of blood flow mechanics, surgical and medical treatment, and design of biomedical devices. This has only been possible through the emergence of advanced technologies in medical imaging, and computers. Since the fundamental principles of CHD are based on CFD, these two terms are quite similar and used interchangeably in this book.

This book is designed to convey information pitched at an appropriate level of knowledge to new students with a diverse background of medicine, physiology, pharmaceutical science, mechanical engineering, and mathematics. This book cannot deliver the entire scope of knowledge that is related to cardiovascular modeling. Rather, it provides an important link between traditional engineering and computational fields with the medical field. For example, an undergraduate who is trained in engineering can selectively read the relevant chapters of the book that provides new knowledge on the computational simulation or medical aspects.

There is a diverse range of CFD books in the market are written with different levels of complexity. To ensure the reader is not overwhelmed by overly advanced mathematical notation and theory, this book describes fundamental mathematical ideas, coupled with sufficient description and backed up with examples. It is intuitive and systematically structured for enhancing the learning process and allowing students to be able to quickly use CFD practically. Nevertheless, a fundamental knowledge of calculus such as basic anti- and differentiation is still required prior to reading the book. With the hands on practice and reading, this book can enable CFD beginners to efficiently perform reliable CFD simulations for the CHD applications.

The book begins with Chap. 1, which prepares the reader for an overview of CHD, its advantages and disadvantages, a preliminary understanding of CHD, and its numerous applications in the cardiovascular system. It aims to initiate curiosity and stimulate creativity for the reader to think of possible solutions to engineering problems related to cardiovascular flow.

In Chap. 2, the human cardiovascular system is presented concisely. By providing the description of the cardiovascular anatomy and physiology at an elementary level, this chapter is useful for readers that have previously not received any form of medical training. An understanding of the cardiovascular anatomy provides a basis for an accurate computational fluid flow setup for geometrical reconstruction based on medical imaging, definition of the relevant anatomical structures, and determining cardiovascular flow boundary conditions.

Medical image reconstruction of the cardiovascular anatomy is covered in Chap. 3, which introduces the reader to medical imaging, geometrical reconstruction, manufacturing or reverse engineering, and Computer-Aided Design (CAD). The book steps through all of the processes from medical image scanning (MRI or CT scans) through to segmentation, and leading to model reconstruction. Important features of each process with informative description of the approach used or available software are given in detail.

Chap. 4 presents the fundamentals of haemodynamics, firstly introducing the properties of blood and then its flow properties. Since blood flow is enclosed by the arterial walls, it is an example of the well-established internal pipe flow theory. This incorporates the laws of fluid dynamics and its interaction with solid surfaces. Characteristic flow features of blood flow in different regions of the circulatory system are presented to exemplify its behaviour.

In Chap. 5, the governing equations of fluid flow, i.e. the Navier-Stokes equations are introduced. Each term in the equations is described in such a way that the reader can appreciate its physical interpretation and impact on the flow. A brief introduction to turbulence and its modelling is also given, which is quite unusual for this type of book. However we feel that since cardiovascular flows through stenosed arteries and some medical devices will exhibit some form of turbulence, it is worth mentioning here. A summary of the different turbulence models is given with guidelines on setting up the different models. A derivation of the equations for solid dynamics is given with an emphasis on the elastic properties of arteries. The discretisation of the continuum equations for the fluid and solid dynamics onto discrete points in a computational domain are shown using the more well-known methods that include the Finite Difference, Finite Volume, and Finite Element Methods. These methods are solved algebraically using both direct and iterative methods. The coupling of the fluid and solid domains produce the Fluid-Structure-Interaction (FSI). This takes place at a shared interface where the fluid and solid meet. Force loadings and structural deformation results are passed from one analysis to the other and the coupling approaches are described. A major difficulty with FSI is the stability and convergence and these issues are discussed.

Description of mesh development for different anatomies are described in Chap. 6. The discretised equations from Chap. 5 are transferred to a corresponding *discrete* set of points in flow field where solutions of the flow variables such as the velocity, pressure, temperature, and deformation are performed. Both static and fluid-structure-interaction meshing are presented. Generation of a quality mesh requires both a good sense of creativity and technical knowledge. Therefore, this chapter provides guidelines to the reader prior to delving into the interesting experiments of meshing different anatomical models. Different types of mesh and its setup for simplified and more advanced geometries are illustrated without presenting overly complicated mathematical algorithms. Furthermore guidelines and best practices are given for developing quality meshes and material presented will give the reader an introduction into mesh creation.

Having laid the groundwork to establish a strong theoretical base, Chap. 7 applies the foundational knowledge and putting them into practice through selected demonstrative applications. The entire process from scans to the final data analysis is shown. The examples include models of a stenosed carotid artery, patient specific carotid bifurcations, stented artery, and cardiac chamber. Post-processing techniques are presented which involves converting the raw data into graphical representations.

Computational simulations increase in complexity when it considers blood flow within elastic wall structures of the cardiovascular system. Chap. 8 presents some FSI cardiovascular applications to provide the reader with an overview of some developments in this emerging field. FSI is being used as a demonstration of practicality of achieving a higher physiological realism in haemodynamics analysis over simple rigid wall assumptions. Applications include atherosclerosis in carotid bifurcation, calcified plaque rupture, aortic aneurysm, as well as coronary artery bypass graft.

The book culminates with Chap. 9 that explores the future trends and more advanced modelling techniques. While the detailed treatment of these advanced techniques are beyond the scope of this book, it is hoped that the chapter will stimulate further enthusiasm in the advancing capability and innovative use of CFD and FEM for studies related to computational haemodynamics.

Jiyuan Tu
Kiao Inthavong
Kelvin Kian Loong Wong

Acknowledgements

The authors are grateful to Springer, particularly Thijs van Vlijmen for reviewing, endorsing, and supporting the idea of publishing this book, and to Sara Germans for assisting in a smooth completion of our project. The authors like to thank research students and colleagues working in the CFD Group at RMIT University for their hard work and contribution. Special thanks are given to Pongpat Thavornpattanapong and Jingliang Dong who have provided worked examples in geometric modelling, meshing and simulations for this book. We would like to thank Dr Sherman Cheung, Prof David Reutens, Prof Thanh Phan, Dr Songlin Ding, Dr Huafeng Li and Mr Nelson Choi for their early contribution of patient-specific geometric reconstruction and CFD simulation, and are grateful to Dr William Yang for his experimental validation of our simulations.

Prof Tu wants to dedicate this book to the memory of his former supervisors, Prof Zhong-Yi Hua, former President of Fudan University, who died in a heart attack, and Prof Shi-Jin Pang who died in a stroke. He would like to express his gratitude to his wife, Xue, and his son, Tian for their understanding and support. Dr Inthavong wishes to acknowledge the immense support, contributions, and fruitful discussions from Laura Frances Portaro; and the constant challenges posed by Antonino Inthavong. He would like to dedicate this book to Samuel August Schmidt for the many nights of online discussions on research, science, and all things academic—until our next discussion. Dr Wong would like to thank his life mentors—Prof Dhanjoo N. Ghista, Prof Jagannath Mazumdar and Prof Derek Abbott for their advice and guidance in his research of computational hemodynamics. He expresses thanks to his wife, Xindan for her understanding and support during the course of his career when this book is being written.

Finally, to those that the authors have failed to mention but have been involved in one way or another, the authors would like to take this opportunity of expressing their deepest appreciation.

Jiyuan Tu
Kiao Inthavong
Kelvin Kian Loong Wong

Contents

Chapter 1
Computational Haemodynamics—An Introduction

1.1 What is Computational Haemodynamics (CHD)

Cardiac diseases remain a major cause of ill health and death in our society and thus advancements in current clinical therapies are of significant importance. Our ageing population and the obesity epidemic, have increased incidences of heart disease giving rise to a new market for therapeutic systems that are computationally based and exclusively dedicated to the treatment of cardiovascular diseases. This coincides with recent advancements in computing and flow visualisation technology, giving greater ease for diagnosis of cardiovascular diseases (Banerjee et al. 2000). As such, there will be an increasing demand for Computational Haemodynamic analysis techniques to further facilitate the prognosis and diagnosis of a failing heart. Haemodynamic parametrical studies have the potential for clinical study of cardiac patients (Tu et al. 2011). The framework can be used to diagnose pre- and post- surgical treatments. In particular, fluid mechanical flow quantification can be adequately documented from simulated cardiovascular flow to aid diagnoses of cardiovascular diseases and to assess cardiac flow analysis. These new approaches to medical imaging provide an alternative to traditional invasive methods of diagnoses such as surgical visualisation, and post-surgical evaluations, to characterise cardiac abnormalities.

Computational Haemodynamics (CHD) is the computer simulation of blood flow in the cardiovascular system. The motion of blood flow can be modelled by using Computational Fluid Dynamics (CFD) and its interaction with blood vessel structures can be studied by combination with the use of Finite Element Method (FEM). This is becoming an emerging research field that may lead to the everyday use of computer modelling techniques for many biomedical engineering applications. The biomedical engineering industry comprises the health, pharmaceutical and medical research fields, whereby interest is increasing in the use of CHD modelling for cardiac flow analysis, effectiveness of stents, and discovery of new phenomena related to physiological flows.

The term "*Computational*" of the CFD terminology refers to the study of the "*fluid dynamics*" represented by the Navier-Stokes equations, which are solved using

J. Tu et al., *Computational Hemodynamics—Theory, Modelling and Applications*,
Biological and Medical Physics, Biomedical Engineering,
DOI 10.1007/978-94-017-9594-4_1

computer programs or commercial software packages. The term "*hameodynamics*" encompasses both the study of biofluids and in this case, blood that is in motion (the biofluid in dynamic mode) both in the natural arteries and in biomechanical devices.

Cardiovascular modelling using CHD is a complex and challenging process. At the front end of the process, medical imaging and computer aided design (CAD) are used to reconstruct the cardiovascular structure. The model geometry is then imported into CFD for blood flow modelling. Due to the interactive effect of vascular compliance and blood flow properties such as pressure and pulsatility, the CHD model also includes structural or FEM-based mechanics for fluid-structure interactions (FSI). FSI for cardiovascular applications solves a system of governing equations of both fluid and solid fields that is based on the mechanics of blood-vessel interaction. Advances in CFD and FEM technologies has been aided by the rapid developments in numerical models to reflect the true physics of flow, and the increase in computing power to perform the immense computational work for solution generation.

Medical imaging, CAD, CFD, and FEM, coupled with some basic medical science knowledge come together for the development of CHD simulations. Expertise in this field requires proficiency in each of these It is expected that this field demands a person who will be proficient from each of the sub-disciplines.

As demonstrated in Fig. 1.1, each sub-discipline discipline is interlinked to each other and does not exist in isolation.

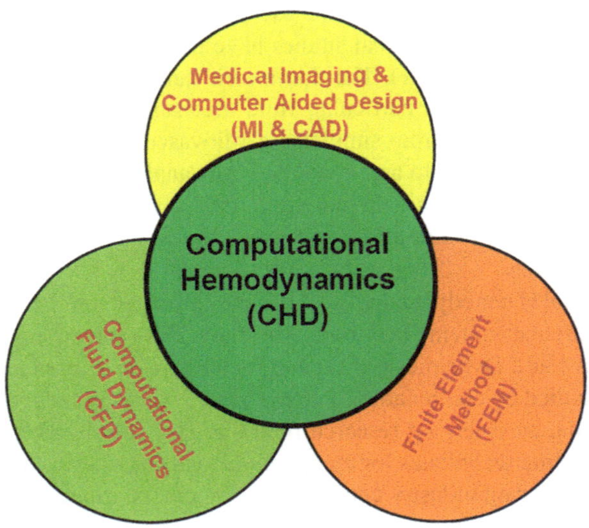

Fig. 1.1 The different disciplines contained within the CHD architecture. The state-of-the-art research CHD architecture involving medical imaging (MI), computer aided design (CAD), computational fluid dynamics (CFD), and finite element method (FEM) components for haemodynamic modelling

1.2 Advantages of CHD

CHD is an emerging field for virtual surgical planning, clinical assessment of vascular diseases and the development of medical devices. The exponential growth in computing power, advancements in technology, and materialisation of interdisciplinary research has seen CFD emerge within the last few decades as a practical tool in modern engineering practice. As such, CHD benefits medical experts in providing support for their diagnosis and clinical treatment of cardiovascular diseases.

Computational modelling is also becoming a staple in research and development (R&D) in practical engineering and product design. The computational results complement experimental and analytical approaches by providing a cost-effective alternative to simulate realistic fluid flows. For example, the visualisation capabilities are highly detailed which can provide vector, contour, and flow streamline plots that describe the physics. Furthermore CHD is a tool that offers the ability to solve a range of complicated flow problems where analytical approaches do not exist.

Finally, another major advantage of CHD is the ability to simulate biofluid flows that are not reproducible in experiments. This is significant in the medical and pharmaceutical fields where invasive methods dealing with human subjects are difficult to undertake. For example, evaluation of performance for medical devices such as heart stent and heart pump. The analysis of cardiac diseases such as atherosclerosis (Fig. 1.2), can also be simulated using CHD, which is indeed much safer and easier to perform than experimenting on a live human subject.

Nevertheless, the suggestion here is that CHD will not replace experimental testing but rather, it will serve as a viable alternative that complements experimental methods. Newly developed models still rely on experimental data as validation for such topics as cardiovascular flows including flow in atherosclerotic arteries and aneurysm, heart stent and ventricular assist devices, and Low-density Lipoprotein

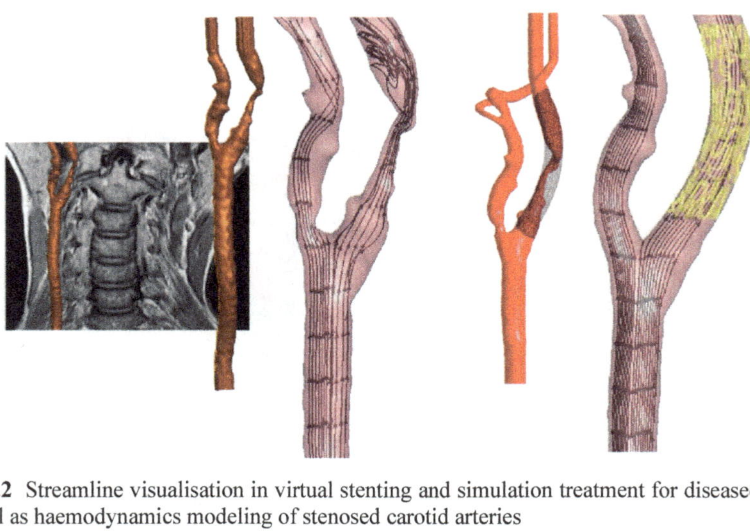

Fig. 1.2 Streamline visualisation in virtual stenting and simulation treatment for diseased artery, as well as haemodynamics modeling of stenosed carotid arteries

(LDL) transportation. By using advanced computer software packages, visualization of numerical solutions using vectors, contours, or animated movies of unsteady blood vessel interaction can have a significant impact on delivering solutions. Some of the aims of this book are to introduce the reader to current research trends and to enable the reader to understand to make the right decisions when setting up CHD models. In particular, new users often encounter incorrect numerically produced flow results that could be wrongly interpreted as acceptable physical phenomena. Numerical results obtained must always be thoroughly examined before they are accepted. Therefore, the new user needs to learn how to properly analyse and judge the computed results.

1.3 Applications in the Cardiovascular System

1.3.1 CHD as a Research Tool

One of the many uses of CHD is to reveal the physical nature of the interactions of fluid around and within objects. Fluid is transported through a domain by many physical processes including dissipation, diffusion, convection, boundary layers and turbulence. CHD can be employed as a *research tool* to perform numerical experiments in order to better understand the physical nature of the fluid dynamics of blood flow. For example it can be used to better understand blood flow within a ventricular assist heart device. Fig. 1.3 shows a snapshot in time of the flow through such a device which reveals the recirculating flow and high wall shear stresses which can result in thrombosis. This example illustrates how CHD can provide detailed visualisation to better understand the observed flow structures and some

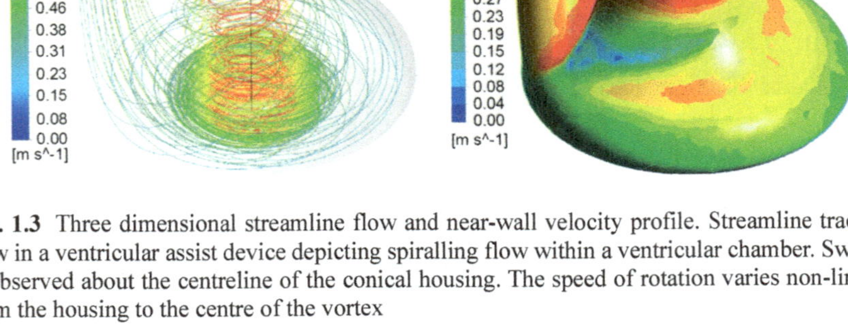

Fig. 1.3 Three dimensional streamline flow and near-wall velocity profile. Streamline traces of flow in a ventricular assist device depicting spiralling flow within a ventricular chamber. Swirling is observed about the centreline of the conical housing. The speed of rotation varies non-linearly from the housing to the centre of the vortex

important physical aspects of the fluid flow, similar to a real laboratory experiment. More importantly, the simulations complement experimental data by providing qualitative comparison and also a means to interpret basic phenomenological aspects of the experimental condition.

Numerical experiments may be performed on problems that are difficult to perform experimentally. This may involve flow through complex bodies or through a porous media, non-Newtonian fluids (blood flow), and moving body problems (heart chamber expansion/contraction). These problems highlight the capabilities of the CHD as a non-invasive technique to study the human cardiovascular system.

1.3.2 CHD as a Training Tool

New users who take a hands-on approach to the medical imaging, CAD, CFD, and FEM disciplines will progress rapidly in understanding CHD, especially through the visualisation tools which work well in conjunction with experimental work. In addition, the use of CHD modelling opens up new teaching methods (virtual surgery, 3D animation), and classes of problems such as human anatomy studies and cardiovascular flow. Research has shown that the use of computational simulations increases learning efficiency and understanding (Wankat 2002) and provides an effective method for novel hands-on learning in combined physical and computational laboratories (Regan and Sheppard 1996).

Traditional medical engineering courses had been limited in terms of strong computer, code development and applications. As multi-purpose commercial codes become more available in research and academic organisations, teaching and learning is evolving in the biomedical engineering field. Nowadays, CFD and FEM software are cornerstones of engineering practice, and many engineers without any post-graduate education are often expected to use it. This has led many engineering and science undergraduate courses to include these disciplines as part of their curriculum.

Furthermore computational models can provide new training methods for medical students by virtual surgery, or virtual anatomy. Virtual anatomy can contribute towards this learning by providing three-dimensional models of the anatomy that can be manipulated visually to convey conceptual ideas. Surgical procedures can be made virtually, and then its effect on fluid flow can be analysed. The surgeon can then make an informed decision on the surgical procedure as well as devising more effective post-surgery recovery plans. One final advantage of the virtual anatomy and surgery, is that communication between medical practitioners and the patient and their family will be improved through visually demonstrating the anatomy and why cardiac surgery is or is not needed.

1.3.3 Examination of Atherosclerosis

Medical image reconstruction of blood vessels has developed rapidly in recent decades. In this section we review some work of atherosclerosis modeling. With the development of modern imaging technology, especially magnetic resonance imaging (MRI) and computed tomography (CT), arterial blood flow in subject-specific physiologic models were quantified (Calcagno et al. 2013; Sanz and Fayad 2008; Silvera et al. 2009). Computational methods to generate velocity and pressure fields in idealized, generic models of the vascular anatomy and physiology was applied by Taylor and Draney (2004). For three-dimensional numerical studies, CFD models of the carotid artery were constructed from MRI or CT images (Campbell et al. 2012) and the prediction for atheroma (Marshall et al. 2004a). The artery and its wall shear stress was quantified using CFD flow equations (Farmakis et al. 2004). Antiga et al. (2002) presented a methodology on bifurcation geometric reconstruction using CT angiography of carotid bifurcations from two patients affected by severe atherosclerotic plaques.

CFD simulations based on MRI plays an important role in evaluating the relationship between local haemodynamics and the prediction for atheroma. This technique can be applied non-invasively at arterial sites where vascular anatomy typically exhibits substantial inter-individual variability. Simulations can demonstrate local mechanical factors occurring in fluid-vessel interactions during the pathogenesis of atherosclerosis. For example, transient simulations that track the blood flow through a complete cardiac cycle can be performed (Fig. 1.4). Results illustrate

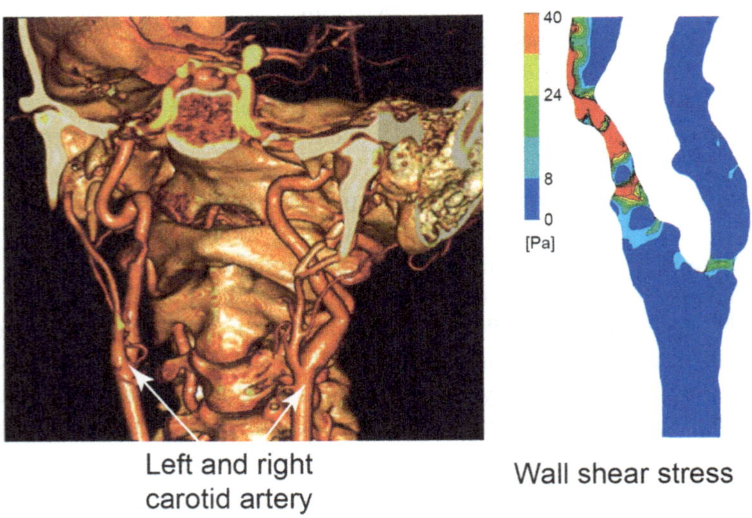

Fig. 1.4 Models of stenosed carotid arteries and CFD simulated time-dependent wall shear stress. Computational fluid dynamics can be used to analyse wall shear stress (*WSS*) distribution in atherosclerotic arteries. The regions of high WSS appears at the stenosis, and can serve as a parameter for grading the severity of atherosclerosis

maximum displacement at the side wall of the bifurcation region. Wall shear stress (WSS) distribution at peak pressure shows the carotid sinus experiences very low WSS, whereas much higher WSS occurs in the interior or exterior carotid arteries where they are stenosed (Tu et al. 2011; Wong et al. 2006). The arterial stenosis introduces higher flow resistance and that regions of high WSS can indicate where stenosis are most severe where percutaneous carotid artery angioplasty and stenting (PTAS) should be implemented.

A study by Schimmer and Malek (2011) where computational haemodynamic characterization of patient-specific carotid bifurcation stenosis is demonstrated for pre- and post- endovascular revascularization shown in Fig. 1.5.

Figure 1.6 shows the CHD modeling results where a series of patients with symptomatic carotid stenosis, their corresponding abnormal flow pattern and WSS can be predicted. By extracting geometries of patient-specific carotid bifurcations pre- and post-percutaneous carotid artery angioplasty and stenting (PTAS), CHD predictions of wall shear rates in diseased and treated carotid arteries can be used for clinical assessment.

In summary, integration of medical imaging with numerical simulation approaches demonstrate effective assessment of stenosis, leading to treatment that best resolves the condition. Furthermore, every successful treatment of the diseased vessel can be saved into a database of records as a future case reference, which can enhance the development of carotid-based treatment strategies.

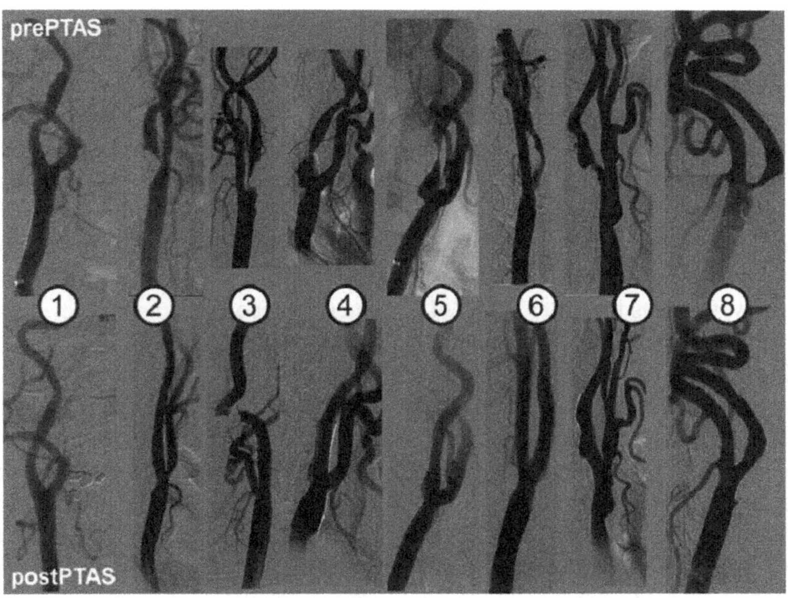

Fig. 1.5 Single plane angiographic imaging of carotid bifurcations (*case study 1–8*) pre- and post-percutaneous carotid artery angioplasty and stenting. The percutaneous carotid artery angioplasty and stenting (*PTAS*) procedure was performed for 8 diseased carotid bifurcations and imaged using angiography. (Schirmer and Malek 2011)

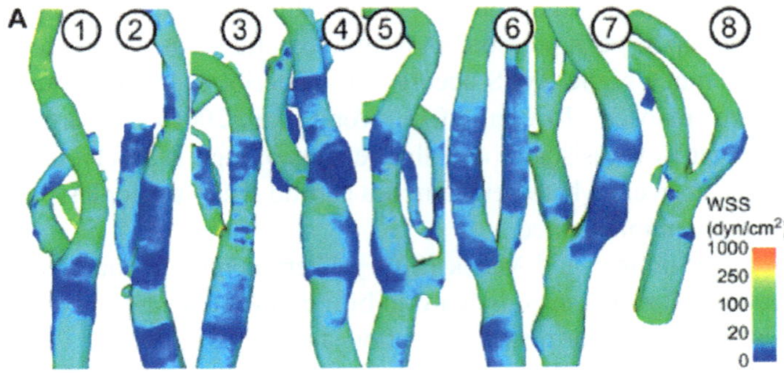

Fig. 1.6 Wall shear stress distribution plots of carotid bifurcations (*case study 1–8*) pre- and post-percutaneous carotid artery angioplasty and stenting (*PTAS*). Contour plots of the wall shear stress (*WSS*) magnitude in dyn/cm^2 averaged over the cardiac cycle before and after carotid artery stenting by PTAS. (Image from Schimmer and Malek 2011)

1.3.4 Plaque Rupture Risk Assessment

Before establishing a numerical simulation to assess plaque mechanical behaviour, the first step is to extract information on plaque tissue morphology and classification, which can be performed by MRI. Coupled with image processing techniques, high-resolution MRI is capable of classifying different tissue types. This forms the basis for determining adverse cardiovascular conditions that exceed the normal threshold for rupture, requiring surgical rectification such as by-pass or stenting.

MR multi-contrast plaque imaging can distinguish plaque components, its morphology, and establish the elastic behaviour of its composites. The mechanical properties extracted from patient-specific plaque condition can be used as clinical data for running the computational modelling of plaque response to stress (Fig. 1.7).

The morphological plaque severity index (MPSI) and computational plaque stress index (CPSI) can then be established (Tang et al. 2005c, 2009). The former index deals with the correlation of plaque morphological characteristics with vulnerability, whereas the latter is linked to numerically computed stress level associated with computationally generated plaque geometry.

Knowledge of diseased vessel structures forms a solid foundation for investigating the pathological development of vascular diseases. This enables effective characterisation of blood flow for patient specific physiological/pathological conditions. This leads to better surgical planning and post-operation therapy that is patient specific and customised.

Plaque distribution and its structural components can be characterised by a three dimensional blood-vessel model with the aim of determining the variable mechanical properties due to changes by the lipid core and calcification contents. Numerical simulations can identify how cap thickness and calcium distribution in lipids interact to influence biomechanical stress.

Geometric models of non-calcified plaque, and partially and fully calcified plaque can also be modelled (Fig. 1.8). Critical stress analysis of a diseased carotid

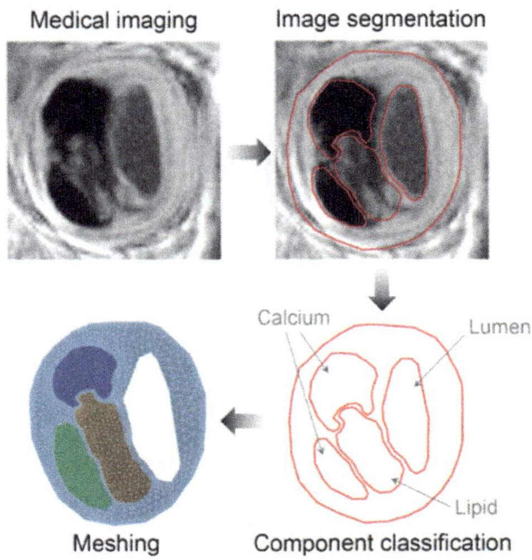

Fig. 1.7 Flow chart of proposed components for plaque vulnerability investigation. The stages of plaque structural modelling which involves medical imaging techniques such as MRI imaging, image segmentation (*discussed later in Chap. 3*), component classification by identifying their intensities, and finally meshing (*discussed later in Chap. 4*)

bifurcation at 80% degree of stenosis (by area) demonstrates the resistance force on the plaque due to calcification. This reduces the maximum principal stress and deformation in the artery. Plaque-rupture analysis based on critical stress of fibrous cap for 3D plaque, (i.e plot of critical stress σ_{cr} versus width of calcification gap d_{cg}) shows that a calcification gap value of > 0.21 mm causes stress levels to exceed 300 kPa leading to plaque rupture.

Modelling of the human atherosclerotic artery with varying degrees of lipid core elasticity, fibrous cap thickness and calcification gap, form the basis of a rupture analysis. The results add new mechanistic insights and methodologically sound data to investigate plaque rupture mechanics.

1.3.5 Preoperative Assessment of Atherosclerotic Arteries

Based on flow and patient specific information a system for decision and evaluation of treatment based on CHD can be performed. By carrying out systematic simulations of diseased vascular systems, appropriate treatment solutions can be formed.

Stenting has been well-established as the most feasible treatment for atherosclerotic lesions. It is becoming an alternative to traditional endarterectomy due to its less invasive nature and high successful rate. However, the placement of stents alters the haemodynamics and coupled with wall movement may lead to the dispersion of late multiple emboli (Richter et al. 1999). The complex structures introduced into the blood flow may enhance biochemical thrombosis cascade (Beythien et al. 1999; Peacock et al. 1995), as well as directly affecting local haemodynamics.

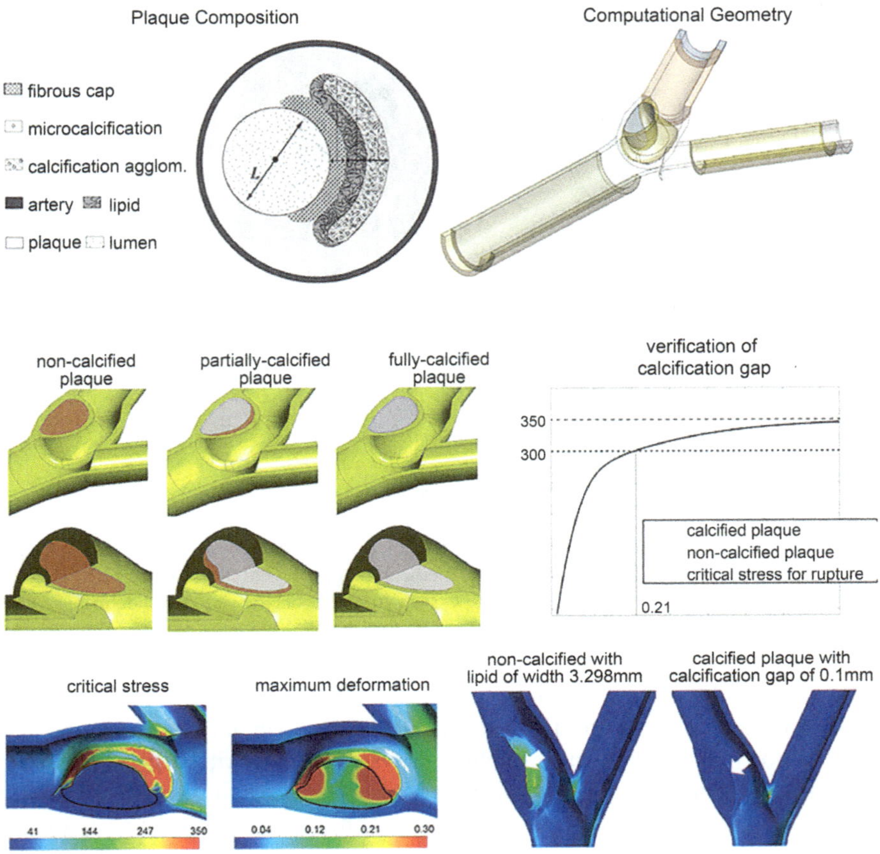

Fig. 1.8 Three-dimensional fluid and structural modelling of plaque. Computational haemodynamics simulation of blood flow through atherosclerotic arteries and structural analysis of calcified plaques with fibrous caps based on fluid structure interaction is implemented

Therefore, understanding the haemodynamics before and after carotid stenting is of paramount importance in determining the need to deploy stents into the diseased vascular regions. This can reduce future procedure-related complications such as restenosis or occlusion that may occur unexpectedly. A simulation-based virtual stenting platform can be implemented by multiple imaging, computational and visualization, as well as flow analysis modules as shown in Fig. 1.9. This provides clinical management, simulation of flow conditions due to cardiovascular diseases, and planning of stent designs.

MRI scans can demonstrate the obliquity of stenosed arteries. Its treatment by a virtual stent design can be produced based on a B-Spline interpolation technique.

The results from simulations of virtual stenting of a stenosed artery is shown in Fig. 1.10 where flow patterns become regular since there is no stenotic section in the artery which interferes with blood flow. Low wall shear stress (of low magnitudes

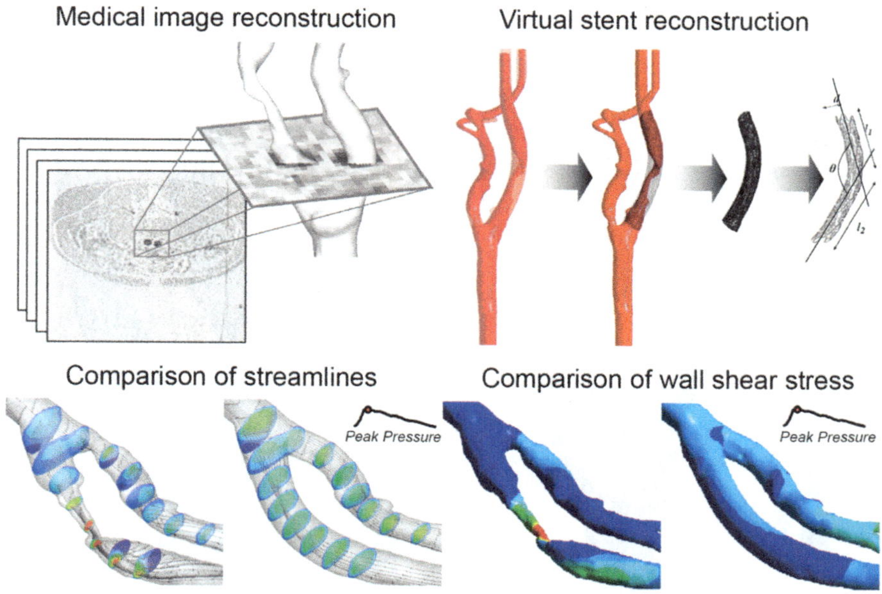

Fig. 1.9 Virtual stent design and analysis procedures. The design of stents via medical image reconstruction, computational modelling and reconstruction, and computational simulation followed by analysis of wall shear stress distribution can determine if stenting is a necessity for the atherosclerotic artery case

of less than 10 Pa) is noticed in the normal carotid artery with simulated stent in place, and this contrasts with high wall shear stress (maximum value at approximately 85 Pa) that is observed in the stenotic carotid artery.

Figure 1.10 also shows that for a normal artery, the wall shear stress distribution is more uniform due to the lower curvature and asymmetry of the carotid sinus and flow divider (at bifurcation point). High wall shear stress values are located at the bifurcation and further downstream. This results from the blood flow having a tendency to flow along the inner walls conforming to curvature of the carotid bifurcation. As a result, low wall shear stress values are constantly formed at the roots and along the outer walls of sinus bulbs. Locations of low wall shear stress are in accordance with the plaque locations of the stenosed model. Therefore, these results affirm that low wall shear stress is related to localization of atherosclerotic lesions (Ku et al. 1985a).

Time-dependent blood flow streamlines through a healthy and stenosed carotid bifurcation are plotted at peak cardiac cycle (Fig. 1.11). Based on the stenosed model, high velocity blood flow occurs at the stenosis due to occlusion caused by the plaque. At the peak cardiac cycle, maximum blood flow is 4.66 m/s. For the virtually stented condition, flow pressure difference between point A and point B of the artery are reduced significantly.

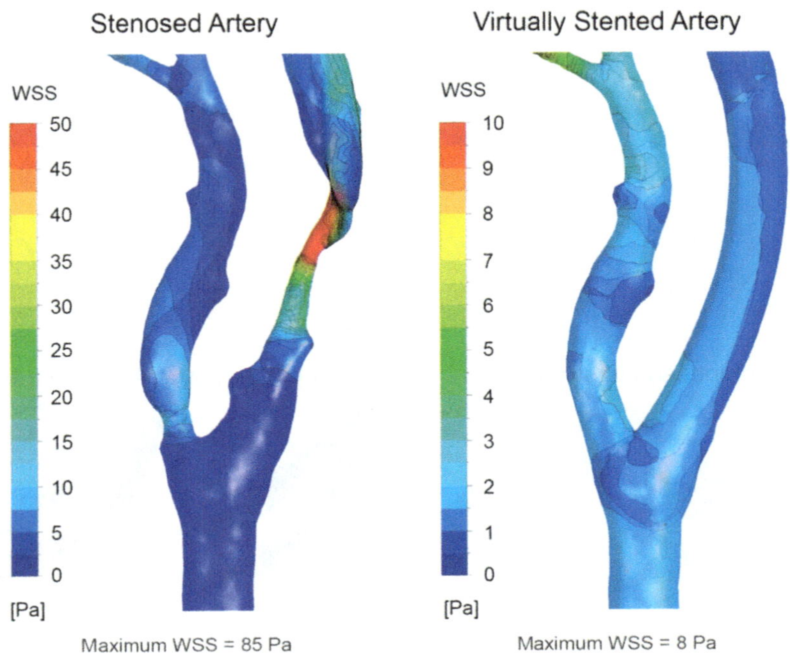

Fig. 1.10 Comparison of the wall shear stress (*WSS*) distribution in the stenosed and the reconstructed normal carotid bifurcation. Comparison of the wall shear stress distribution in the stenosed and the reconstructed healthy carotid bifurcations demonstrate the difference between diseased and reconstructed normal arteries. WSS is higher in the stenosed region (*left*) than in the reconstructed healthy carotid models (*right*). Reverse engineering can dilate the stenosis and restore the lumen to normal diameter and distribute the WSS more evenly throughout the artery. It is observed that a more uniform WSS distribution is found throughout the reconstructed healthy arterial regions, which confirms superior haemodynamics performance as compared to the stenosed artery

1.3.6 Surgical Treatment of Atherosclerotic Arteries

The coronary stent device is an expandable tube-like device that is inserted into a vessel to treat localized stenosis in the arteries. With the rapid development of medical technology, stenting has been a feasible intervention method to prevent stroke or ischemia[1] caused by atherosclerotic arteries during the last decades. The total global market for intravascular stenting, including in vitro cardio intervention device and related consumables has been estimated at US $ 6.12 billion per annum, with a market growth potential of 3.4 % per year (Greenemeier 2008).

Drug-eluting stents (DESs) have decreased restenosis rates compared to bare metal stents (BMSs). For example, DESs have restenosis rates in the range of 6–8 % (Leon et al. 2009; Stone et al. 2011). Low WSS is correlated to neointimal hyper-

[1] *Ischemia* is typically caused by damage or dysfunction of blood vessels resulting in restriction of blood supply and causing a shortage of oxygen and glucose to keep tissues alive.

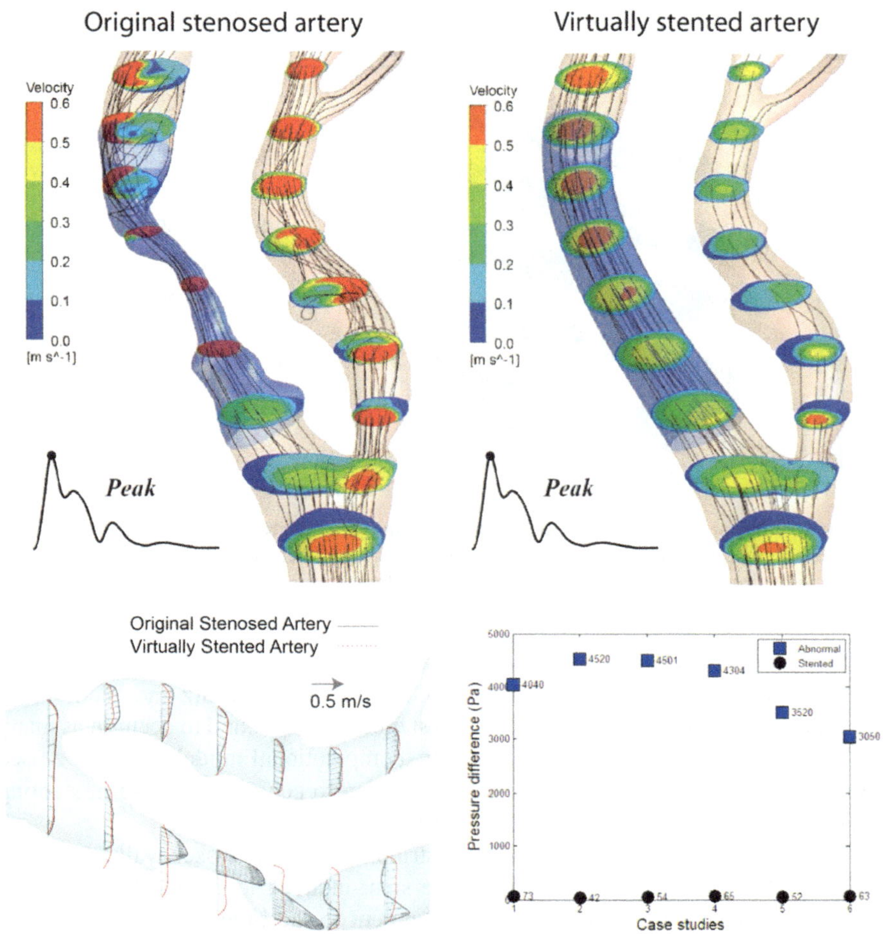

Fig. 1.11 Comparison of velocity profiles for pre- and virtually stented arteries. Velocity profile and pressure differences across a stenosed region (*based on 6 case samples*) for pre- and virtually reconstructed arteries can be used to decide the suitability of stenting for the atherosclerotic arteries

plasia, leading to restenosis (LaDisa et al. 2005b; Liu and Goldman 2001), and also with areas of inhibited endothelial cell migration onto the stent surfaces (Sprague et al. 2000).

However, the presence of a non-biocompatible device inside an artery can lead to inflammation and also influence the fluid dynamic behaviour in regions next to the arterial wall. Parts of the stent struts protruding into the lumen may induce formation of vortices and stagnation zones which affect WSS. Other issues exist for the patient after stent implantation such as vessel dissection and elastic recoil. These effects depend on parameters such as stent configuration, its global length, the delivery system, the struts dimension, shape, and spacing (Tominaga et al. 1992). Stents

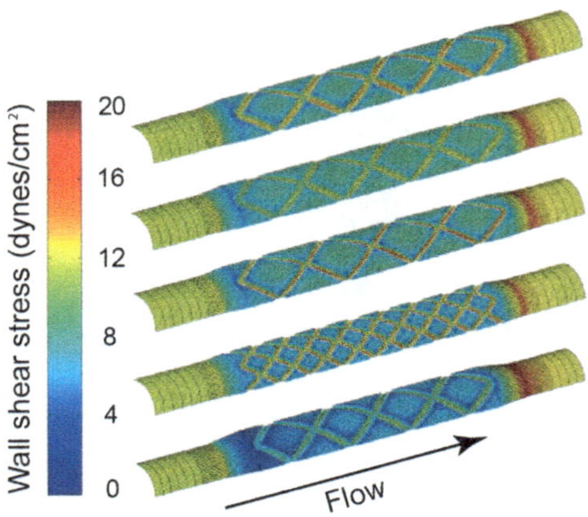

Number of struts	Width (mm)	Thickness (mm)	Deployment ratio
4	0.329	0.096	1.1-1
4	0.329	0.056	1.1-1
4	0.197	0.096	1.1-1
8	0.197	0.096	1.1-1
4	0.329	0.096	1.2-1

Fig. 1.12 Analysis of WSS on the arterial tissue in the stented arteries. Wall shear stress (*WSS*) analysis can be carried out based on various stent design parameters such as number, width, and thickness of struts and stent-to-artery diameter ratio. (Image from LaDisa et al. 2004)

that are poorly matched to patient-specific vessels may pose vascular complications and lead to restenosis after stenting (Gordon et al. 1994; Kuntz and Baim 1993; Rogers and Edelman 1995). The in-stent restenosis rate is found to occur in as many as 20–50% of stented vessels. Fortunately, computational modelling can be used to improve the stability of stent-vessel interaction to achieve a long term stenting performance without restenosis from the engineering point of view.

Since stent geometry affects WSS distribution on the vessel wall, designing an ideal stent structure that can minimize stent-induced low wall shear stress is desired. Thinner struts and with more alignment to the blood flow direction minimises the occurrence of low WSS (LaDisa et al. 2004, 2005a). Moreover, CHD can also be used to compare various commercial stent designs (Duraiswamy et al. 2009; Gundert et al. 2011; He et al. 2005; Murphy and Boyle 2010) as shown in Figs. 1.12, 1.13, and 1.14.

Sometimes, during stent implantation, fitting correctly to the geometry of the vessel does not occur. The stent may be deployed and expanded uniformly with a poor curvature fit along the wall of the curvilinear artery. In the case of asymmetric distribution of the plaque along the arterial axis, the stent is forced to bend against the artery wall, causing the sharp edge of the stent to cut through the artery wall. Over time, this may lead to platelet activation and thrombosis, which can result in restenosis of the artery. Fragments of the formed thrombus may travel downstream of the carotid artery to block the blood flow in micro vessels, and thereby causing stroke. Therefore, it is important to fully assess the need of stent implantation based on medical expert opinion.

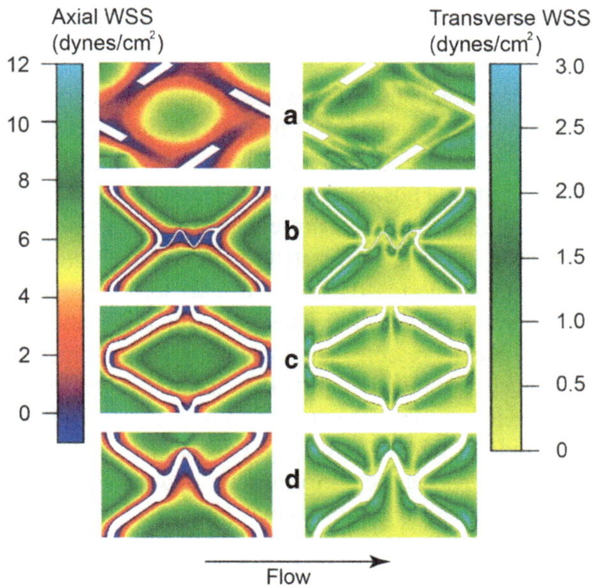

Fig. 1.13 Analysis of WSS on the arterial tissue in the stented arteries. Computational modelling can be extended to analysis of commercial stents—Wallstent, Bx-Velocity, Aurora, and NIR (Image from Duraiswamy et al. 2009). Other parameters of analysis include (**a**) Normalized effective WSS, (**b**) normalized average axial WSS, (**c**) normalized average transverse WSS, and (**d**) ratio of normalized axial WSS to transverse WSS are presented for analysis

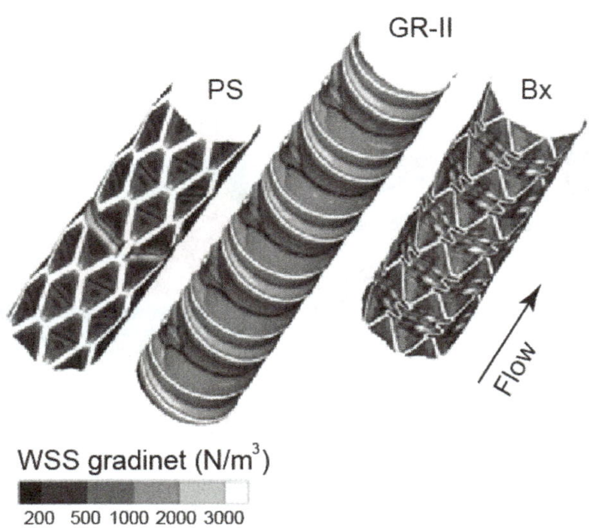

Fig. 1.14 Analysis of WSS and WSSG on the arterial tissue in the stented arteries. In the study by Murphy and Boyle (2010), the wall shear stress gradient (*WSSG*) was used. Three commercial stents: Palmaz–Schatz (*PS*) stent, Gianturco–Roubin II (*GR-II*) stent, and Cordis Bx-Velocity (*Bx*) stent, are analysed based on the mean WSSG value. The Bx stent shows a 12 % higher value than that for the PS stent due to regions of WSSG exceeding 2000 N/m³ on the uneven prolapse near the S-connectors of the Bx stent

1.3.7 Preoperative Assessment of Aneurysm

An aneurysm is an occlusion in the arteries caused by a blood-filled sac occurring in the wall of the blood vessel (see later in Chap. 2). Using CHD the flow field on an untreated and stented aneurysmal artery can be characterized. Streamline tracings can visualize the strength and position of vortices that are present in the aneurysmal sac. In Fig. 1.15, a simulated model is presented to show stenting inside a diseased artery at the aneurysm, and to demonstrate the interference with blood flow before and after stent treatment. The reason for stent insertion inside the aneurysm is to exclude it from the systemic blood circulation, so that the aneurysm will gradually shrink (due to reduced pressure) and eventually become insignificantly small. While this may induce stagnation and possible thrombosis formation after stenting, the chance of rupture is low since the aneurysm does not grow in size. Aneurysms can form on either side of the artery wall and so simulations of different types of diseased arterial geometries can be performed for understanding the effect of stent treatment. Three dimensional stent models may be created based on patient-specific aneurysm to enhance research and clinical value.

The simulated results in Fig. 1.15 demonstrate that stenting causes a reduction in pressure, velocity, vorticity, and shear rate. The reduced pressure exerted by blood on the aneurysmal sac will decrease the risk of rupture. However, a lower volume of flow into the sac increases the viscosity of blood in the aneurysm (Kim et al. 2010a). A reduced vorticity in the sac can be seen in the stented aneurysm which corresponds

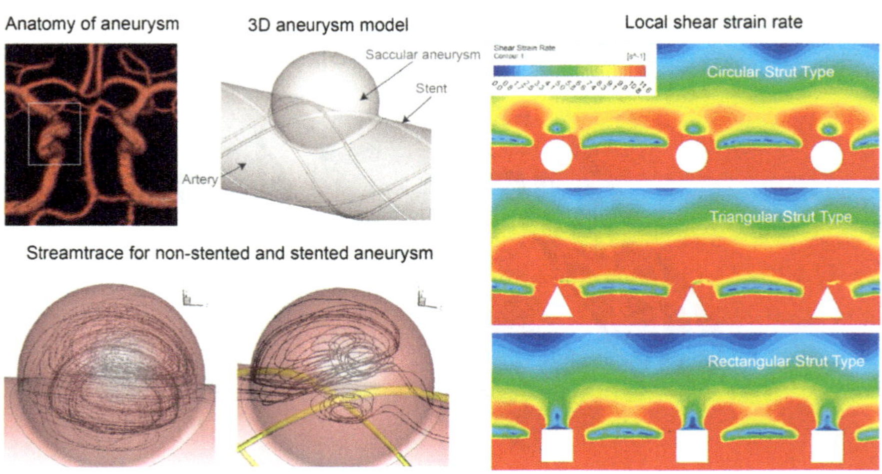

Fig. 1.15 Saccular aneurysmal stent modelling and analysis. Idealistic modelling is implemented and streamline tracing shows that the stent limits flow into the aneurysm. Different stent struts are modelled to demonstrate the effect on local shear strain rate of blood

to a lower fluid shear stress and shear strain rate. It is noted that high shear stress is necessary for preventing platelet-dependent thrombosis (Sukavaneshvar et al. 2000b). Moreover, reduced blood into the aneurysm also means flow stagnation and the induction of thrombosis increases. These undesirable conditions all contribute towards aneurysm rupture (Liou and Liou 2004b). Therefore, the type of stents deployed must have sufficient porosity to minimise aneurysmal rupture but prevent platelet aggregation (Kim et al. 2010a). The results shown can enable medical experts to evaluate the effectiveness of stent designs and their corresponding porosities in prevention of aneurysm dilation leading to rupture.

1.3.8 Assessment of Medical Devices

Congestive heart failure at its end stage leaves limited therapeutic options for cardiac patients. In such situations, mechanical assist therapy such as the ventricular assist device can be used to assist the cardiovascular system via means of artificial heart pump. Since its first implantation by Rose et al. (1954) and its first clinical application by Hall et al. (1964), research on improving its safety and mechanical reliability has been pursued. Clinical trial by Randomized Evaluation of Mechanical Assistance for the Treatment of Congestive Heart Failure (REMATCH) is a large-scale assessment of ventricular assist device efficacy in recent years. The purpose of REMATCH is to conduct a series of randomized clinical trials and evaluate the ventricular assist device with reference to optimal medical management in the case of congestive heart failure patients who are ineligible for transplant. The study demonstrated that ventricular assist device is a superior surgical remedy when compared to drug therapy (Rose et al. 1999). It has the capability to restore haemodynamics and the nutritional status of the blood circulatory system and reverses the abnormalities of the heart caused by heart failure syndrome (Jessup 2001; Kherani et al. 2004). However, one of the primary concerns by most patients is the associated financial costs with this therapy (Bieniarz and Delgado 2007; Hernandez et al. 2008).

A Spiral Vortex Ventricular Assist Device (SV-VAD) supports cardiac patients with refractory heart failure. Unfortunately, thrombus formation and risk of stroke due to flow complications may lead to aggravated conditions. The haemodynamics of a continuous flow in the ventricular chamber of a SV-VAD can be analyzed using CHD. Particle image velocimetry and laser Doppler anemometry can measure cross-sectional averaged axial and tangential velocities; which are used for validating the simulated spiral flow in a transparent acrylic SV-VAD replica. The relationship between swirling flow and blood cell damage can be established by evaluating flow effect on thrombosis due to high shear stress (Fig. 1.16). Flow stagnation within the SV-VAD should be insignificant and its low shear stress should minimize hemolysis (Wong et al. 2010a).

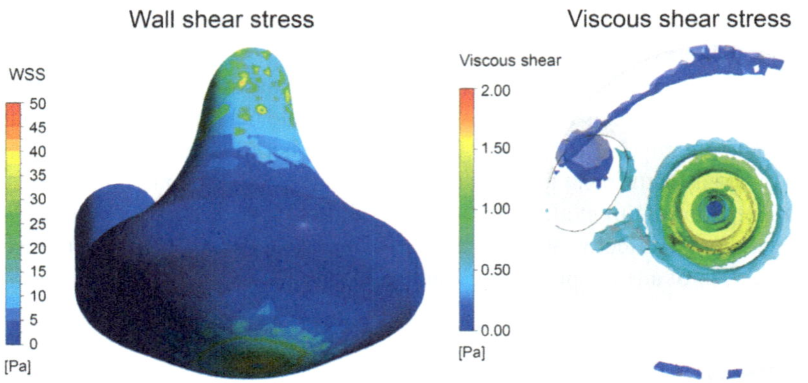

Fig. 1.16 Flow in ventricular assist device analysed using plots of wall and viscous shear stress profiles. Wall shear stress of blood at the near wall surface of the device is demonstrated to have low orders of magnitudes. The iso-surface shear stress plot in the device chamber pertains to low viscous shear stress that is experienced by the blood. These results relate to minimal risk of erythrocyte membrane rupture and hemolysis

1.4 Summary

A computational approach to study the human cardiovascular system and blood flow has been largely driven by the growth in computing power, advancements in technology, and materialisation of interdisciplinary research. Many advantages were presented in this chapter. For example the ability to simulate blood flows in cardiovascular structures after implantation of medical devices that are difficult to reproduce experimentally. Surgical treatment of human atherosclerotic arteries can be invasive and the CHD approach can provide a cleaner alternative to plan surgical procedures and examine feasibilities of medical device implantation. CHD can also be used as an educational and research tool in many biomedical applications such as assessment of cardiovascular diseases, R&D of medical cardiac devices, and assistance to surgical procedures to the heart.

Using CHD for biomedical applications is multi-disciplinary, incorporating medical imaging, computer aided design, computational fluid dynamics, and finite element method modelling. As such, CHD requires medical engineering trainees to obtain knowledge subsets from each discipline or field, which is the objective of this book. By providing the necessary background material for an understanding of the architecture of CHD, the reader can attain proficiency in program codes and its successful operation. In the next chapter, we begin by presenting the anatomy and physiology of the human cardiovascular system that serves as a base for developing the computational haemodynamics simulation settings. The primary aim is to summarise the important features of cardiovascular blood flow and how it is incorporated into a computational model.

1.5 Review Questions

1. For which research fields is CHD most relevant to?
2. What specific examples can you think of where CHD is being used within the biofluid engineering applications?
3. What are some of the advantages of using CHD for the analysis or evaluation of cardiac patients?
4. What are the limitations and disadvantages of using CHD?
5. What are some of the biomedical devices that can be assessed by CHD?
6. What CFD measurements can be obtained to assess flow in diseased arteries?
7. The bio-medical science field is turning to CHD to resolve flows within vascular systems. What advantages does CHD hold over experiments in obtaining these numerical results?
8. How can CHD influence the necessity of implanting a stent in a stenosed artery?

Chapter 2
The Human Cardiovascular System

2.1 Introduction

Before delving into the computational methods of CHD, this chapter provides a preliminary understanding of the circulatory system from a physiological and functional perspective, as well as related terminologies that will be used in the succeeding chapters. This includes descriptions, locations, geometry, and naming conventions for the relevant anatomy. This facilitates the process of computational reconstruction of the anatomy related to cardiovascular modelling introduced in Chaps. 3 and 4 and establish physiologically correct boundary conditions discussed in Chap. 5. The models arising from this process are linked to haemodynamic analyses that are discussed in Chaps. 7 and 8. Knowledge of the anatomy and its functions provides a smooth transition towards understanding cardiovascular function, and its CHD modelling requirements, while also stimulating interest in the reader after having established the background knowledge.

2.1.1 Functions of the Circulatory System

The circulatory system comprises the cardiovascular system that transports blood, and the lymphatic system that distributes lymph throughout the body. theIt circulatory system, which is a network of blood vessels that transports nutrients in the form of amino acids, electrolytes, lymph (fluid containing white blood cells), hormones, and oxygenated blood to tissues or organs in the human body. This maintains homeostasis, the immune system, and stabilizes body temperature and pH levels.

The human cardiovascular network is a closed loop system that enables the transport of oxygenated blood to the tissues and organs of the human body and the de-oxygenated blood to the respiratory organs. The heart pumps approximately 5 L of blood through the cardiovascular network to vital organs of the human body, providing nutrients and oxygen that are needed, and then transporting the waste products and harmful chemicals away from them. The cardiovascular system

J. Tu et al., *Computational Hemodynamics—Theory, Modelling and Applications*,
Biological and Medical Physics, Biomedical Engineering,
DOI 10.1007/978-94-017-9594-4_2

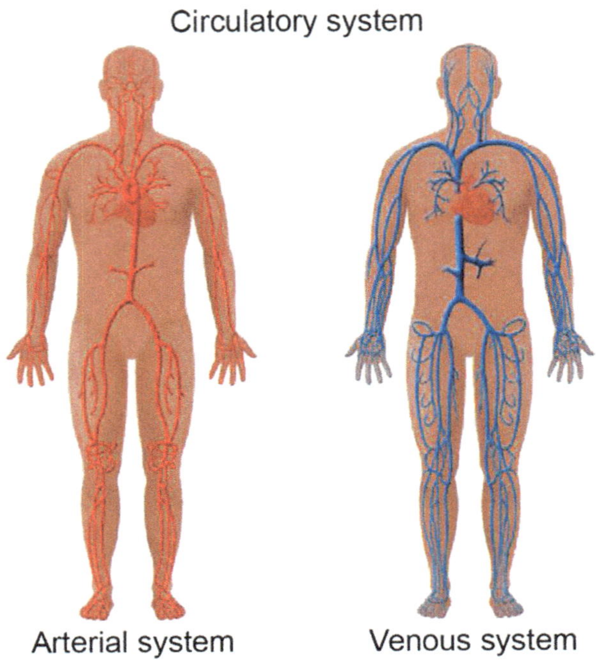

Fig. 2.1 Vasculature of the human body. Circulation is based on the vascular network system of arteries and veins that is distributed throughout the entire body. Oxygenated blood (*red*) is supplied by the aorta to the rest of the human body by an intricate network of small arteries and arterioles. Deoxygenated blood (*blue*) from these organs is transported via the venules, veins and finally the vena cava back to the heart for oxygenation in the respiratory system

comprises the pulmonary system; the coronary system; and the systemic system. It is common to combine both the pulmonary system with the coronary system to form a system known as the cardiopulmonary system.

The cardiovascular system is an inter-connection of arteries and veins that branch at multiple levels to reach all parts of the human body via an intricate network of vessels (Fig. 2.1). Functionally there are the arterial and venous networks that supply oxygenated and return de-oxygenated blood respectively. Blood circulation is achieved by the vascular system such that oxygenated blood (represented in red) is transported from the heart via the aorta to all parts of the body. The de-oxygenated blood (represented in blue) results after oxygen is depleted from the blood and used by the cells to produce energy and waste within the body. As illustrated, oxygenation and de-oxygenation occurs at designated tissues and organs in the system of vascular networks. In CHD modelling we focus primarily on this system. Cardiac health issues arise when this system fails transport the blood through the network.

The lymphatic system is made up of lymph vessels and nodes forming an open network of intricate tubes to transport lymph throughout the body, thus acting as the secondary circulatory system. As it moves through the body it maintains fluid levels in the body and filters out bacteria by collects waste products and disposes them through certain organs in the body such as the bladder, bowel, lungs, and skin. A major difference between the cardiovascular and the lymphatic system is that the latter does not have a pump (e.g. heart) to move the lymph and instead relies on muscle contractions and gravity to move it through the body.

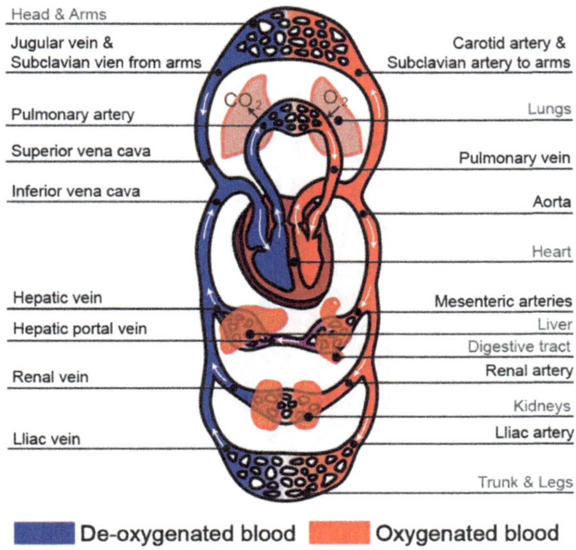

Fig. 2.2 The cardiovascular system. Gas exchange takes place in the lungs that oxygenate the blood by diffusion of oxygen into the blood stream through the alveoli. The oxygenated blood is then returned via the vena cava back to the heart, and is then transported around the body to the organs. To move the blood around the body through the organs and lungs, the heart acts as the central pump for the entire cardiovascular system

2.1.2 Organization of the Cardiovascular System

The cardiovascular system can be sub-categorised further into the pulmonary, systemic circulation, and coronary circulation (Fig. 2.2) based on the routes that the blood takes (i.e. circulates). The *pulmonary circulation*, as its name suggests, involves blood transfer to and from the heart and lung. In this circulation, deoxygenated blood is transported to the lungs and freshly oxygenated blood is taken back to the left side of the heart. The *systemic circulation*, deals with the whole body and the heart whereby oxygenated blood is circulated to the tissues and deoxygenated blood returns to the right side of the heart. The *coronary circulation* involves the blood that circulates in the heart itself.

2.1.2.1 Pulmonary Circulation

When blood is deprived of oxygen and contains waste materials, such as carbon dioxide the de-oxygenated blood enters via two large veins called the *vena cavae* into the right atrium of the heart (lower chamber), which then contracts (systole) and pushes the blood into the right ventricle (upper chamber) via the tricuspid valve (right atrioventricular valve). The right ventricle then contracts forcing the fluid out via the pulmonary semilunar valve into the pulmonary artery and into the lungs, whereby gas exchange occurs (Fig. 2.3).

The oxygenated blood is then returned into the left atrium via the pulmonary vein, which is pumped into the left ventricle of the heart and expelled through the aorta, the largest artery in the body (to withstand the high pressures), to the other portions of the body. There are a series of valves within the heart and within the

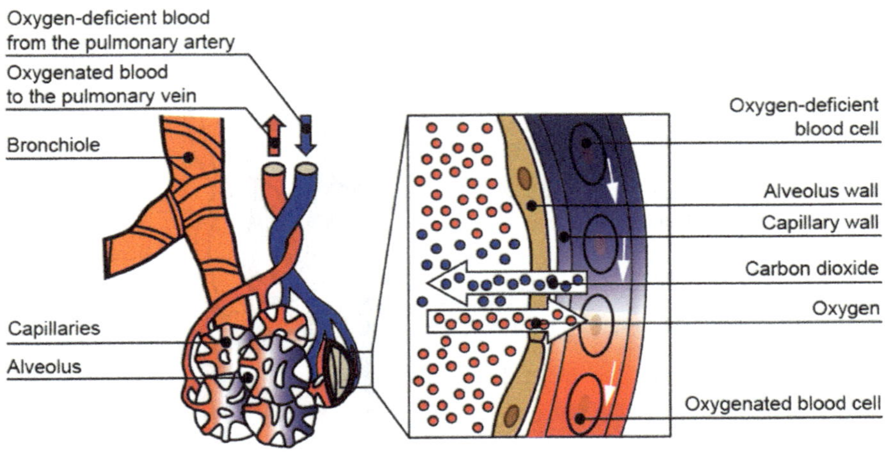

Fig. 2.3 Oxygen and carbon dioxide gas exchange. During breathing, gas exchange takes place in the alveoli where oxygen is passed into the bloodstream while carbon dioxide and other waste products are expelled. In the pulmonary circulatory system, the pulmonary artery is responsible for transporting de-oxygenated blood away from the heart to the lungs. The pulmonary vein transports oxygenated blood back to the heart. The oxygenation and de-oxygenation processes take place simultaneously at the alveolar-capillary interface in the deep lung airways

veins around the body that prevent backflow from occurring by sealing off the vessels when the heart is expanding (diastole), causing a lower pressure upstream.

2.1.2.2 Systemic Circulation

High pressure is exerted on the blood as it is squeezed out of the aorta in the left ventricle to pass through to the whole body. Within this flow, the blood will absorb nutrients attained from the digestion, which is then used to provide fuel for energy and storage. The waste are taken away through the liver and then expelled from the body. The vessel walls are smooth enough to allow for ease of flow and strong enough to withstand the high pressure of the flow. Eventually, the de-oxygenated blood carrying the waste product is returned to the heart. Due to greater distances to be transported, the left ventricle muscle is stronger and provides the blood with sufficient pressure to circulate it further.

2.1.2.3 Coronary Circulation

Oxygenated blood circulates through the heart via the coronary arteries, while deoxygenated blood is taken away to the lungs through the cardiac veins. The coronary circulatory system is very much like the systemic circulatory system in the sense that blood is supplied to the heart but via the coronary arteries. The heart and the coronary arteries are analogous to the organs of the human body and the cardiovascular network of arteries.

2.2 Physiology of the Cardiovascular System

2.2.1 Anatomy of the Heart

The heart supplies oxygenated blood to the rest of the body and then transports the de-oxygenated blood to the respiratory system for oxygen replenishment (Fig. 2.4). The heart comprises the right ventricle and left atrium, separated by a partition septum. Each half consists of two chambers; a thin-walled atrium and a thick-walled ventricle. The atria receive blood from the veins, while the ventricles pump blood out of the heart and through the circulatory system. The right atrium is the upper right chamber of the heart collecting de-oxygenated blood from the vena cava, and then passes it via the tricuspid valves. This goes into the right ventricle for pumping into the lungs through the pulmonary valve and via the pulmonary artery for oxygenation. The oxygenated blood returns via the pulmonary vein into the left atrium, which then pumps the blood into the left ventricle through the mitral valve. The left ventricle is the strongest chamber of the heart that supplies the oxygenated blood to the rest of the human body via the aorta after passing through the aortic valves. The valves in the tricuspid,

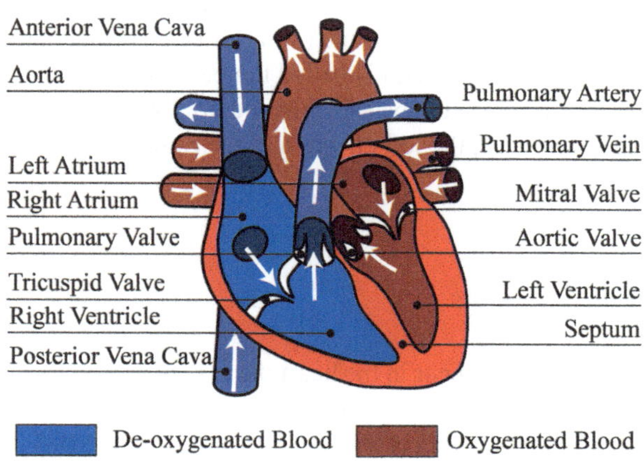

Fig. 2.4 Cardiovascular circulation of the heart. The heart comprises the *left and right atria*, which are responsible for collecting de-oxygenated and oxygenated blood from the vena cava and pulmonary vein respectively. The *right ventricle* is a heart chamber that pumps de-oxygenated blood to the respiratory system, and *left ventricle* is the most muscular chamber that pumps the oxygenated blood to all parts of the body. Heart valves are present at the connections of the atria and ventricles, as well as the pulmonary artery and aorta to achieve a single flow direction circuit. The *white arrows* indicate the direction of blood flow

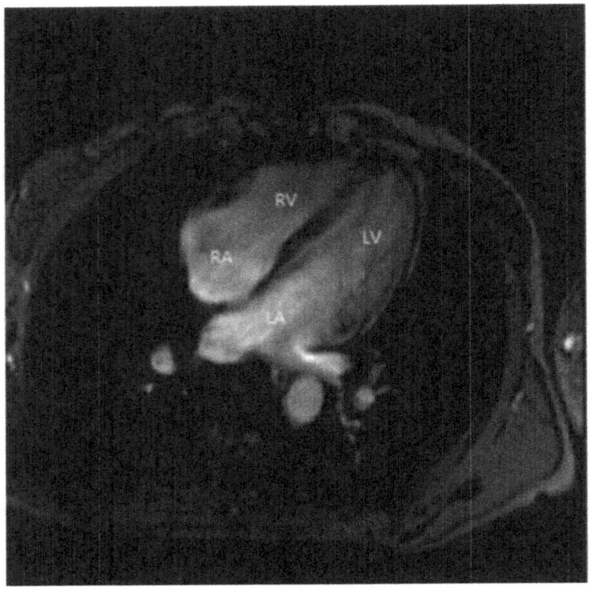

Fig. 2.5 A magnetic resonance showing how the heart is positioned in a human. The *image* shows a slice through all the four heart chambers during ventricular filling. The image is seen from the feet so the *right side* in the picture is the subjects *left side*. The MR acquisitions were performed at St. Olavs University Hospital, Trondheim. (Image from Dahl 2012)

pulmonary, mitral and aortic valves of the heart ensure that blood flows in the circulatory system effectively without reversing its direction in the circuit.

The magnetic resonance image in Fig. 2.5 shows a slice through all four heart chambers during ventricular filling. The subject is lying supine and the image is seen from the feet which mean that the right side in the picture is the subjects left side. Fig. 2.2 shows the heart situated in the chest cavity in between the two lungs and behind and slightly left of the sternum (breastbone). The narrow end of the heart is called the apex and the broad end is called the base. The apex is pointing downwards and to the left, whereas the base is directed upwards and to the right. The left ventricle, the right ventricle, the left atrium and the right atrium (RA) are indicated in the scanned image.

2.2.2 *Cardiac Cycle*[1]

The sequence of events that occur in the heart during one heart beat is called the cardiac cycle. The events occur nearly simultaneously for both sides of the heart. The typical resting heart rate in adults is 60–90 beats/min (bpm). A physically fit person has a lower heart rate compared to an inactive person. Each heart beat is commonly divided into two main phases: systole and diastole. Systole and diastole are synonymous with contraction and relaxation of a heart muscle, respectively. Both the

[1] Contribution from Dahl, S.K. (2012) Numerical Simulations of Blood Flow in the Left Side of the Heart. PhD Thesis, Norwegian University of Science and Technology.

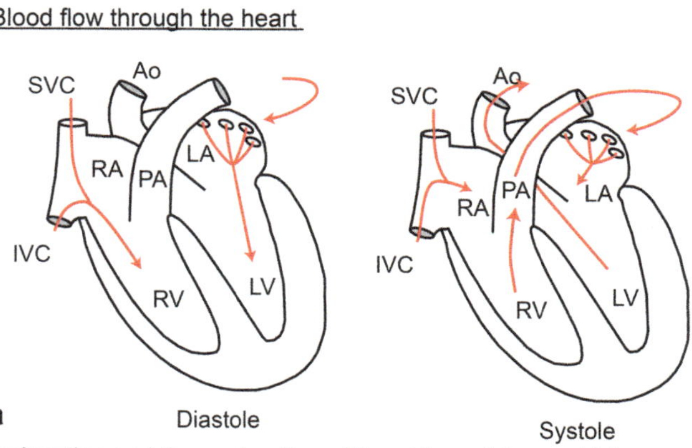

a

b

relaxation and the contraction of the atria and the ventricles

	ventricular diastole		*ventricular systole*
ventricles	relax		contract
atria	relax	contract	relax

Fig. 2.6 **a** Blood flow through the heart during ventricular diastole and systole (*adapted from Klabunde (2005)*, cvphysiology.com). **b** Correspondence between the relaxation and the contraction of the atria and the ventricles with respect to the ventricular diastole and systole

atria and the ventricles go through these two stages every heartbeat, but the terms diastole and systole alone, often refer to the ventricular stages. Figure 2.6 shows how the blood travels through the heart during ventricular diastole and systole and the correspondence between the relaxation and the contraction of the chambers with respect to the ventricular phases.

2.2.3 Physiology of the Aorta

The aorta, which extends from the left ventricle in the upward direction and then channels down towards the abdomen, is the largest and strongest artery in the human body. Oxygenated blood is transported via this artery to the body organs through the systemic circulation. Anatomically, the entire aorta is made up of three main segments: Ascending aorta; Aortic arch; and the Descending aorta (that comprises the Thoracic aorta and the Abdominal aorta), labelled in Fig. 2.7.

The aorta is a heterogeneous combination of smooth muscle, nerves, intimal cells, endothelial cells, fibroblast-like cells, and a complex extracellular matrix. Its wall is made up of several layers—the tunica adventitia, tunica media, and tunica intima, which are mainly composed of collagen giving it stability by helping to anchor it to nearby organs. Once blood is squeezed out from the left ventricle, it transports the high pressure and pulsatile blood to the rest of the body. Being distensible

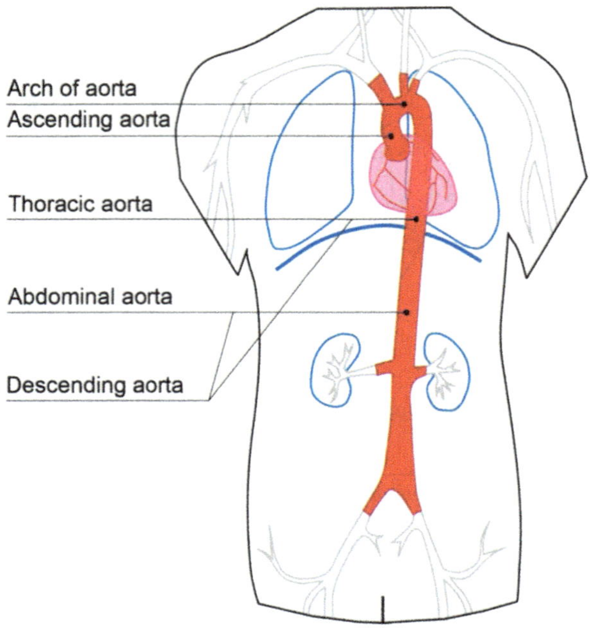

Fig. 2.7 Segments of the aorta. The components of the aorta are the ascending aorta, aortic arch, and the descending aorta (*that comprises the thoracic aorta, and the abdominal aorta*). The aorta is the largest and strongest artery in the human body and the main channel through which blood is pumped from the left ventricle. A network of vessels is further branched from the aorta to distribute blood to the vital organs

and elastic, the blood pressure decreases in strength and becomes less pulsatile from the aorta to arteries and to capillaries.

The blood spreads from the aorta down to the rest of the arteries which disperse through the body in a branching pattern. This gives rise to the term the arterial tree to describe the branching pattern of all the arteries in the body. The blood travels through the arteries in a pulsatile manner. Reflected waves rebound at bifurcations back to the semilunar valves and the aorta, which create a dicrotic notch in the aortic pressure waveform when they push onto the aortic semilunar valve. This can be visualised in the cardiac cycle profile (Fig. 2.8) which shows a small dip that coincides with the aortic valve closure. This dip is immediately followed by a short rise, referred to as the dicrotic wave, then declines gradually.

As a body ages, the artery stiffens and causes the pulse wave to circulate faster and the reflected waves return to the heart at a higher speed before the semilunar valve closes, and resulting in higher blood pressure. Determining the pulse wave velocity via invasive or non-invasive techniques can assess the arterial stiffness, which is related to the degree of the disease.

2.2.4 Physiology of the Carotid Bifurcation

The carotid bifurcation, which includes the Common Carotid Artery (CCA), External Carotid Artery (ECA) and Internal Carotid Artery (ICA), transports oxygenated

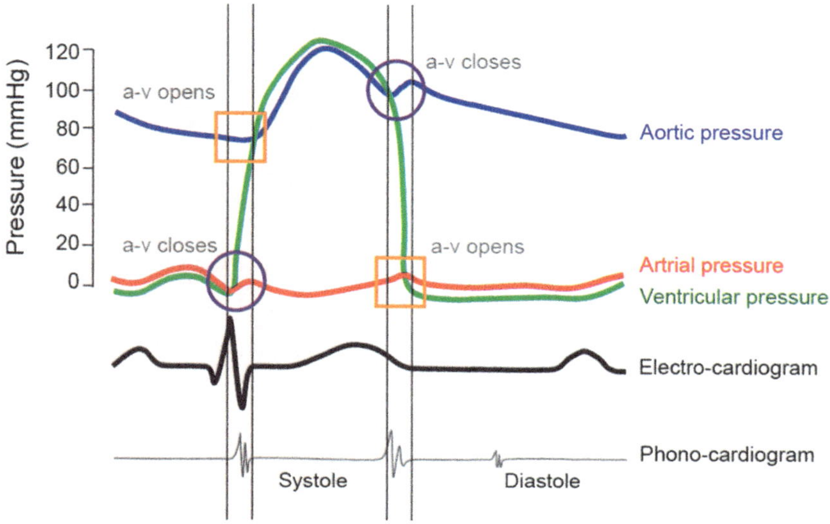

Fig. 2.8 Cardiac cycle events occuring in the *left ventricle*. The *square symbols* represent the aortic valve (a-v) opening while the *circle symbols* represent closing

blood to the head and neck regions (Fig. 2.9). The CCA is the main channel for supplying this oxygenated blood. It commonly exists as a bifurcation from which the ECA and ICA are originated.

The left and right CCA branches out from the aortic arch in the thoracic region and the brachiocephalic artery respectively (Fig. 2.10). This artery, which is also

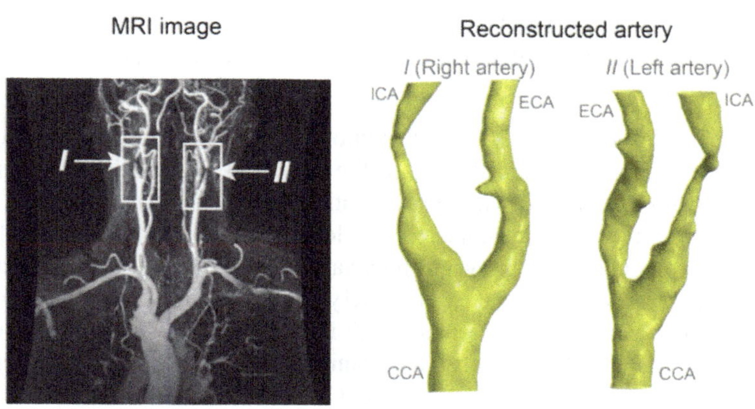

Fig. 2.9 Components of the carotid arterial network. The carotid bifurcation comprises the Common Carotid Artery (*CCA*), the Internal Carotid Artery (*ICA*) and the External Carotid Artery (*ECA*) at the neck and the head. The carotid artery serves the function of supplying oxygenated blood to the cerebral region

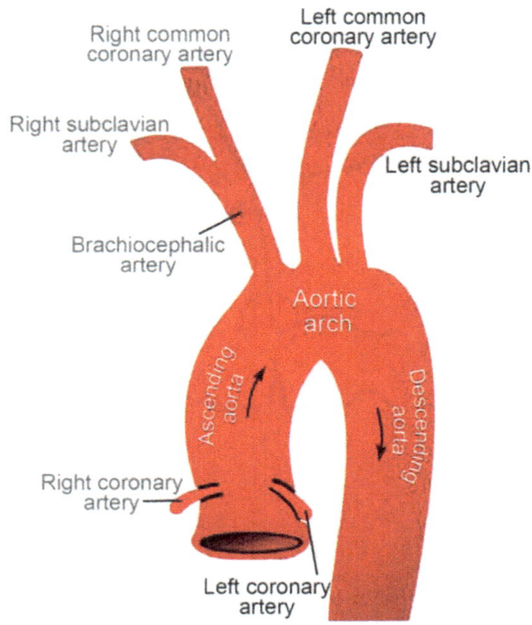

Fig. 2.10 Components of the aortic arch. The *right and left* common coronary arteries (*CCA*) branch from the aortic arch. The *right CCA* extends from the brachiocephalic artery

known as the trunk or innominate artery, extends from the first branch of the aortic arch, and divides into the right common carotid artery and the right subclavian artery. It connects to the mediastinum that supplies blood to the head and neck as well as the right arm.

The ICA and ECA branches supply blood to different organs through arteries at their downstream. In particular, the brain and eyes located at the downstream of ICA branch are the most active and important part of the human body, which consume a high quantity of oxygen and require a high volume of blood supply per unit time. Therefore the blood flow volume through the ICA branch is greater than that of the ECA.

Common Carotid Artery (CCA) The Common Carotid Artery ascends through the superior mediastinum[2] anterolaterally in the neck and lies medial to the jugular vein[3]. The two CCA are not symmetrical, with the left artery having greater length than the right artery. This accounts for the longer path from the aortic arch. The carotid artery, jugular vein, and vagus nerve are enclosed in the carotid sheath. The CCA bifurcates into the internal carotid artery and the external carotid artery at the superior border of the thyroid cartilage with inter-individual variations in terms of angle of bifurcation and asymmetry. The diameter of the CCA in adults ranges from 0.2–0.8 cm with an average value of 0.7 cm (Xu 2002).

---

[2] The *mediastinum* is the central compartment in the thorax that contains a group of structures that includes the heart, the esophagus, the trachea, and the lymph nodes of the central chest.

[3] The *jugular vein* is part of the venous network of vessels that bring deoxygenated blood from the head back to the heart via the superior vena cava (see Fig. 2.2).

External Carotid Artery (ECA) The External Carotid Artery is typically smaller than the ICA, is located anteromedial to the ICA and ascends posterolaterally. The branches of the ECA are: the superior thyroid; ascending pharyngeal; lingual; facial occipital; posterior auricular; superficial temporal; and the internal maxillary artery.

Internal Carotid Artery (ICA) The Internal Carotid Artery is the larger of the two bifurcation branches, comprises four main segments: the cervical, petrous, cavernous, and the cerebral. The cervical section originates at the CCA bifurcation and ends up near the base of the skull. Anatomically, the ICA runs deep to the sternocleidomastoid muscle, and it typically extends posterolateral to the ECA and courses medially as it ascends towards the neck region. The ICA is slightly dilated at its origin. This dilated bulb is known as the carotid sinus.

2.2.5 Physiology of the Coronary Arteries

Coronary arteries form a network of micro-scale vessels within the heart that is responsible for the transport of oxygenated blood to the myocardium (the muscle tissue of the heart that is made of thick contractile muscle cells responsible for the contractile pumping) to enable heart pumping. The cardiac veins then deliver the de-oxygenated blood away from the myocardium (Fig. 2.11).

The coronary artery network includes the epicardial coronary arteries which are found on the surface of the heart. These arteries regulate blood supply to the myocardium however their relatively narrow size make them susceptible to atherosclerosis, and blockages leading to angina or a heart attack. During the systolic event, ventricular myocardium contracts and high intraventricular pressures compresses

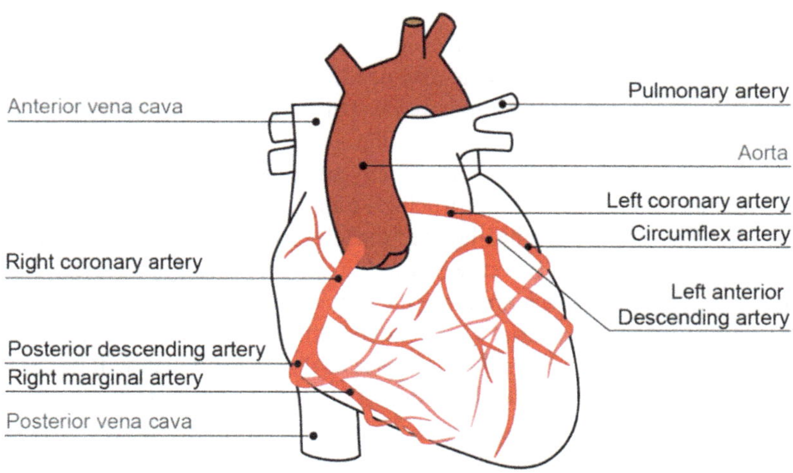

Fig. 2.11 Anatomical structure of the coronary arterial tree. The branches of the coronary arterial tree, as labelled in *red text*, supply blood to the heart muscles. Obstruction to the coronary arteries blocks blood from reaching the cardiac muscles and resulting in myocardial ischemia since they are the only vital source of blood supply to the myocardium

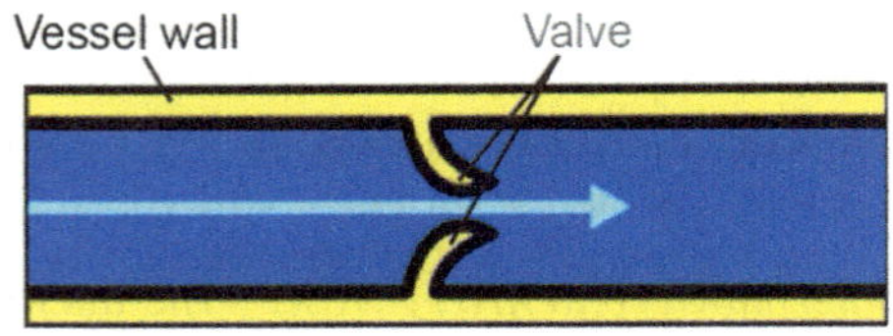

Fig. 2.12 Schematic of venous blood vessel. Valves are present along the blood vessels allowing only a one-way flow and hence prevents flow reversal

the arteries, while the epicardial vessels stay patent. During the diastolic phase, the intraventricular pressures decreases and myocardial perfusion occurs.

Coronary arteries are vital to the functions of the heart. Myocardial ischemia is a condition of the heart whereby oxygenated blood is unable to be supplied adequately. During atherosclerosis, blood flow in coronary arteries is obstructed, and intense chest pain known as myocardial angina occurs, which results in heart failure. Prolonged ischemic conditions cause the myocardium to weaken from hypoxia during myocardial infarction[4].

The left and right coronary arteries branch out from the left and right aortic sinus respectively, which is near the aortic root just above the aortic valve. The posterior descending artery or posterior inter-ventricular artery determines the coronary dominance of the heart.

2.2.6 Physiology of the Vascular Network

The relatively large arteries branch into a network of arterioles that become a system of capillaries, which enables maximum transfer of nutrients between the blood and the vital organs. Capillaries have micro-thin walls that facilitate nutrient transfer and at downstream flows, they merge to become venules, which are further restructured into larger vessels to form veins. Veins are notably smaller than arteries in terms of diameter and strength as they need not withstand high pressures and do not transfer as much materials. In addition, valves are present along the vessel to prevent flow reversal (Fig. 2.12).

Blood flow characteristics such as velocity and pressure, as well as the total area of the vascular structure along the structural transition from artery to vein is shown in Fig. 2.13. The artery is a multi-layer of muscles and connective tissue that serves the purpose of transporting blood (Refer to Fig. 2.1). The wall of an artery typically comprises three layers: the intima, media and adventitia. To minimize flow resistance, the internal of the arterial wall is lined with a smooth and thin endothelium. Damage to the endothelium results in disease of the artery, which will be described in Sects. 2.3.1 and 2.3.5. Next to the endothelium is the basement membrane, which allows diffusion of nutrients and waste. At the next level, the elastic layers absorb energy from the blood to minimize arterial deformation and provide structural rigidity to maintain the flow channel. The elastic properties of the artery declines as it

---

[4] Myocardial infarction is the medical term for a heart attack which is a result of blockage in the cardiovascular network that prevents blood flow to the heart a long enough time that part of the heart muscle becomes damaged or dies.

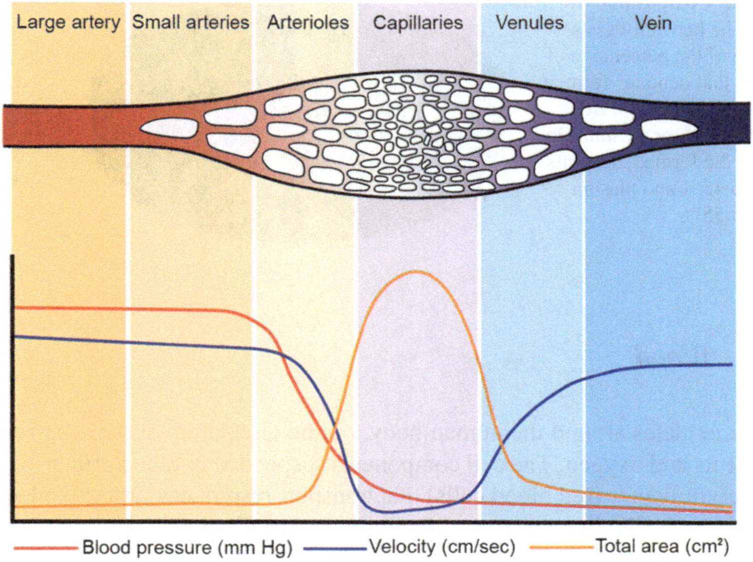

Fig. 2.13 Variation of flow characteristics for an artery-to-vein connection. The flow velocity and pressure of blood flow within the vessels of the systemic circulation as it undergoes structural transition from being an artery to a vein demonstrates that the blood pressure and velocity is inversely correlated to the total effective area presented by the vessels

branches away from the heart due to the reduction in pressure. However, aging and arterial disease can lead to the loss of this elasticity, accounting for increased resistance to blood flow that can lead to hypertension (Figs. 2.14 and 2.15).

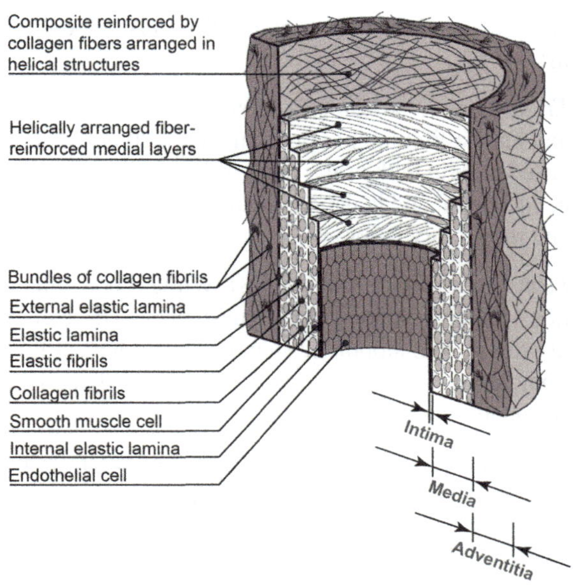

Fig. 2.14 Dissection of the multi-layer tissues that constitute the artery wall. The components of the artery consist of the endothelium, basement membrane, elastic layers, smooth muscle, and connective tissue in a configuration that provides elasticity to the vessel and low resistance to flow through it

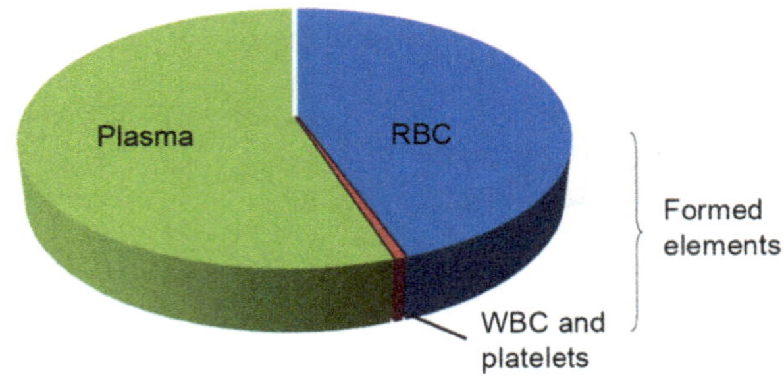

Fig. 2.15 Composition of blood. The hemacrotit is a measure of the percentage of blood is that contains formed elements (*the RBC, WBC, and platelets*), or plasma. On average the formed elements make up 45 % and plasma make up 55 %

2.2.7 Blood

Blood circulates around the human body via the circulatory system, carrying with it nutrients and oxygen. The cell components suspended in blood primarily consists of the erythrocytes (red blood cells) that transport respiratory gases; the leukocytes (white blood cells) that enhances immunity against diseases; and the thrombocytes (platelets) that provide the mechanism for blood clotting. The proportion of blood occupied by red blood cells is normally about 45 %. White blood cells and platelets make up a very small portion of the blood and therefore are often included in this percentage. The remaining proportion, approximately 55 %, is made up of plasma. The hematocrit is a measure of the proportions of blood that are either cellular (RBC) or liquid (plasma). For example normal haematocrit is approximately 48 % for men and 38 % for women. This represents the percentage of blood by volume that contains red blood cells.

Plasma is composed predominantly of water (making up to 93 % by volume), proteins, glucose, mineral ions, hormones and carbon dioxide. It serves as a form of protein reserve in the body, maintains the electrolyte level by intravascular osmosis, and prevents infection or blood disorders. Blood plasma has a density of 1025 kg/m^3.

Erythrocytes (Red Blood Cells) Red blood cells are oxygen carriers of the blood and have a lifetime of approximately 120 days. The hemoglobin, which is an iron-based protein with molar mass of approximately 65,700 g/mol., gives the red blood cell its color, and it also facilitates oxygenation of the blood and transport of oxygen and respiratory gases.

Leukocytes (White Blood Cells) White blood cells protect the body against infection and non-biocompatible substances. The cells are manufactured in the bone marrow (a hematopoietic stem cell), and have a lifespan of approximately 3 days. They are then distributed to the human body and exist in both the circulation and lymphatic systems.

Thrombocytes (Platelets) Platelets are micro-scale cell fragments of approximately 2 μm in diameter and have a lifespan of approximately 7 days. Platelets are circulated in the blood and are responsible for hemostasis, which leads to the release of thread fibers to enable blood clotting.

Low quantity of platelets, which is a condition known as thrombocytopenia, results in excessive bleeding. On the other hand, high quantity of platelets can cause thrombocytosis results in blood clotin which blocks blood vessels and increase the risk of stroke, myocardial infarction (heart attack), or pulmonary embolism. If the carotid arteries are blocked, then cerebral thrombosis occurs. This abnormality is known as thrombocytopathy.

2.3 Disease of the Cardiovascular System

2.3.1 Atherosclerosis

Atherosclerosis is a pathological process that takes place in the major arteries and is the underlying cause of heart attacks, stroke and peripheral artery disease. In an atherosclerotic artery, the lumen is narrowed and the arterial wall is stiffened by the build-up of fatty plaque beneath the endothelium (Fig. 2.16). During atherosclerosis, monocyte adherence to the endothelium is one of the earliest steps in lesion development. A monocyte entering the arterial wall at a junction between the endothelial cells consume excess Low Density Lipoprotein (LDL) that is present in the arterial wall, and then transform into macrophages. The macrophages then consume more LDL and then oxidize to become foam cells. The accumulation of foam cells can damage the arterial wall increasing its permeability and the rate of plaque build-up. When the plaque ruptures and a thrombus blocks the blood flow path, the tissue suffers from a lack of oxygen supply, which results in myocardial ischemia and lead to myocardial infarction (heart attack) or angina pectoris (chest pain from an ischemia).

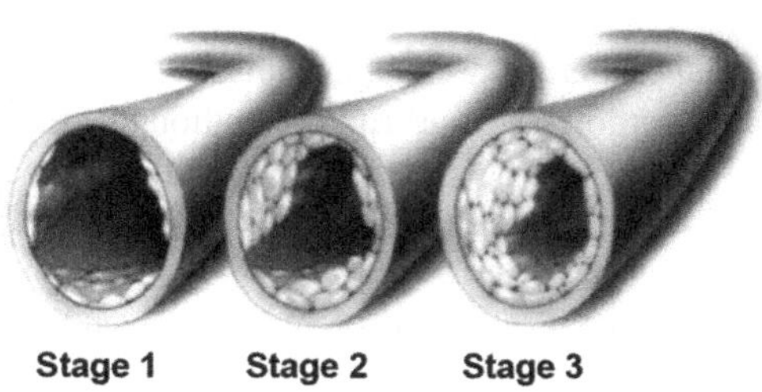

Fig. 2.16 Narrowing of lumen based on build-up of plaque over time. The localization of plaque and its growth takes place over time and results in lumen narrowing. This further aggravates the growth of plaque. When plaque ruptures, the platelets of the blood tend to adhere to the cytokines at the arterial wall at the injury site to form a clot that blocks the flow completely. Eventually, a complete blockage results to obstruct the entire blood supply

In some cases, the arterial wall can remodel[5] itself by increasing its external diameter to accommodate a plaque without narrowing the lumen, producing angiographic silent plaques detectable only by intravascular ultrasound. In the majority of cases where this does not occur, the resulting stenosis reduces blood flow to some degree. Due to a pathological remodelling of the arteries, a relationship exists between localization of atherosclerotic lesions and haemodynamics.

The preferred surgical intervention for atherosclerosis is balloon angioplasty, which is quick and involves minimal risk. It involves a small catheter with an inflatable balloon inserted into the femoral artery and guided to the stenosis by angiography. Inflation of the balloon then crushes the stenosis. Healing entails proliferation of smooth muscle cells followed by the deposition of new collagen (Davies 1996) thus thickening the cap and reducing the risk of thrombus formation, as well as pushing the plaque into the wall thereby increasing blood flow.

2.3.2 Calcification of Lesions in Plaque

Atherosclerosis constitutes a high number of deaths related to cardiovascular diseases in developed countries. It is a chronic systemic disease, frequently leading to vascular morbidity and premature mortality. Atherosclerosis causes the thickening and hardening of arteries by lining the arterial wall with fatty deposits, which results in plaque build-up or atheroma. Three main components (the atheroma, cholesterol, and calcification) form an atheromatous plaque.

Although atherosclerosis is systemic, plaque rupture is local and leads to acute cardiac syndromes such as ischemia and myocardial infarction or cerebrovascular events. Plaque material and structural characteristics are important factors in the natural progression of the disease and may have important clinical predictive value.

Atherosclerotic stenoses are the main cause of stable angina. Plaque rupture can lead to the formation of a thrombus that blocks blood flow to the heart leading to unstable angina or myocardial infarction. Plaques with large atheromatous cores, thin fibromuscular caps and inflamed caps are particularly vulnerable regardless of the degree of stenosis. There is some speculation that wall shear stress leads to luminal thinning and promotes plaque rupture although there are also questions about whether high haemodynamic shear alone would disrupt a stenotic plaque as haemodynamic stresses are usually much smaller than mechanical stresses imposed by blood and pulse pressures.

Calcified lesions most likely represent atherosclerosis at later stages of remodelling and may reflect more stable lesions (O'Rourke et al. 2000). However, early stages of atherosclerosis that do not contain calcium deposits may be more prone to rupture with subsequent occurrence of acute events (Schuijf et al. 2007).

Non- or less-invasive imaging can identify flow-limiting coronary stenosis (Meijboom et al. 2008), detect plaque components, measure atherosclerotic plaque burden

[5] Artery remodelling is a permanent change in the walls in response to long-term inflammation or obstruction in the vessel walls.

and its response to treatment, and differentiate stable plaques from those that are prone to rupture (Kitagawa et al. 2009; Takumi et al. 2007). Non-invasive imaging modalities include computed tomography (Harada et al. 2010) and magnetic resonance imaging (Helft et al. 2002; Yuan et al. 2001), while invasive imaging includes intravascular ultrasound modality (Chandran et al. 2003; Fuessl et al. 2001). Of these technologies, MRI provides the greatest intrinsic contrast between soft tissue structures. In particular, it has been shown to accurately document atherosclerotic plaque composition (Worthley et al. 2000b) and arterial wall remodelling (Worthley et al. 2000a).

Plaques can be characterized into three types based on histology analysis (Stary 2003): calcified and non-calcified plaques; and mixed plaques which refer to lesions with non-calcified and calcified components within a single lesion (Refer to a and b of Fig. 2.17). The presence of calcification in lipid, based on observation agglomerate of calcium clusters, occurs in some plaques (Refer to c and d of Fig. 2.17). Calcium content increases in patients with acute coronary syndrome (Hodgson et al. 1993). Large lipid core and calcified areas (defined as $>10\,\%$ of the plaque area each) and thin-cap fibroatheroma have been found to be associated with positive vascular remodeling (Burke et al. 2002; Varnava et al. 2002). Regardless of

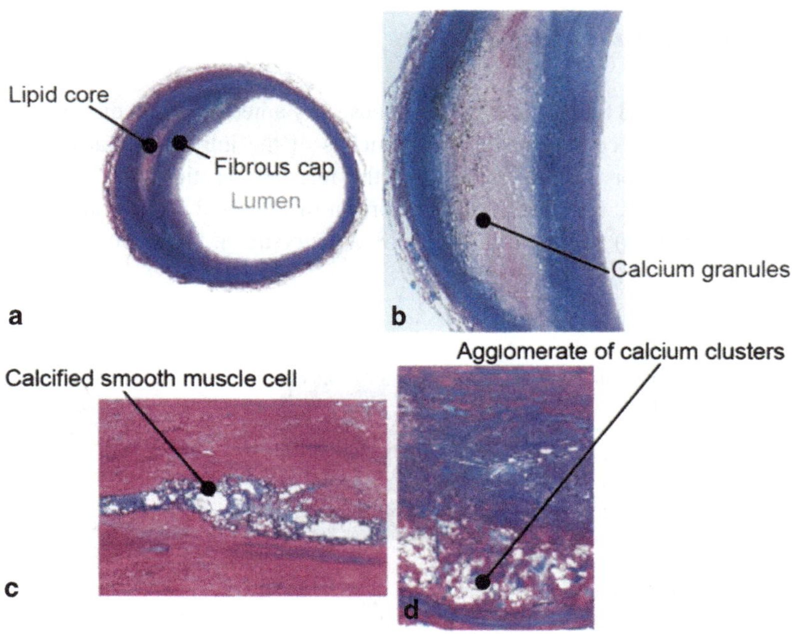

Fig. 2.17 Histological observation of plaque composites. **a** Lipid distribution in a crescent formation within the plaque can cause protrusion into the lumen and tend to obstruct distal arteries after plaque rupture. **b** Calcification can be observed by *white granules* embedded in lipid. **c** A *zoom-in view* of calcified smooth muscles cells reveals the agglomerate of calcium clusters within the plaque. **d** Calcification agglomerates are also present in the lesion adjacent to the elastic lamina. (Revised from images by Stary 2003)

the mechanisms of calcium formation involved, histology shows that calcium is a common but variable component in advanced atherosclerotic plaques.

Plaque rupture is dependent on biomechanical events acting on the fibrous cap such as haemodynamic shear stresses (Gertz and Roberts 1990), turbulent pressure fluctuations (Loree et al. 1991), cyclic variation of intraluminal pressure and maximum principal stress by the pulsatile blood pressure (Loree et al. 1992; Richardson et al. 1989b). In particular, large eccentric lipid cores are of mechanical disadvantage since circumferential tensile stresses are configured in such a way that fibrous caps have a tendency to rupture most of the time (Cheng et al. 1993a). This gives rise to the relationship between plaque rupture and the critical stress acting on the fibrous cap.

Autopsies of patients that are diagnosed of cardiac ischemia showed that the level of macrophages is high, smooth muscle cells are reduced, the proportion of crescentic acellular mass for a lipid core is significant, and the fibrous cap is thin (Fayad and Fuster 2001; Moreno et al. 1994b). For plaque rupture, a 65 μm thick fibrous cap with an infiltrate of macrophages is defined as the threshold after histological analysis (Burke et al. 1997). This can give guidance to critical risk analysis of plaque condition.

2.3.3 Aneurysm

Aneurysm is a blood filled dilation that is caused by atherosclerotic disease of blood vessel walls and is common near the branches of the intracranial arteries (especially at the anterior cerebral artery, and the internal carotid artery), and also at the abdominal aortic artery (distal to the origin of the renal arteries at the infrarenal abdominal aorta) as shown in Fig. 2.18. Aneurysms at the cerebral arteries are

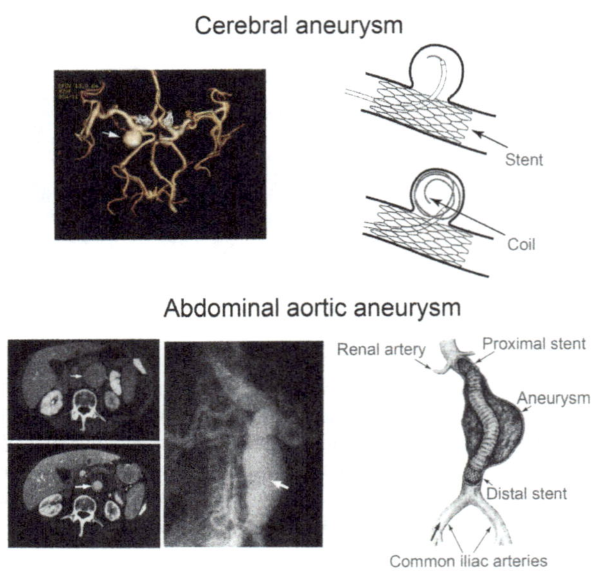

Fig. 2.18 Cerebral and abdominal aortic aneurysms. Aneurismal dilatations can occur at different arteries and locations. For a cerebral aneurysm, stenting is performed at the aneurysm neck to support coiling within a wide-neck aneurysm and to prevent the coil from falling into the artery. For abdominal aortic aneurismal stenting, the placement of an endovascular stent graft within an aortic aneurysm is via a catheter which is inserted up to the abdominal aorta in order to release the stent. (Adapted from images by Moore 1996)

commonly known as cerebral, intracranial or brain aneurysms, discussed in depth in Chap. 7. In the abdominal artery the aneurysm is known as Abdominal Aortic Aneurysm (AAA) which is a localized dilation of the abdominal aorta by more than half of its original diameter.

Aneurismal conditions can also be hereditary resulting in congenital aneurysm, which deteriorates the vascular wall structure. With aging, aneurysms increase in size resulting in the rupture and initiation of bleeding within the brain. This causes a medical condition known as stroke that can result in death or disability.

Aneurysms are classified as saccular or fusiform is based on their geometry (spherical- or spindle- shape) and dimension, which reaches up to approximately 5–20 mm in diameter. A saccular aneurysm is spherically shaped, affects a short length of the vessel, and typically occurs in the cerebral arteries. A fusiform aneurysm, which is typically spindle-shaped, is variable in dimension and is affected by arterial length. They typically affect a longer length of the ascending and transverse aortic arch, the abdominal aorta.

An aneurysm can grow large and rupture leading to severe haemorrhage, and other complications, death. Nearly 8.9% of the population above 65 years old is diagnosed with this disease. As it normally remains asymptomatic, surgical intervention or follow a wait-and-watch strategy is preferred for vascular surgeons if the rupture risk is being assessed at a low level. Currently, the maximum diameter criterion is widely used in clinical practice. When the diameter exceeds 55 mm or the expansion of the aneurysm is greater than 10 mm/year, a surgery is recommended (Brady et al. 2004).

Treatment of aneuryms can be made by inserting stents into the arteries. The affected arterial region is generally treated using angioplasty with stents or by open surgery that involves inserting a clip across the aneurysm in order to prevent blood from entering the aneurysmal bulge and aggravating its dilation (Fig. 2.18). For AAA stenting, an endovascular stent graft is delivered to the aortic aneurismal site via a catheter and a stent graft is released, and positioned to reinforce the weak arterial section caused by the aneurysm. This channels the blood through it without exerting pressure on the aneurismal bulge (Fig. 2.18).

2.3.4 *Thrombosis*

A thrombus is formed on a blood-contacting surface or in flowing blood which blocks circulation at a micro-level. We discuss two possible causes of thrombosis. When a calcified cap within a plaque is unable to withstand stresses by the flowing blood exerted on its wall it ruptures and exposes the interior of the artery to the blood stream. This causes blood to coagulate about the rupture and form blood clots. The blood clot gives rise to a thrombus, which essentially adds a further blockage to the already narrow channel -this condition is referred to as thrombosis. Should the blockage be severe, it can cut off the blood supply through the artery. If this occurs in the coronary artery, the result is a heart attack. If the thrombosis occurs in the cerebral artery, it can trigger a transient ischemic attack or a stroke. A blockage in other non-major arteries can result in pain in the areas of the body deprived of oxygen, which can lead to cell death.

Medical devices such as heart stents or artificial heart valves used for cardiovascular treatments are complicated by thrombosis formation and fatal haemorrhage. These issues are partially linked with the bio-fluid dynamics phenomena within the heart and arteries. For example blood cell damage leading to trauma of the erythrocytes is caused by excessive turbulence; and blood clot formation incurred by stagnant flow. Hemolysis or rupture of erythrocyte membrane due to high shear flow also aggravates platelet activation. Furthermore, platelets in a high shear flow have a higher chance of activation, leading to thrombus formation. At the same time, high shear stress is necessary for preventing the activated platelets from adhering to the artificial surface of an implanted medical device and forming platelet aggregation. It should be noted that a layer of adsorbed plasma protein accumulates onto an artificial surfaces rapidly when it comes into contact with blood, and this forms an interface for the platelets. Therefore, platelet activation and adhesion have to be balanced in terms of viscous shear stress, which significantly complicates the design of an artificial device.

2.3.5 Stroke

Haemorrhagic stroke develops when bleeding is present within the brain of a patient diagnosed with aneurysm in the cerebral arteries. It is a common cause of death by cardiovascular disease and affects 15 % of stroke patients in the world's population (AIHW Board 2008). Its onset has been linked to diabetes, obesity, alcoholism, tobacco use and copper deficiency (Valencia et al. 2008). Patients who die from such a medical condition exhibit copper deficiency of approximately a quarter quantity of a normal person. Tobacco and alcohol intake diminishes the copper content within the body, resulting in the progression of aneurismal growth (Utter and Rossmann 2007).

As mentioned in Sect. 2.3.4, a blood clot can travel to the cerebral artery within the brain of a patient and block the blood supply to result in ischemic stroke. In a similar manner, when plaque in the atherosclerotic region of a carotid artery ruptures, debris of the vulnerable plaque may result. This debris flows with the blood and blocks the narrow vessels downstream of the artery (Fig. 2.19). Another means of a complete arterial occlusion is by tearing of the tunica intima at the arterial wall which results in an arterial dissection of the carotid arteries, causing blood clots to enter between the inner and outer layers of the vessel. Ischemic and haemorrhagic stroke conditions account for 85 % and 15 % of cases respectively.

Blockage of cerebral arteries in the brain can give rise to a condition known as cerebral thrombosis. This can cut off blood supply to the brain in the micro-scale arteries. When the smaller diseased vessels rupture and bleed, cerebral haemorrhage occurs. Both these events are known as cerebrovascular disease of which the most common is stroke. A stroke results in numbness or paralysis, affecting speech, swallowing, vision, body balance and coordination. Stroke is a growing epidemic

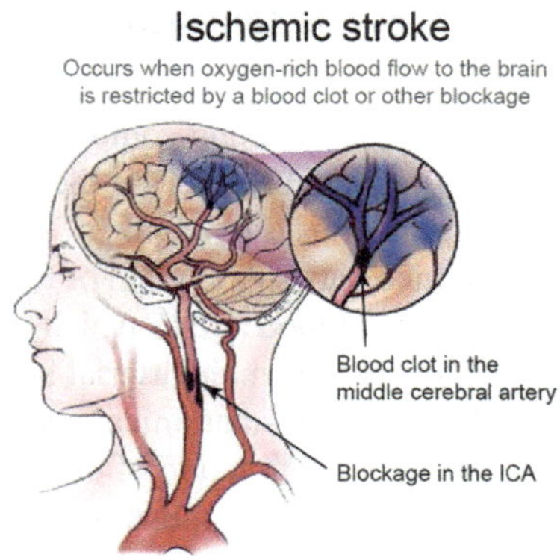

Fig. 2.19 Genesis of ischemic stroke: A blood clot in the blood stream can be lodged in a cerebral artery. This undesirable item can be debris of a ruptured plaque. The ruptured plaque components can travel downstream to the arteries in the brain region and initiating a cut-off of blood supply in the cerebral region resulting in stroke

in our society and there exists significant motivation to study its cause and medical remedies.

As discussed previously, atherosclerosis is one of the well-known causes of stroke (Ross 1999). The blockage of a carotid bifurcation is demonstrated in Fig. 2.19. In the regions of disturbed blood flow, which are found around carotid bifurcation, there is an increased chance of atheroma deposition. The worst scenario occurs when such atherosclerotic lesions aggravate and lead to health complication such as stroke or heart attacks.

2.4 Summary

This chapter gives an overview of the components and functions of the circulation system, which comprises the cardiovascular system, lymphatic system, the pulmonary, the systemic and the coronary circulations. Major physiological components of these systems such as the heart, aorta, carotid arteries as well as the coronary arteries and vascular network are provided. This chapter also presents common cardiovascular health problems related to the disease of the heart.

The primary aim of this chapter is to summarize the important features of blood circulatory flow by presenting the anatomy and physiology of the cardiovascular system. It by no means is a replacement for a comprehensive anatomy and physiology study. It does however aim to provide an understanding of the anatomy and physiology of the cardiovascular system, and the mechanics of blood flow. Furthermore, variations in the physiology and diseases of the blood vessels should be taken into account as it influences many settings for the computational model. The next chapter presents techniques in reconstructing the blood vessels from medical imaging scans, which is the first step in a CHD analysis.

2.5 Review Questions

1. What are the main components of the circulation system?
2. Describe the cycle for the transport of oxygenated and deoxygenated blood around the human body.
3. Which particular vessel of the venous system is responsible for delivering oxygenated blood and to which organ does it transport the blood to?
4. What are the main arterial branches of the carotid bifurcation?
5. What are the anatomical components of the aorta?
6. What is the function of the coronary arteries?
7. What are the composites of blood?
8. What are the common cardiovascular diseases?
9. What is the cause of thrombosis?
10. What are the types of calcified lesions that can occur in a plaque?
11. How is aneurysm caused?
12. Describe the cause of ischemic stroke.
13. What are the common causes of stroke?

Chapter 3
Geometric Model Reconstruction

3.1 Introduction

Computational reconstruction of the human cardiovascular structures can be divided into four stages: image acquisition, data conversion, segmentation and surface reconstruction. The development of a model first begins with medical imaging of the anatomy which can be obtained from various sources, yet all provide essentially similar information. This includes a 3D matrix (or series of 2D matrices) of volume elements (voxels), in which tissues and structures are distinguished by differences in brightness or greyscale. Two dimensional slices contain pixel data; while a voxel is the three dimensional analogy of a pixel where the third dimension is the spatial distance between each slice.

Visualizing and reconstructing morphological structures from scanned images is an area of active research. This chapter provides an overview of contemporary methods for image processing, edge detection, and surface and volume definitions, for both clinical and research images of cardiovascular structures. Specifically, computer algorithms are developed for the delineation of anatomical structures and other regions of interest. This includes image segmentation which is central in model reconstruction and many other biomedical imaging applications of anatomical structure, and pathologies (e.g. stenosis, and related cardiovascular disease identification).

3.2 Medical Image Acquistion

To identify the cardiovascular geometry various medical imaging modalities can be used. In this section we provide an introduction to Magnetic Resonance Imaging (MRI), or Computed Tomography (CT) imaging modalities, and describe specific

J. Tu et al., *Computational Hemodynamics—Theory, Modelling and Applications*,
Biological and Medical Physics, Biomedical Engineering,
DOI 10.1007/978-94-017-9594-4_3

issues relevant to each modality and the anatomical regions for which segmentation methods can be applied. New techniques are continuously emerging through rapid advances in technology and these techniques are covered in peer reviewed journals such as *Current Cardiovascular Imaging Reports, IEEE Transactions on Medical Imaging, Medical Image Analysis*, that are dedicated to medical imaging.

MRI scans use magnetic fields, and radio waves to obtain cross-sectional images of the body. During a scan an electric current passes through coils of wires to generate a magnetic field. Hydrogen protons of water molecules inside the body that normally spin in random directions are then aligned with the magnetic field. A short burst of tuned radio waves is sent through the body which momentarily changes the quantum state of the hydrogen protons (e.g. flips the spin of the proton). When the radio wave stops, the proton returns to its original orientation and in doing so echo's its own radio signal that a scanner detects and deciphers into images. This means that different tissue structures produce different pulse sequences, leading to contrast changes for a number of tissue parameters. In addition, anatomical and physiological variation between subjects requires different pulse sequences to achieve the correct contrast.

CT scans use multiple x-rays taken at thin cross-sections in the region of interest along the person's body forming slices (like slicing a loaf of bread). During a scan x-ray beams consisting of photons are absorbed or redirected (i.e. scattered) by material in the body which reduces the strength of the x-ray beam. Electronic detectors collect the x-ray information from each cross-section and send them to a computer that combines them into a single image. CT scans produce images with resolutions equal to or better than MRI. However soft tissue contrast in CT is not as good as in MRI, but is superior for imaging bone and bone tumors. Since CT scanners use x-rays which are a form of ionising radiation, its cumulative use has associated risks that are unavoidable and therefore CT scans are only performed where the benefit of the examination outweighs any potential risks. Both MRI and CT are non-invasive techniques however a contrast dye is sometimes injected into the body via one of the veins during the scan, referred to as an MR- or CT- angiograph. The contrast dye highlights the circulatory pathway and allows detection of the coronary arteries on the x-ray images. This type of scanning is referred to as *angiography*.

The scanned images produced are in a variety of formats but the most generic and common format is DICOM which contain both image data and patient information. These images are a series of stacked 2D pixels separated by a slice thickness, and when combined provide 3D volumetric data. The images contain pixels of greyscale values with respect to different organs or tissues. These values represent a mapping (Fig. 3.1) of the linear x-ray attenuation coefficient in CT or a measure of the radio-density in MRI. In medical imaging a Hounsfield scale with units of HU is used, while in image processing the greyscale is numbered between 0 and 255.

In a complete MRI/CT scan a set of 2D cross-sectional images, each separated by a thickness is produced. When the set of images are collated together, 3D volumetric information can be obtained. Each 2D image is made up of an array of pixels—the smallest element of the image. When the slice thickness between two images is considered, these pixels become voxels (*volume of pixels* Fig. 3.2). The

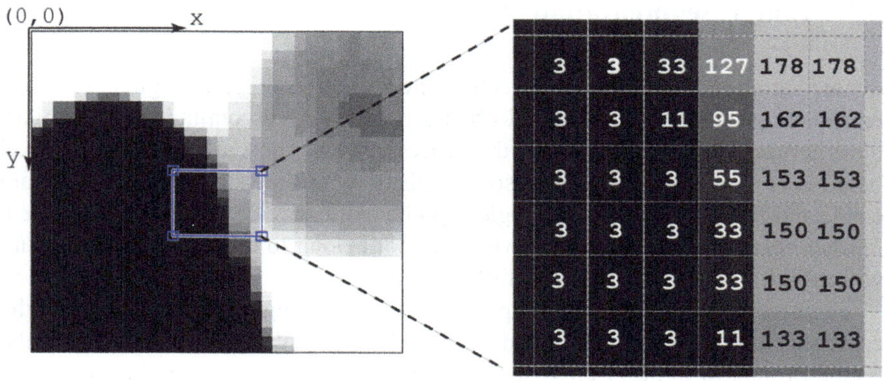

Fig. 3.1 A greyscale image represented digitally by row (x) and column (y), and the greyscale function $f(x, y)$ applied over each pixel

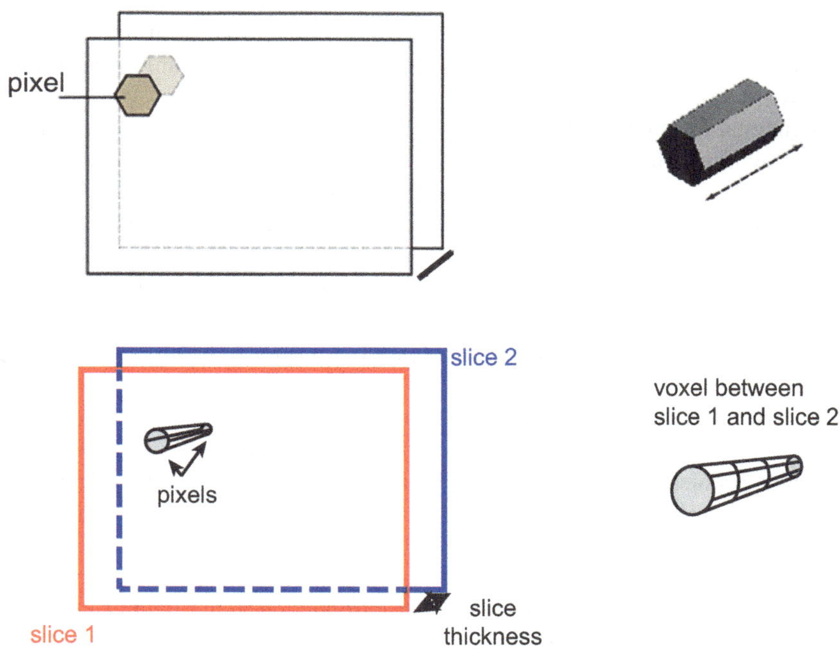

Fig. 3.2 Representation of a pixel (2D) and a voxel (3D)

spatial resolution of the scan is related to the voxel volume which is dependent on the field of view, pixel size, and the slice thickness where a greater resolution will allow better distinction between anatomical structures. The digital representation of an image can therefore be represented as matrices which are simply in the form of a 2D or 3D array.

3.3 Image Segmentation

Segmentation is the process of extracting one or more contiguous region of interest representing individual anatomical objects based on a discontinuity or a similarity criterion. More precisely, it is the process of assigning a label to every pixel based on a criterion in an image so that pixels with the same label and that are connected forming a contiguous region can be extracted, identified or categorised. Furthermore neighbouring pixels which do not belong to the anatomical structure will exhibit a pixel value outside of the criterion.

Segmentation can be manually performed by selecting individual pixels on the cross-sectional slices or automated. In manual segmentation the user selects the region of interest in every cross-sectional image in a set of scanned images which can have up to 1000 images. Therefore this procedure is time-consuming and also lends itself to inter-observer and intra-observer variability. Fully automated or semi-automated segmentation for monochrome images are generally based on discontinuities or similarities within the image based on greyscale values (referred to as intensity). For example, discontinuities at edges in an image can be identified based on abrupt changes in the intensity (greyscale level). Similarity in the intensity is used to extract a region which exhibits similar properties according to a set of predefined criteria.

Segmentation methods have been explored for many years producing a large number of algorithms dependent on the specific application, imaging modality, and other factors. For example segmentation of carotid arteries involves handling of outliers, feature point detection, and additional user interaction. However segmentation remains a challenging problem to overcome the increasing number of anatomical structures of interest, large variations in the properties within the images, and imaging artefacts such as noise, partial volume effects, and motion-blur. Therefore there is no single algorithm that can produce sufficient results for all types of medical images.

A segmentation issue related with MR or CT imaging is intensity inhomogeneity artifact (Condon et al. 1987; Simmons et al. 1994; Sled and Pike 1998), which can cause problems with algorithms that assume a constant intensity value for a tissue class. Performing a pre-filtering operation can remove the inhomogeneity by assuming mean tissue intensity for each tissue class is spatially varying within a certain range.

This section provides the reader with an introduction to the field of image segmentation and a description of some of algorithms among the many that exist in the literature. For further readings, there are general reviews on the segmentation of MR images (Bezdek et al. 1993; Clarke et al. 1995; Liang 1993; Peters et al. 1993), comparisons of different segmentation methods for MR images (Clarke et al. 1993; Hall et al. 1992; Vaidyanathan et al. 1995), segmentation of CT images (Sivewright and Elliott 1994), while CT segmentation applications of the abdominal aortic aneurysms (Juhan et al. 1997), carotid artery (Zhu et al. 2013) and segmentation of the heart (Ecabert et al. 2008), and cerebral artery (Manniesing et al. 2008).

3.3.1 Segmentation Approaches

A region within an image can be defined by its pixel properties (e.g. greyscale intensity), boundary (edge) or its interior. Therefore segmentation approaches can be broadly categorised into the following:

- *Pixel based (Thresholding)*: each pixel is labelled based on its grayscale values that represent intensity from the scans.
- *Edge based*: detects edge pixels to form a boundary containing the region of interest
- *Region based*: considers pixel greyscale levels from neighbouring pixels by including similar neighboring pixels (region growing).

Pixel based methods are the simplest and easiest approach to implement, however they lack contextual information and fail in scans with high inhomogeneity through a single region. Edge based methods are the next simplest approach and are efficient on scans of anatomical structures that have clearly defined boundaries such as the artery. A common problem however is that noise or occlusions can cause false or missed edge detection. Region based methods are the most complete but complex methods since regions of interest includes more pixel categorization than edges. Furthermore region growing techniques are useful in noisy images where edges are difficult to detect.

3.3.2 Threshold Segmentation

Image segmentation is defined as the partitioning of an image area or volume into non-overlapping, connected regions which are homogeneous with respect to some characteristic such as intensity (Pham et al. 2000). The simplest form of segmentation is the selection of pixels in 2D (or voxels in 3D) based on criteria typically its greyscale level, referred to as thresholding. This uses either global or local information to select only those pixels within a greyscale range (threshold) and a binary function is applied as,

$$g(x) = \begin{cases} 1 & \text{if } \min_x(f(x)) \leq \theta \leq \max_x(f(x)) \\ 0 & \end{cases} \tag{3.1}$$

where θ is the selected greyscale value or a range of values and $f(x)$ is the greyscale values of the images. The output is a binary image where each pixel is coloured as black or white, depending on a pixel's label of 1 or 0. It is most effective for images containing different structures that have uniform but contrasting intensities to other quantifiable features or structures. The segmentation problem becomes one of selecting the proper value for the threshold, θ. In many cases θ is chosen manually by the user, by trying a range of values and observing the range that best identifies the anatomical structure of interest.

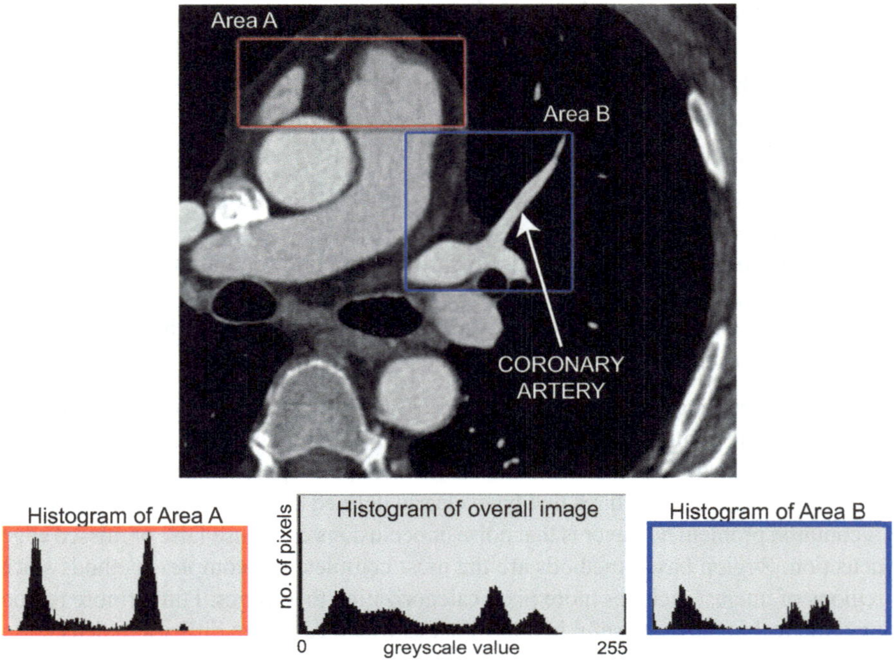

Fig. 3.3 Histogram analysis for a scanned image of the coronary artery. Local histograms can also be defined by selecting the local area of interest. The horizontal axis of the histogram represents the greyscale values that range from 0 (*black*) to 255 (*white*). The vertical axis represents the number of pixels that contain the greyscale value

A useful method to determine the appropriate θ is to analyze the histograms of the set of scanned images that are to be segmented. A histogram plots the occurrence (or number of pixels) of all greyscale intensities (0–255). By selecting sub-areas of the image for histogram analysis, the correct θ can be found. Figure 3.3 shows a scanned image of the coronary artery with histogram analysis. An overall histogram shows a range of peaks as the image contains a collection of different anatomical structures. By selecting a sub-area namely Area A and Area B, the local histogram reveals the dominant threshold pixel values that represents the structures within the selected sub-area. By looking at the histogram for a specific image a viewer can determine the greyscale distribution at a glance.

When a greyscale range is defined, boundaries are traced for all pixels within the range in the image. Greyscale threshold works well when an image that has uniform regions and contrasting background. Typically selected pixels from a given threshold criteria will not usually constitute a single connected component. This would lead to the initial selection being subdivided into smaller sub-selections. Each separate threshold selection can be treated as a 'mask' which can be used in Boolean operations. Figure 3.4 shows the result of a single threshold criterion applied to a scanned coronary artery image. The criterion sets all pixel values greater than θ converted to white; while all pixels less than θ are converted to black. A number of thresholding algorithms exist and the following review papers that

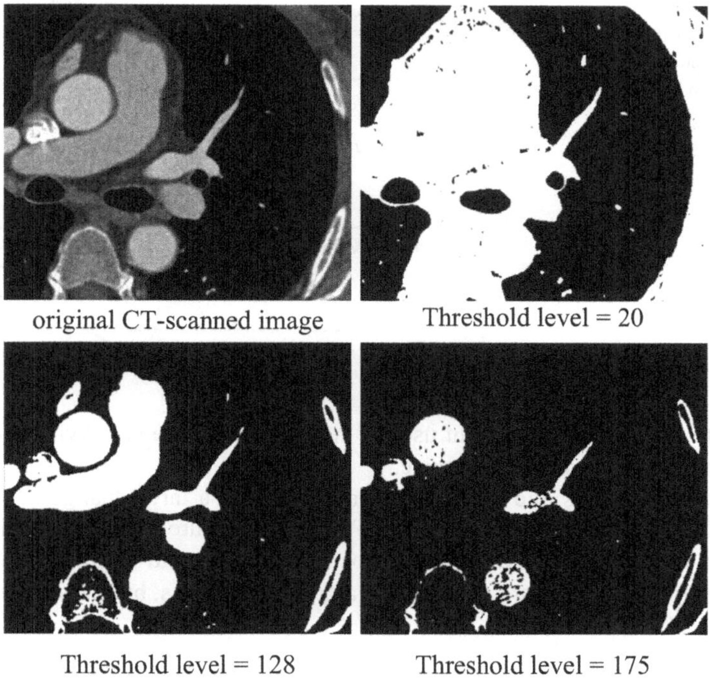

Fig. 3.4 A simple threshold criterion applied to the scanned coronary artery image. Greyscale values of the pixels range from 0 to 255 where 0 is typically set as *black* and 255 is set as *white*

summarise threshold algorithms are suggested, (Glasbey 1993; Le et al. 1990; Sahoo et al. 1988; Sezgin and Sankur 2004).

3.3.3 Edge Based Segmentation

Threshold algorithms are highly dependent on the defined criteria. A narrow or stringent criterion may lead to loss of pixels that exists in the region of interest, while conversely a loose criterion may lead to inclusion of non-required regions. A scanned object, such as the artery passageway exhibits a change in its structure between the inner blood flow, the lumen wall, and the surrounding tissue, and this is reflected through the greyscale levels on the image. If the boundaries of an individual structure can be detected, then the enclosed region containing the structure, can be separated and segmented from the scan. This boundary detection is a form of edge based segmentation techniques.

Edge-based segmentation represents a large group of methods based on the idea that an edge is defined as an area in an image where the intensity changes rapidly. This is achieved by applying a filter over the image to detect the rapid change in pixel values and thereby labelling the pixel as part of an edge or non-edge. Segmentation is performed by allocating a single label category of all non-edge pixels which are contiguous. Edge detection of an image significantly reduces the amount

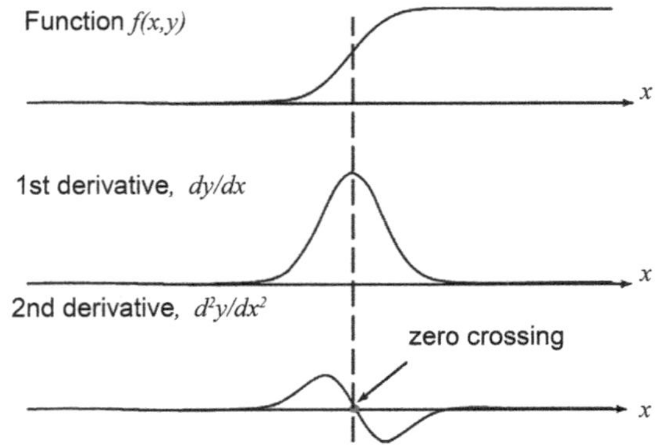

Fig. 3.5 First-order and Second-order derivatives of a function applied to an edge shown by the jump in intensity

of data to analyse and filters out useless information, while preserving the important structural properties in an image.

Ideally, an edge is one that has a discontinuity or a large change in greyscale intensity between neighbouring pixels. The salient feature of edge detection algorithms is the use of a derivative operator to determine and amplify the intensity change (Fig. 3.5). There are a number of algorithms for this, but these may be classified as either:

- **derivative based**—where the first derivative of the intensity change is greater in magnitude than a specified threshold or
- **gradient based**—to find regions where the second derivative of the intensity has a zero crossing (i.e. a point where the sign of a function changes from positive to negative or vice-versa and represented by a crossing of an axis with zero value). In gradient based edge detection a gradient of consecutive pixels is taken in both x and y direction.

From a graphical point of view, consider an edge which has a one-dimensional shape of a ramp and is depicted with a jump in intensity (Fig. 3.5). Taking the first derivative shows a maximum at the centre of the edge of the original function. This is the derivative based method which identifies a pixel as an edge location if the value of dy/dx exceeds some threshold. Since edges have higher pixel intensity values than its neighbours, you can set a threshold and compare the dy/dx value to find those values that exceed the threshold to detect edges. Taking the second derivative gives a zero at the maximum or turning point of the first derivative. As a result the gradient based method involves finding an edge based on the zero crossing in the second derivative curve. Taking the first and second order on each individual pixel of an image can be computationally intensive and hence impractical. A kernel (or mask), which is a small matrix itself, can be used to pass over an image to obtain an approximate derivative value.

First-order derivative (derivative based) algorithms include the **Sobel, Roberts**, and the **Prewitt** edge detectors that perform a 2-D spatial gradient measurement on an image. It is used to find the approximate absolute gradient magnitude at each point in an input grayscale image. The Prewitt detector can detect more edges but produces somewhat noisier results compared with the Sobel detector. For

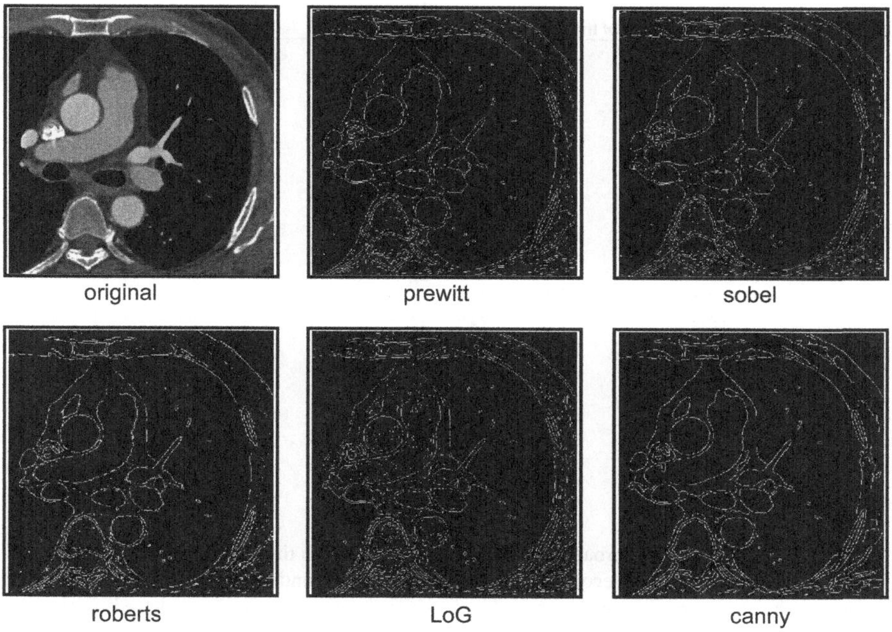

Fig. 3.6 A DICOM slice image of a coronary artery in the axial plane is shown with different edge detection algorithms applied

gradient based edge detection, the second order derivatives may be obtained by the **Laplacian of a Gaussian (LoG)** detector which applies a Laplacian of Gaussian filter to look for zero crossings (Marr and Hildreth 1980); or the **Canny** edge detector (Canny 1986) which determines the local gradient maxima of the image based on the derivative of a Gaussian filter. If two thresholds are used, then strong and weak edges can be detected, which makes the method more robust in the presence of noise. Each of these edge detectors are applied to the scanned coronary artery used earlier (Fig. 3.6).

Common problems arising from edge detection are primarily due to the absence of an edge where a real border actually exists. In addition, the presence of noise, fake, and weak edges will also have a negative influence on the algorithm. To overcome this, detected edges are connected to build up the border into an edge chain which will remove fake and weak edges. It should be noted that edge detection techniques are typically used in conjunction with region-based technique for complete segmentation.

3.3.4 Region Based Segmentation

In Sect. 3.3.2 regions were identified through threshold values based on the intensity values of the pixels while in Sect. 3.3.3 the segmentation process involved finding edge boundaries between regions based on pixel differences. In this section,

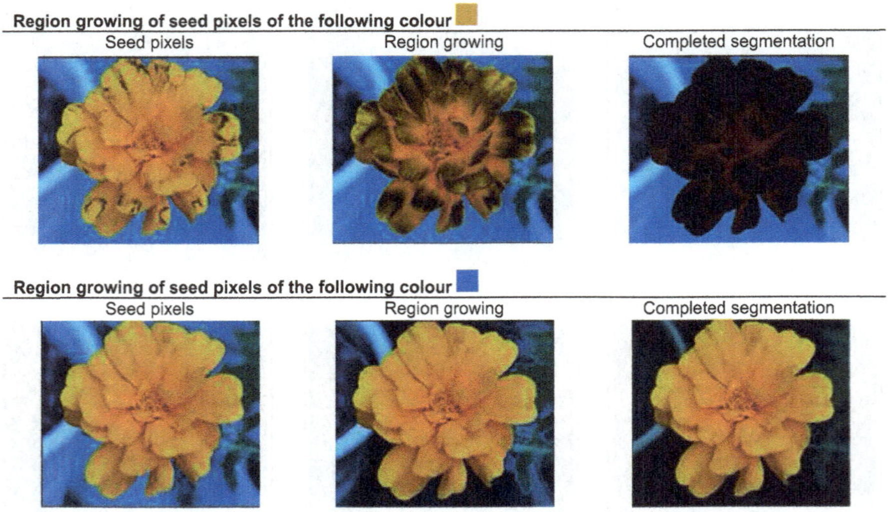

Fig. 3.7 Region growing approach using a single criterion each time, firstly by selecting the *yellow* flower petal colour, and secondly by selecting the background *blue* colour. The first criterion segments the flower petals, as shown in *black*, while the second criterion is based on the *blue* background pixels

region based segmentation is presented which involves finding the region of interest directly. Unlike the edge based methods that detect boundaries, region-based segmentation determines the regions directly working from the inside and growing outwards, instead of outside in. The difficulties lie in the criteria for region classification which are typically more difficult than edge detection methods. In addition it generally can't find objects that span multiple disconnected regions.

Region growing is a group of pixels or sub-regions collected together with some predefined criteria, based on greyscale intensity and/or edge information, in order to form a larger region. The process begins by manually defining *seed* pixels in a way that the local group of seed pixels grows by appending to each all pixels connected to the initial seed pixel that satisfies some predefined criteria (greyscale, texture, colour or shape) are included, thus the region grows. Such criteria include specific range of grey level (average intensity or variance), colour, or shape. Region growing is advantageous over edge detection in noisy images, where edges are difficult to detect.

An example of region growing is shown in Fig. 3.7 on a yellow flower with a blue background. To demonstrate the region growing we firstly define the seed pixels as those pixels matching a shade of yellow. The selected pixels are shown as a black colour where the initially grouped seeded pixels grow with connected neighbouring pixels until the entire flower petals are selected. In the second criterion, the background is segmented based on a selection of blue seed pixels. The region growing is able to be constrained within the blue background only.

The predefined criteria wield significant influence on the output. For example specifying pixels too similar gives good homogeneity and coherent regions, but it lends itself to *over-segmentation* where the region suffers from being smaller than the actual object and probably won't span over separate objects. More relaxed criteria can produce larger regions that fill the entire object but may lead to leaks across the boundaries of those objects and fill multiple objects unnecessarily. Selecting the criteria for DICOM images typically involves a threshold range on the pixel intensity, or to apply an average intensity over a neighbourhood around each pixel. For multiple regions, seed points should be selected within each of the desired regions since the seeded points grow through neighbouring pixels that are similar, and is expected to end at boundaries. The use of a histogram (Fig. 3.3) in this situation can help reveal the appropriate range of greyscale values that should be selected as seed pixels. Further readings regarding region growing segmentation can be found in Haralick and Shapiro (1985), Gibbs et al. (1996), and Kallergi et al. (1992).

Region splitting is an alternate method to region growing which begins by dividing up the whole image into disjointed sub-regions and then merges similar regions together. The splitting criteria continues as long as the properties of a newly split pair continue to differ from those of the original region by more than a threshold. Recursively splitting an image into smaller regions becomes more efficient than region growing which involves recursively merging individual pixels to produce larger coherent regions. The main problem with region splitting is deciding where to make the partition. One technique is to subdivide the image into smaller quandrants building a quadtree structure. This involves looking at an area of interest to determine if the region satisfies a set criterion. If so, then the area retains its region, if not, then the area is split into four equal sub-areas and each subarea is reconsidered. This process continues until no further splitting occurs. In the worst case the areas become so small that they are actually a single pixel. One can also allow merging of two adjacent regions if each has similar characteristics and satisfies the set criteria. Further reading can be found in Manousakas et al. (1998), Tremeau and Borel (1997) and Goldstein et al. (2010) (Fig. 3.8).

The **watershed algorithm** concept comes from the field of topography, referring to the division of landscape into water catchment basins or areas first proposed by Meyer and Beucher (1990b). The grey scale image is seen as a topological surface, where each pixel is a point situated at some height or altitude as a function of its grey level. Thus the image is visualised in 3D as a function of *x, y, g* defined by spatial coordinates (*x, y*), and greyscale values (*g*). A white colour (grey level 255) is taken as the maximum altitude and black colour (grey level 0) the minimum. Thus darkest pixels will exhibit deep regions, known as catchment basins, which are local minima. These are separated by ridges in the topology called watershed lines or simply watersheds (Fig. 3.9).

Two approaches may be used to extract the regions of interest. The first is to find catchment basins, then their enclosing watershed lines by taking a set complement. This is analogous to the topological surface being filled with water. The catchment basins fill up with water starting at these local minima. As the flood level rises,

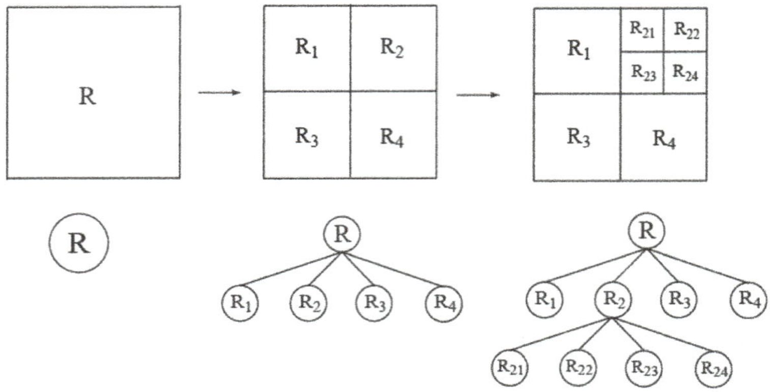

Fig. 3.8 A single region in an image that is progressively split into quad-regions when adjacent regions do not satisfies a set criteria

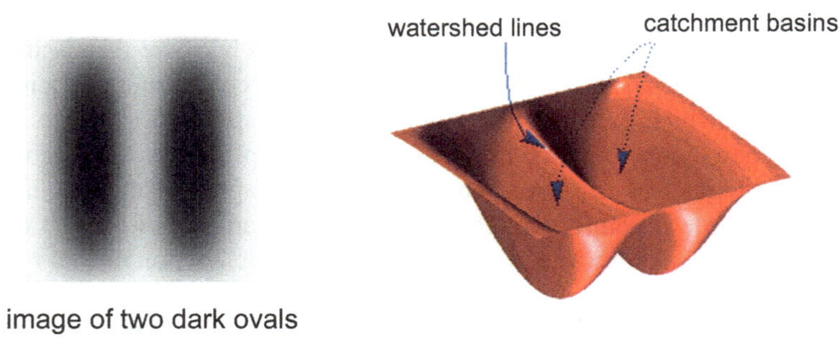

Fig. 3.9 Conceptual idea for watershed segmentation (Image adapted from *mathworks.com.au*)

boundaries between adjacent segments (catchment regions) erode and those segments merge. The other approach is to compute a complete partition of the image into basins, and subsequently find the watersheds by boundary detection (Fig. 3.10).

This section introduced the reader to some common algorithms for segmentation but is certainly not limited to these alone. For those interested in the field of image processing, the authors suggest the following textbooks and review articles that provide greater detail, Demirkaya et al. (2009); González and Woods (2011); Suri et al. (2002, 2005); Umbaugh (2005).

3.3.5 Using Specialised Medical Software

Segmentation algorithms presented in the previous section is a sample of what exists in an active research field. Published algorithms in the literature allow computer scientists, capable of advanced programming skills, to program and implement the

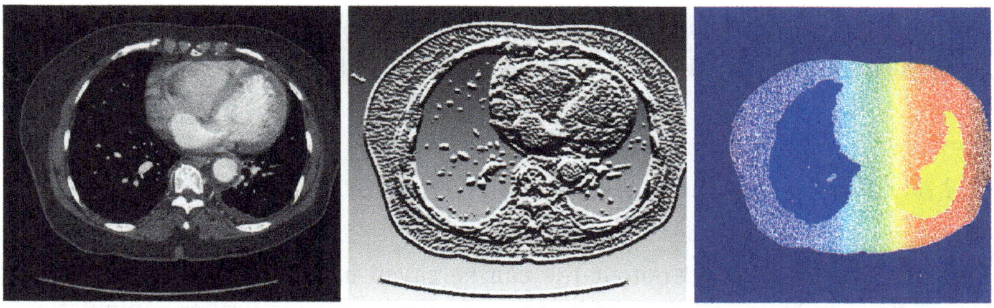

Fig. 3.10 A slice CT-scan of the aorta with the watershed segmentation algorithm applied

algorithms using computer languages such as C, C++, and Java. For most engineers and medical practitioners this step is not feasible without such skills. Fortunately a number of commercial and free specialised medical software exists, that converts the CT/MRI images and applies segmentation algorithms with an easy to use Graphical User Interface (GUI. A sample of what may be out in the public domain brief description of free open source software is given below:

Open Source/Freeware

- **CVIPTools** is a UNIX/Win32-based package and contains a collection of C and C++ computer imaging tools that includes edge/line detection, segmentation, and many other functions. (www.ee.siue.edu/CVIPtools)
- **Fiji/ImageJ** is a Java-based image processing package that uses additional plugins for a variety of functionalities including segmentation algorithms (*pacific.mpi-cbg.de/wiki/index.php/Fiji).*
- **GemIdent** is an interactive program that is designed for colour segmentation in images with few colours, and the objects of interest look alike with small variation (*gemident.com*).
- **ITK-SNAP** (www.itksnap.org) is an interactive software application that allows users to navigate three-dimensional medical images, manually delineate anatomical regions of interest, and perform automatic image segmentation.
- **Megawave 2** is made up of C library modules, that contains original algorithms written by researchers and is run using Unix/Linux (*megawave.cmla.ens-cachan.fr*).
- **MITK & 3Dmed** is made up of C++ library for integrated medical image processing, segmentation, and registration algorithms (www.mitk.net/download.htm).
- **Slicer** has a GUI that allows manual and automatic segmentation, registration, and three-dimensional visualization. It is a modular platform which means that it allows addition of new modules (www.slicer.org).
- **VXL** (*vxl.sourceforge.net*) is a collection of C++ libraries designed for computer vision research and implementation. The acronym stands for Vision-X-Library where the letter 'X' represents another letter related to the application. For ex-

ample the letter 'G' in place of the 'X' gives VGL which is the geometry library files and the letter 'I' gives VIL which is an image processing library.

3.3.6 Surface and Volume Reconstruction

After segmentation a 3D output file can be saved under different CAD formats which need to have certain conditions to be mesh-able for further computational fluid dynamics analysis. For example the outputted model may only contain a wire-frame model which lacks geometric topological relationships such as vertex, lines, faces, and the topology of the face perimeters or volume shells. A generic work flow is shown as a flowchart in Fig. 3.11. After segmentation the output file can be directly read into a CFD meshing software or additional topological data can be included into the model. An advantage of including the topological data is that the file becomes compatible with CFD software. During the addition of topological

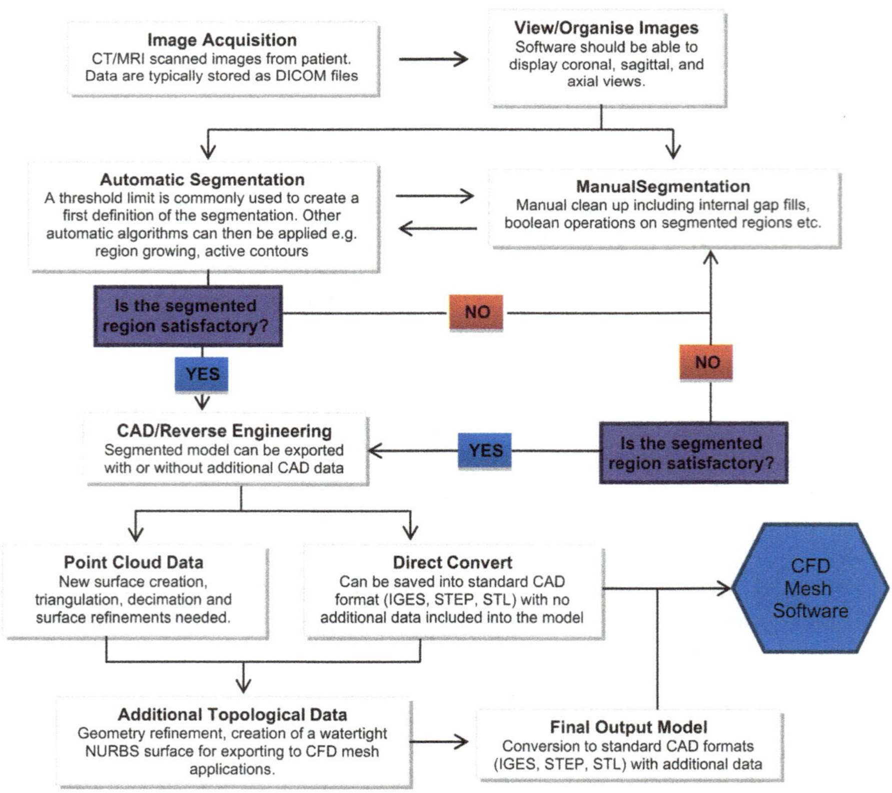

Fig. 3.11 Flow chart showing the segmentation process to extract and export a region of interest into a 3D computer model commonly referred to as a Computer-Aided-Drawing (CAD) model

data, the geometry can be mathematically described with a Non-Uniform Rational B-Spline (NURBS) for simple integration into a CFD-mesh software. This involves patching or overlaying a water-tight surface which holds point, line, and face data that are interconnected and relational. This is important since some CAD models will not be water-tight. The decision to directly output or apply a water-tight surface is dependent on the software, since a lot of CAD processing is self-contained and can be performed within the software itself.

It is important for the reader to appreciate the structure of 3D CAD models and its relationship to CFD meshing. The exported model should be in a format that can be read by the CFD meshing software. The format of the model serves as an intermediary between the segmentation software and the receiving CFD/CAD software. Therefore there is a reliance on the file format adhering to standards to allow smooth data exchange. However this is not always the case and there are variations in CAD formats among different software vendors. This causes loss of data and therefore loss of geometry details, leading to what is known as a '*dirty*' CAD geometry. Such *dirty* geometries exhibit stray points, non-intersecting lines, cross-overs, and slivers in the geometry. When this occurs, the user needs to clean up the geometry which can be performed within the CFD meshing program or external CAD-specific software. This is another reason for ensuring that the segmented model contains a water-tight surface.

The output file can be set as standard 3D CAD formats such as STL, IGES (Initial Graphics Exchange Specification), or STEP. As an example, the data structure for the above mentioned CAD formats is given below:

- *STL* or stereolithography, files are commonly used for rapid proto-typing manufacturing. It contains only surface geometry of a three dimensional object only. The surface is made up of small triangulated faces, called facets. Each facet is represented by three vertices (corners of the triangle) and its perpendicular direction. user. If too many triangles are created, the STL file size can become unmanageable. If too few triangles are created, curved areas are not properly defined and a cylinder begins to look like a hexagon (see example below).
- *IGES* (Initial Graphics Exchange Specification) is a platform independent format used to allow digital translation between CAD software. It contains two main data entities: geometry (to describe shapes such as curves, surfaces, solids and relationships) and non-geometry (to describe graphical or computational data such as colour, mass, time etc). The format is standardised to allow file interchange between different vendors. Unfortunately, CAD vendors have their own interpretation of the IGES standard, which renders some the information useless to others. Additionally, the size of an IGES file can be typically larger than other formats because of the two data entities it stores. Further information on IGES format can be found in www.steptools.com.
- *STEP*, is the acronym for Standard for The Exchange of Product model data which is a proposed International Standards Organization (ISO) standard for electronic data exchange. It is also known as the product data exchange using STEP (PDES) in the United States.

3.4 Examples

Reconstruction of the different cardiovascular structures can be performed using the generic flow process in Fig. 3.11. Segmentation algorithms form the core step in extracting the region of interest. However instead of presenting the same procedures in this section, alternative methods are presented.

3.4.1 *Abdominal Bifurcation*

The DICOM images of a CT scan from a 61-year-old female patient is shown in Fig. 3.12 in three planes of view: coronal, axial, and sagittal. The original CT scan images comprised of 955 contiguous slices with a slice increment of 0.625 mm. The scan image had a resolution of 512 × 512 pixels, with a pixel size representing 0.703 mm.

Thresholding segmentation is first applied by defining the minimum and maximum pixel greyscale values based on the Hounsfield scale (set to 168 and 596 respectively—see Sect. 3.2) to extract the artery passageway. The thresholding segmentation extracts the pixels within the defined limits and is coloured in green in Fig. 3.13. However this will also include additional side artery branches and other tissue that exhibit pixels within the threshold range. In such a circumstance, manual operation is needed to remove these additional pixels that are not part of the region of interest.

The final segmented volume is then extracted from the threshold segmentation. A preliminary three-dimensional surface geometry is reconstructed based on the exported CAD file as an *.stl* format. During the CAD conversion, a large amount of noise is present and the preliminary 3D geometry requires further refinements. The preliminary model and the refined model are shown in Fig. 3.14.

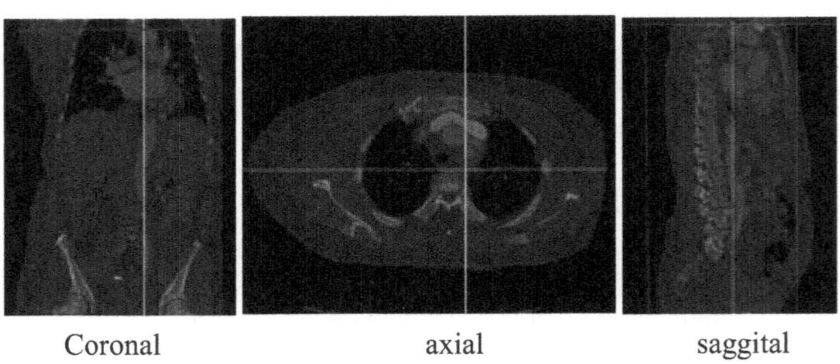

Fig. 3.12 Orientation assignment of the imported scanned images

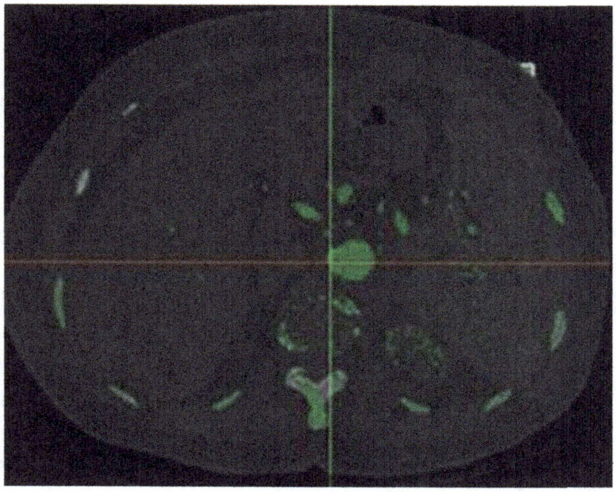

Fig. 3.13 Thresholding of interested abdominal aorta on one selected slice using the Hounsfield scale within the range of 168–596

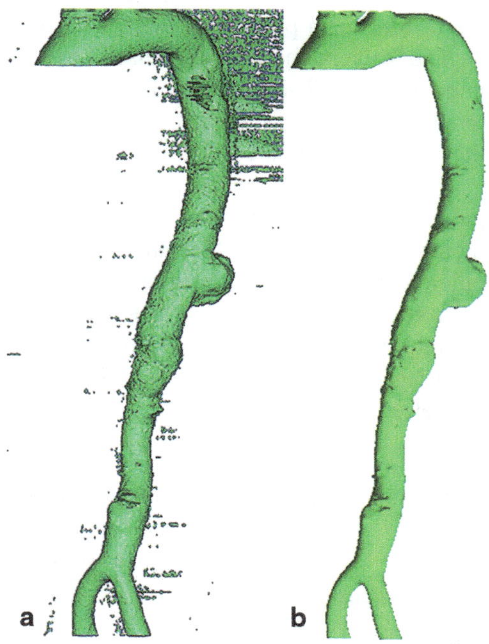

Fig. 3.14 Calculated three-dimensional geometry of the (**a**) preliminary model with the presence of noise, and (**b**) refined model removing unwanted voxels

3.4.2 *Left Atrium*[1]

MRI recordings of the left atrium and each of the pulmonary vessel stubs were acquired from a 25-year old healthy male on a Siemens Avanto 1,5 T system. A temporal resolution of 26 ms was achieved, resulting in 40 frames during one cardiac cycle. Images were acquired in the four-chamber and the short-axis orientation. A total of 14 slices with a slice thickness of 5 mm were needed in both orientations to cover the atrium and the mitral plane.

In-vivo measurements of the flow velocity through the mitral orifice and the mass flow rate through each of the venous inlets were performed by velocity phase mapping scans. The MR images were analyzed using the freely available software Segment v1.8 R1172. The anatomical geometry of the atrium was reconstructed using the automatic segmentation tool to outline the endocardial border in the short-axis view. By combining both image stacks (short-axis and four chamber view), the proposed contour in every short-axis slice was evaluated using the intersection points to the four chamber view. If necessary, the contours were refined using the manual dynamic contour editing tool.

The MRI showed that the subject had the typical pulmonary vein pattern consisting of four main pulmonary veins with four separate ostia. The location and inlet angle of each pulmonary vein relative to the left atrium were determined with fairly good accuracy. The veins are denoted left or right according to their position relative to the center of the body, and superior (upper) or inferior (lower) according to their distance from the head.

Hence, the name of the four common pulmonary veins are the right superior pulmonary vein (RSPV), left superior pulmonary vein (LSPV), right inferior pulmonary vein (RIPV) and left inferior pulmonary vein (LIPV). The orifice of each pulmonary vein was reconstructed by measuring the major and minor diameter of each vein in their respective MR scan (see Table 3.1). The measured values fall within clinical reported data (Fung 1993; Popel and Johnson 2005). Figure 3.15 shows an elliptic pulmonary vein orifice, the two lines indicate the minor and major diameters.

Table 3.1 Diameters and areas of pulmonary veins extracted from MRI scans

	Left superior pulmonary vein	Left inferior pulmonary vein	Right superior pulmonary vein	Right inferior pulmonary vein
Minor diameter [mm]	11.9	10.6	12.1	16.1
Major diameter [mm]	15.4	15.8	17.0	16.1
Area [mm^2]	113.8	131.2	156.4	203.1

[1] This section has been contributed to by **S. K. Dahl** in Dahl, S. K., Thomassen, E., Hellevik, L. R., Skallerud, B. (2012b). Impact of Pulmonary Venous Locations on the Intra-Atrial Flow and the Mitral Valve Plane Velocity Profile. Cardiovascular Engineering and Technology 3:269–281.

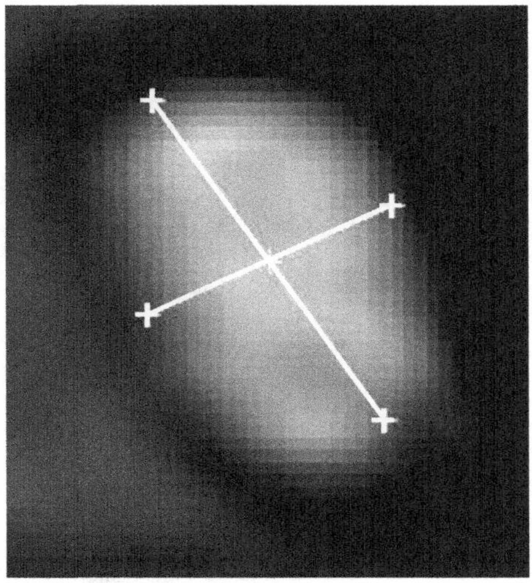

Fig. 3.15 A MR image showing an elliptic pulmonary vein orifice, the *two lines* indicate the minor and major diameters

The segmentation data was used to build a subject-specific atrial geometry. The first model, denoted c_A, has four PVs located in the anatomically correct positions as adopted from the MR recordings. The geometry of c_A is shown in Fig. 3.16a, and b. There is a close proximity between the ostia of the right pulmonary veins (RPVs) and between the ostia of the left pulmonary veins (LPVs), this is in accordance with findings in the literature (Bernaschi et al. 2013; Fung 1993; Popel and Johnson 2005). In general, and also in our subject, the entry locations of the left pulmonary veins are closer to the mitral plane than the right pulmonary veins (Fig. 3.16a). The left atrium appendage (LAA) lies adjacent to the ostium of the left superior pulmonary vein and is indicated with an arrow in Fig. 3.16b.

In order to examine the impact of the venous entry locations, two additional models were constructed; c_B and c_C. There are several possible entry locations for the pulmonary veins. We chose to keep the angle fixed between the pulmonary vein trunks, as seen from an atrial view, while the pulmonary veins vertical distance to the mitral valve plane were modified. The geometries of the atrial chambers were identical with c_A. Regarding the distance between the veins and the mitral valve, only studies on "The Isthmus Line" were found. The Isthmus Line is the distance between the lower border of the left inferior pulmonary vein ostium and mitral annulus. Schmidt et al. (2006) measured this line to be 28.5 ± 6 mm (range 17.3–40.5 mm), whereas Westerhof et al. (2010) found a mean value of 35 ± 7 mm (range 23–50 mm). The two studies illustrate the wide dispersion among subjects. In our subject this line was measured to be 30 mm. Based on the findings in the literature and in our subject we chose the following: In c_B, the left pulmonary veins are moved up to the same level as the right pulmonary veins, the isthmus line is then in accordance with the maximum distance measured in Schmidt et al. (2006). We assume the interpatient variability also applies for the right pulmonary veins, and thus in c_C the right pulmonary veins were therefore moved down and located at

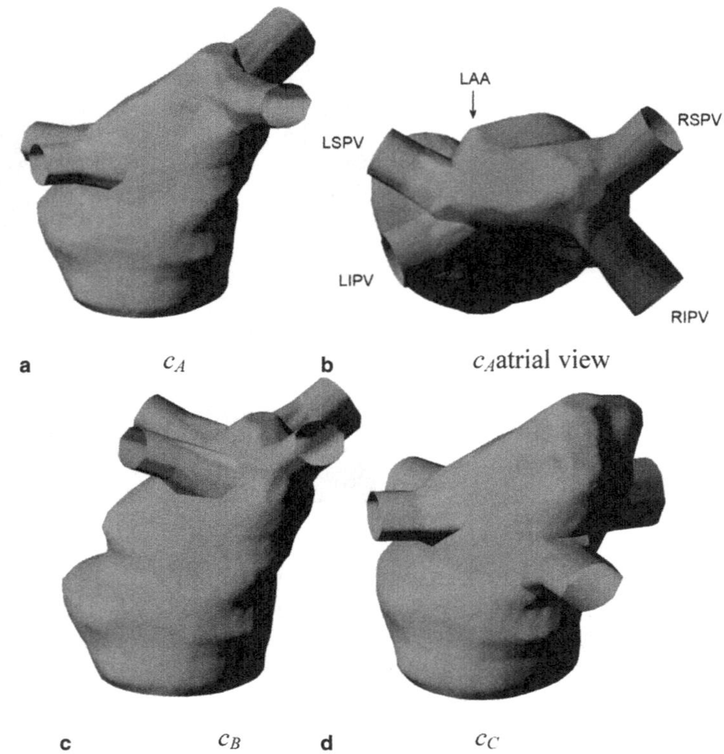

Fig. 3.16 Geometry of C_A from (**a**) an inferior, and (**b**) atrial view, whereas C_B and C_C are shown in (**c**) and (**d**), respectively. The name of the four veins and the location of the LAA are indicated in (**b**)

the same vertical position as the left pulmonary veins in cA. The geometry of c_A, is shown in Fig. 3.16a and b, whereas Fig. 3.16c and d, show c_B and c_C, respectively.

3.4.3 Left Ventricle[2]

In this example real-time 3D echocardiography (RT3DE) is used to record moving structures of the left ventricle during a cardiac cycle. The 3D echocardiography left ventricle volume of a 30 year-old female volunteer was acquired using a Vivid E9 scanner using a 3V matrix probe with a centre frequency 2.4 MHz (GE Healthcare

[2] This section has been contributed to by **S. K. Dahl**, in Dahl, S. K., Fagerholt, E., Kiss, G., Prot, V., Amundsen, B., Hellevik, L. R., Skallerud, B. (2011). 3D moving boundary conditions for heart CFD simulations—from echocardiographic recordings to discretised surface. MekIT'11: Sixth National Conference on Computational Mechanics:33–46.

Vingmed, Horten, Norway). The volume was acquired during apnea over 4 heart cycles, from the apical window, in harmonic mode, one QRS triggered sub-volume acquired per heart cycle. The frame rate was 27 per cycle.

The endocardial border was generated using the AutoLVQ tool (Hansegard et al. 2009), EchoPAC workstation (version BT 11), GE Vingmed Ultrasound, Hortem, Norway. AutoLVQ represents the left ventricle boundary as a deformable model and relies on 3D energy minimization for evolving it. A combination of internal, external and temporal forces ensures shape continuity, while adapting the model to a particular 3D echo recording. The endocardial contour process was initialized by manual positioning of the apex and the mitral valve attachment points in a long-axis view (e.g. four chamber), both at end-diastole and end-systole. After manual selection, the endocardial border is automatically generated throughout the cardiac cycle. The proposed contour was then evaluated in both short and long-axis cut-planes of the 3D volume. If deemed necessary the border can be further refined by adding additional attractor points that pull the model towards the endocardium. In this case, the border was adjusted by placing a limited number of attractors. The papillary muscles and major trabeculae were included in the left ventricle cavity both in diastole and systole.

The segmented endocardial left ventricle wall derived from RT3DE consisted of closed three dimensional surface meshes for 27 different time steps. Together they describe the inner ventricular wall movement. The algorithms were constructed in such a way that user intervention was necessary only in the first frame, i.e. start of systole.

To create our computational transient 3D model we have to include a model of the mitral valve and the left ventricular outflow tract into the segmented left ventricular surface mesh for all time frames. The mitral valve and left ventricle outflow tract model was reconstructed from the same RT3DE data as the ventricle. Figure 3.17 shows the geometry of the complete 3D model at start- and end-systole.

To illustrate the correlation between the model and the echocardiographic recordings, the model was realigned with the original RT3DE data as shown in Fig. 3.18. The left ventricle was placed back at the same position as when it was segmented in AutoLVQ software (Fig. 3.19).

3.5 Summary

Reconstruction of the cardiovascular anatomy begins with obtaining the relevant data from medical imaging. While a number of imaging modalities exist this chapter presented CT and MRI based modes. Regardless technique, the scanned images typically form 2D contiguous slices that are separated by a known interval distance. The reconstruction of this data from 2D to 3D requires extraction of the region of interest in what is known as the segmentation process. A number of algorithms have been developed by researchers and biomedical scientists which can be accessed through either open-source/free software or commercial software. Access

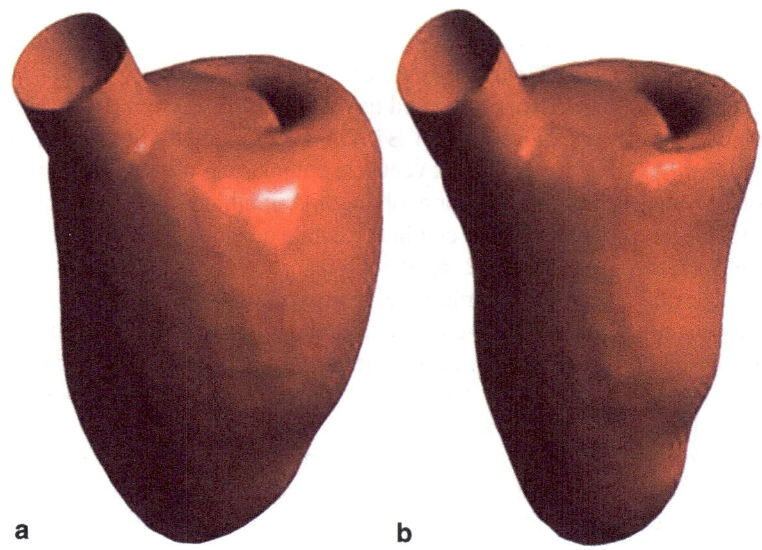

Fig. 3.17 (**a**) Geometry of the complete 3D model at (**a**) beginning of systole stage and (**b**) end of systole

Fig. 3.18 3D model realigned with the original RT3DE data at peak systole

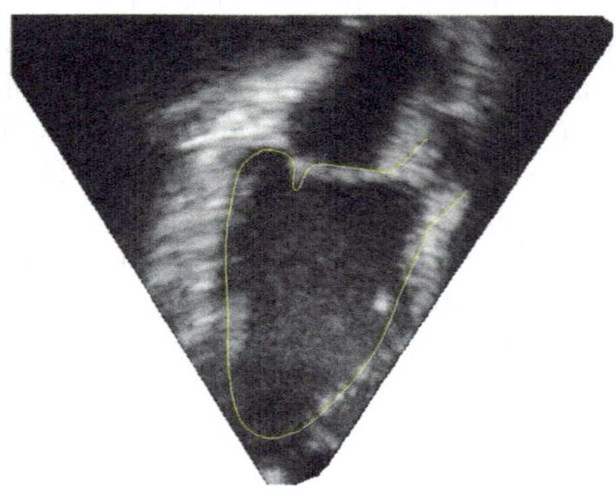

to the algorithms is available either through a graphical user interface or command line text. After segmentation, the extracted region typically needs to be 'cleaned' for any noise or artificial regions that are not needed. Sometimes the model may need further processing in CAD software. An important step before moving to a CHD mesh model is to ensure that the segmented model has a water-tight surface. This chapter aimed to bring the reader up to speed with techniques of reconstruction that cross over into different disciplines including biomedical imaging, manufacturing/ reverse engineering, and CAD/drafting fields. The content presented is an overview of some theory and its application and is certainly not an exhaustive description what exists for the disciplines mentioned. However it is hoped that the reader has gained an appreciation for the cross-discipline nature of CHD research.

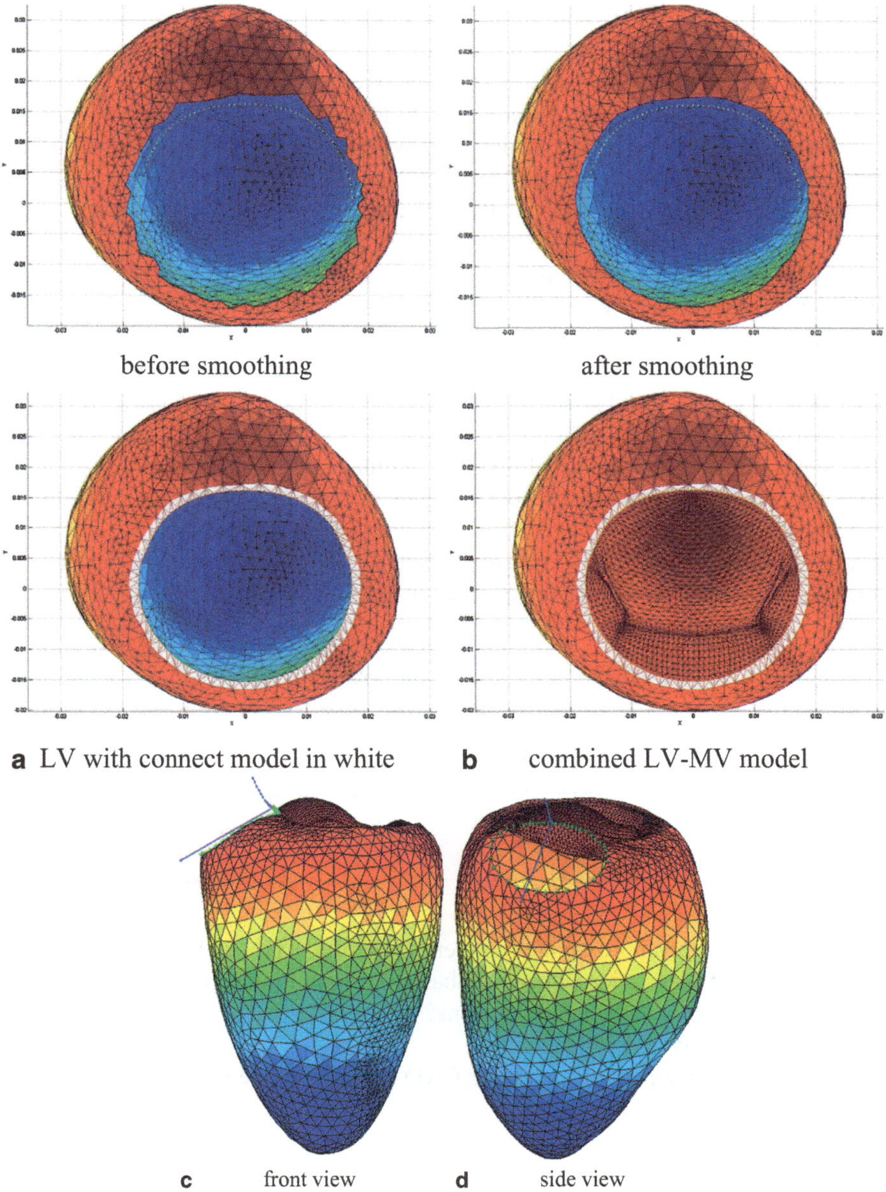

Fig. 3.19 Left ventricle (LV) edge before and after smoothing. The *dotted line* is the contour of the mitral valve (MV) annulus. The manually positioned MV on the LV base in the start frame, i.e. start systole. (**a**) illustrates the LV model with its manually generated connect model in *white* and the dotted MV annulus. (**b**) Combined LV-MV model from the top and the side, respectively. (**c**) and (**d**) show the LV model with the smoothed edge (*green dots*). The *blue line* is the vector used to obtain the cutting plane. The tracing of the aorto-mitral-curtain used in the construction of the LVOT is illustrated with *blue dots*

3.6 Review Questions

1. Why are the scanned images from CT referred to as *'slices'*?
2. What is the significance of water content in the body for MRI scans?
3. For the following anatomy, which scan (MRI or CT) would you use: soft tissue, bone structure, mucous walls.
4. For the respiratory airway, what are the main issues to consider when using MRI and CT scans?
5. What is the Hounsfield scale?
6. What is the difference between a pixel and a voxel?
7. What does manual segmentation involve?
8. What is the problem with using a fully automated segmentation?
9. What is the threshold greyscale range? What are you looking to segment if you had a range of 2000–3072 HU?
10. What is the main property of pixels in an image where an edge occurs?
11. Why do images that use the LoG and Canny edge detection schemes need to be smoothed?
12. What does the term '*zero crossing*' mean?
13. The Roberts edge detection algorithm uses the following mask:

Prewitt, G_x

-1	0
0	1

Prewitt, G_y

0	-1
1	0

What would the approximate first derivative of a pixels intensity be?

14. Give a brief description of a region-based segmentation algorithm. How does it differ from edge detection algorithms?
15. What is a 'dirty' CAD geometry?
16. Why is it problematic in accessing CAD models between different software?

Chapter 4
Fundamentals of Haemodynamics

4.1 Introduction

Haemodynamics is the study of the properties of blood and how it flows. Its physical principles can be described by the same fluid flow equations that derive from classical fluid dynamics however closer inspection of the dynamic nature of blood reveals its complex nature. It is a liquid tissue that under fluid motion necessitates principles of modern fluid dynamics such as rheology and fluid structure interactions to describe its flow behaviour. This chapter introduces both classical principles and modern aspects of fluid mechanics, while Chap. 5 introduces principles of fluid structure interactions.

4.2 Fluid Properties of Blood

Blood is comprised of formed elements (40–55 %) suspended in plasma (45–60 %) (Fig. 4.1). The formed elements are mostly made up of red blood cells called erythrocytes (approx. 99.3 %) > 0.3 % of the formed elements are made up of white blood cells (leukocytes), and platelets (thrombocytes). Plasma is mainly comprised of water and functions to dissolve and transport solutes such as proteins, nutrients, organic waste and ions.

The main fluid properties of blood that govern its flow behavior are viscosity and density. The **viscosity** of a fluid is material-dependent and describes the fluid's resistance to deformation under shear force. Physically, this means that the property of viscosity is a measure of any liquid's resistance to flow. It describes the internal friction between fluid particles as they slide past one another. A fluid with high viscosity resists motion because of high internal friction, while a fluid with low viscosity flows easily with little internal friction, e.g. the flow of honey has a high viscosity compared with the low viscosity of the flow of water.

In a homogeneous fluid, such as water, viscosity remains constant when a change in velocity occurs. Fluids exhibiting constant viscosity during changes in shear rate

J. Tu et al., *Computational Hemodynamics—Theory, Modelling and Applications*,
Biological and Medical Physics, Biomedical Engineering,
DOI 10.1007/978-94-017-9594-4_4

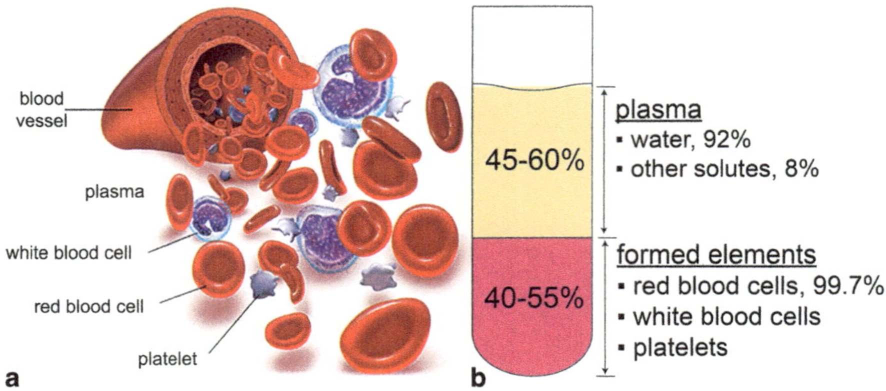

Fig. 4.1 Blood is a liquid tissue made up of blood cells and platelets

(e.g. change in velocity) are known as a Newtonian fluid. Blood is a heterogeneous fluid containing blood cells that change in shape and movement within the plasma during velocity changes. This leads to a variable viscosity making blood a non-Newtonian fluid (this is discussed in more detail in the next section). In many haemodynamic studies, blood viscosity is assumed constant (typically 3.2×10^{-3} Pa.s) because only small changes in the blood are found throughout a given study and this change is negligible.

The viscosity of a fluid is quantified by the dynamic and kinematic viscosity. The dynamic viscosity, μ and has standard units of Newton seconds per square metre, $[Nsm^{-2}]$ or kilograms per meter per second $[kgm^{(-1)}s^{(-1)}]$. The kinematic viscosity, ν, is the ratio of dynamic viscosity to the fluid density $\nu = \mu / \rho$ which has units metres squared per second $[m^2 s^{(-1)}]$.

The **density** of blood is approximately 1060 kg/m^3 at 37 °C and is related the volume of red blood cells in blood. Only to a minor extent is the density influenced by other plasma solutes (Kenner 1989). Comparatively the density of water is 998 kg/m^3 at 20 °C and the density of air is 1.22 kg/m^3 at standard atmospheric pressure and temperature.

In general the density of a fluid changes with temperature and pressure, but this is typically small for liquids and therefore is often assumed constant despite any variation in temperature and pressure. To exemplify this, we consider a fluid's compressibility which is the fractional change in volume (m^3) per unit increase in pressure or temperature. A volume change will alter its density (since its mass remains the same); however for a liquid this is typically involves a change of 0.001 m^3 per 1 kPa of pressure change and 0.00001 m^3/1 °C change in temperature. Such small volume changes have a small effect on its overall density.

4.3 Viscosity of Blood

In general the viscosity of blood is a function of plasma viscosity (which is relatively constant) and viscosity of the formed elements (not constant). When the velocity is constant, blood behaves like a Newtonian fluid since the blood cells do not deform. But this is not the case for large velocity changes where blood cells deform and move in different proximity to the vessel walls, both of which influence its viscosity. Viscosity of blood is influenced mainly by the red blood cells and to a lesser extent, blood vessel diameter, shear rate, and temperature.

Red blood cells occupy 99.7 % of all formed elements, and are a major determinant for the difference between plasma and blood viscosity. The remaining volumes contain white blood cells and platelets which contribute 0.16% and 0.12 % of the total formed elements, respectively (Popel and Johnson 2005). The percentage of whole blood occupied by the formed elements is called the *haematocrit*. Since red blood cells occupy almost the entire formed elements, the haematocrit is commonly reported as the volume of packed red cells. In adult males the average haematocrit value is 46 (range: 40–54), while for females it is 42 (range: 37–47).

The viscosity of plasma is $\mu = 1.2 \times 10^{-3}\,\mathrm{kg \cdot m^{-1}s^{-1}}$ and this increases when red blood cells are considered. The size, shape, and flexibility of red blood cells are influencing factors to its viscosity. The viscosity of whole blood at hematocrit of 45 % is approximately $\mu = 3.2 \times 10^{-3}\,\mathrm{kgm^{-1}s^{-1}}$ about 4–5 times of water. A change in haematocrit means that the volume occupied by the red bloods changes and this has a significant effect on viscosity; one unit increase in haematocrit can cause up to a 4 % increase in blood viscosity (Baskurt et al. 2007). Figure 4.2a highlights the influence of haematocrit levels on viscosity which shows a non-linear relationship.

When exposed to increasing shear, the viscosity decreases. The rate of shear (i.e. rate of deformation) of a fluid flowing between the centre of a blood vessel and its walls is defined by

$$\gamma = \frac{u}{h} \tag{4.1}$$

where γ is the shear rate, measured in reciprocal seconds; u is the velocity of the fluid; and h is the distance from the vessel wall. Increased shear rates occur when there are high velocities in the vessels. This causes red blood cells to move towards the centre of the vessel and orient in the flow direction, reducing its viscosity. For low velocities, the viscosity is highest as the red blood cells become more evenly distributed throughout the vessel. For extremely low velocities, red blood cells can aggregate, which increases the viscosity dramatically. The influence of shear rate is shown in Fig. 4.2b.

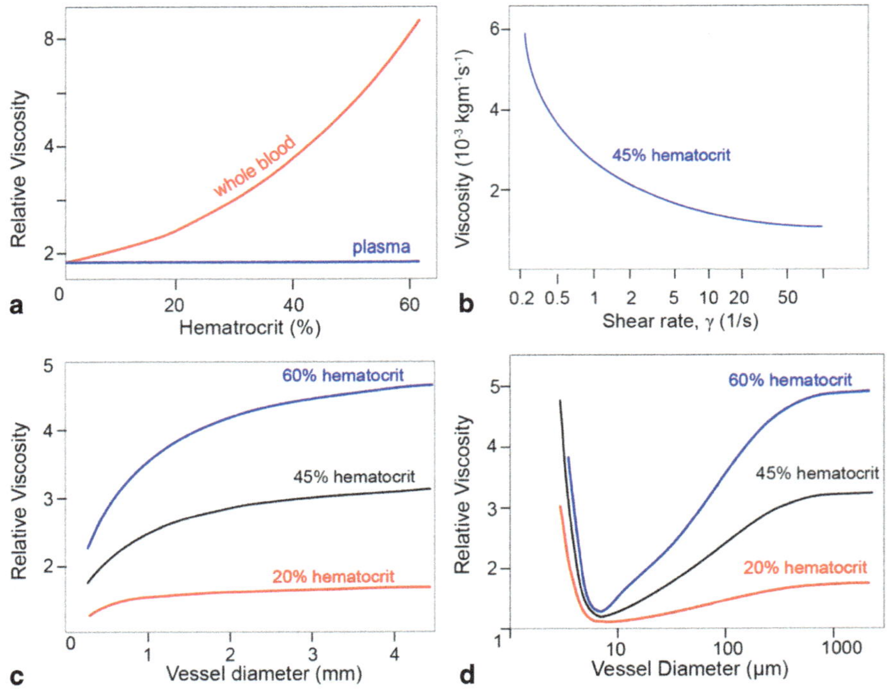

Fig. 4.2 Relative viscosity is taken as the ratio of the viscosity of blood to that of water at 37 °C, $\mu_{water} \approx 0.7 \times 10^{-3} kgm^{-1}s^{-1}$ **a** Relative viscosity as a function of haematocrit fraction. **b** Relative viscosity as a function of the vessel diameter for different haematocrit fractions. **c** Dynamic viscosity as a function shear rate for blood with 48 % haematocrit fraction. Data for figures from (Bernaschi et al. 2013; Pries et al. 1992; Westerhof et al. 2010)

Blood flows continuously through the network of vessels in the cardiovascular system. These vessels vary in size from the smallest vessels in the capillary network to the large heart chambers. The viscosity is also influenced by vessel diameter due to shear stresses caused by its walls. The viscosity decreases gradually for a vessel diameter decreasing from 3 mm down to 1 mm (Fig. 4.2c). In these vessel diameters, the effects on red blood cells are not significant and the viscosity remains relatively constant. Blood can then be considered *as a homogeneous fluid with Newtonian properties in larger arteries.*

When the vessel diameter is in the micron range, there is a rapid decrease in the viscosity with decreasing vessel diameter from 1000 μm down to 10 μm. In these small vessels, the red blood cells move towards the centre of the vessel, leaving the plasma to remain close to the walls. This reduces the blood viscosity and this effect called the "Fahraeus-Lindqvist effect" extends down to vessels with diameters of about 7 μm (Fig. 4.2c). At the smallest vessels the inner diameter is about the same size as the red blood cells, ranging from 3 to 8 μm (Popel and Johnson 2005; Pries et al. 1992). When blood flows through these vessels, red blood cells deform considerably causing microcirculation and increasing viscosity as the vessel diameter approaches the theoretical minimum blood cell diameter of 2.7 μm.

Finally like all fluids, (e.g. water) viscosity is dependent on temperature. A decrease of 1 °C in temperature yields approximately a 2% increase in the viscosity of blood. Thus in colder regions of the body such as our fingertips, the viscosity of blood is higher than in that found in the brain. Furthermore in a case of hypothermia where the temperature is very low, the blood viscosity increases impeding on blood flow.

Representing blood as a non-Newtonian fluid can be achieved by using the blood apparent viscosity power-law,

$$\mu = ke^{T_0/T}S^{n-1} \tag{4.2}$$

where $k=0.00622$ [kg s^{n-2}/m] is the consistency index, $n=0.7$ is the powerlaw index, T [K]and T_0 [K] are the local and reference temperatures, respectively, and S is the shear rate.

4.4 Clinical Relevance of Blood Viscosity

Haematocrit levels and its corresponding viscosity provide insight to the health implications of a patient. Abnormally high haematocrit levels of 60–70% are called polycythaemia where there is increased oxygen content and increased viscosity. This leads to an increased resistance to blood flow, straining the heart to work harder and can impair organ perfusion. Similarly athletes that perform blood doping by injecting packed red blood cells, or illicit products such as erythropoietin (EPO), increase their haematocrit to increase oxygen delivery. This elevates blood viscosity and increases the workload on the heart. Conversely low haematocrit <39% leads to anaemia which produces decreased oxygen and decreased viscosity. Some common causes include trauma leading to loss of blood, bone marrow disorders, dehydration, lung disease and the abuse of certain drugs.

Increased viscosity has been linked with major cardiovascular risk factors including high blood pressure, cholesterol, obesity, smoking, and age (Lowe et al. 1997; Sloop 1996). Blood vessels affected by these cardiovascular risk factors hinder natural blood flow. When blood flow slows down, viscosity increases as a result of more cell-to-cell and protein-to-cell adhesions. Blood cells start to bind together, forming clumps that thicken the blood, potentially clotting the vessels. Since viscosity describes the thickness or stickiness of a fluid, it also describes the resistance of blood to flow. This leads to a measure of how hard the heart has to work to delivery blood to the body, a key measure of cardiovascular performance.

4.5 Blood Flow Properties

Blood flow through the cardiovascular network of vessels is essentially a form of internal flow since the fluid is wall-bounded. Within this class of flow, a number of flow properties exist including velocity, pressure, and shear stress. Furthermore the blood flow can be steady or unsteady, laminar or turbulent, or in linear or curved arteries. Each of these fluid dynamics properties is discussed.

4.5.1 Shear Force

Unlike solids, a fluid cannot resist an applied force and instead it reacts by deforming continuously under the action of the force, while its viscosity tries to resist the deformation. When blood is moving across the non-moving artery, a resultant shear stress occurs due to the fluid particles moving past one another at different velocities, causing a friction-like effect. If its velocity is the same, and all fluid particles move together then no shear stress is found. To understand shear stress we consider a 3D rectangular fluid element next to a stationary vessel boundary wall shown in Fig. 4.3. The fluid velocity is zero at the wall surface, but increases in velocity away from the wall. Each layer of fluid moves faster than the one below it, creating friction between them. This friction acts as a force resisting the motion, which is the viscosity.

Experimental observations made by Sir Isaac Newton showed that for certain fluids under laminar flows, the shear force is proportional to the velocity u, and the area A, and inversely proportional to their distance from the wall y,

$$F = \mu A \frac{u}{y} \tag{4.3}$$

where the proportionality factor is the dynamic viscosity μ. Therefore fluids that behave according to Eq. 4.3 are known as Newtonian fluids. Earlier we saw that for

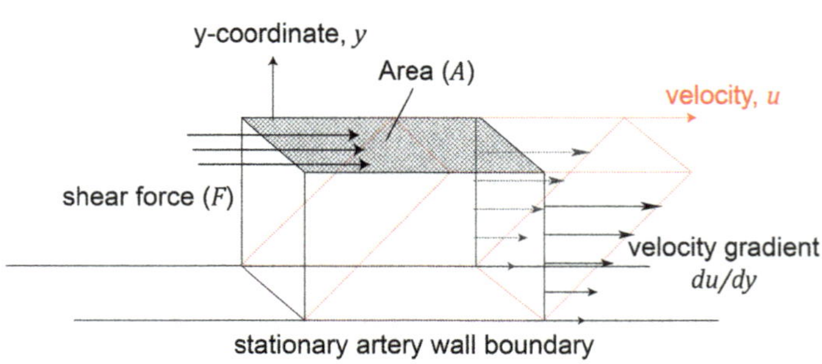

Fig. 4.3 Shear stress and velocity gradient arising from an applied force deforming a rectangular fluid element

blood flow in large arteries, its behaviour is indeed Newtonian. The ratio u/y is the average velocity change over the distance y and in differential terms this is written as a derivative, du/dy which represents the fluid velocity in the direction perpendicular to the fluid element. Furthermore stress is defined as a force per unit area, and therefore the shear stress, denoted by by tau, τ is defined as

$$\tau = \frac{F}{A} = \mu \frac{du}{dy} \tag{4.4}$$

Equation 4.4 is known as Newton's law of viscosity. Sometimes a negative sign is used so that $\tau = -\mu du / dy$ if the coordinate system uses the origin point as zero at the centre of the pipe which corrects the equation to allow a positive shear stress for a velocity gradient that is consequently negative, i.e. $du / dy < 0$.

4.5.2 Pressure Force

Pressure in fluid flow is a type of stress similar to shear stress but the force acts inwards and perpendicular to the surface of a fluid body. It is defined as a force per unit area with standard unit of Pascal which is a Newton per square meter. Pressure values in haemodynamics are normally reported as gauge pressure which is its value relative to local atmospheric pressure (atmospheric pressure = 101325 Pa). This means that gauge pressure is zero when its pressure is equal to atmospheric. Figure 4.4 is a schematic demonstrating the pressure force that is exerted on the surface of an arbitrary fluid body and a rectangular fluid element. For the rectangular fluid element, each of its six faces has a pressure force that acts normal and inward on the face.

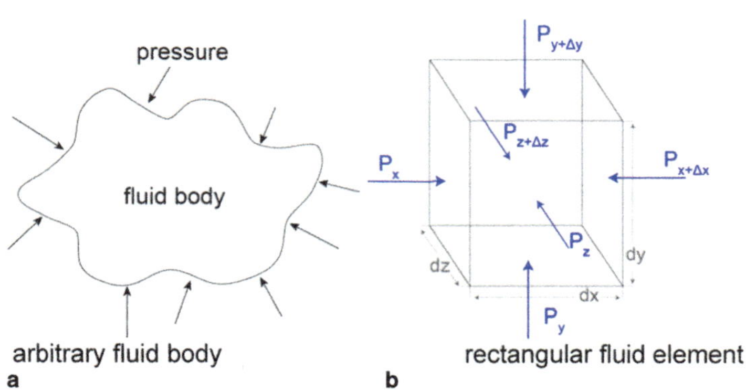

Fig. 4.4 Pressure force acting on an arbitrary body and a rectangular fluid element **a** arbitrary fluid body, **b** rectangular fluid element

Pressure is defined as a force per unit area, and if we consider the pressure force in the x-direction then

$$P_x = \frac{F_x}{A_x} = \frac{F_x}{dy \cdot dz} \qquad \text{which gives} \qquad F_x = P_x(dy \cdot dz)$$

The change from P_x *to* $P_{x+\Delta x}$ is based on the rate of change of P in the x-direction multiplied by the distance of which the rate of change takes place. This is given as

$$P_{x+\Delta x} = P_x + \frac{dP}{dx}dx \quad \text{and since} \quad F_x = P_x(dy \cdot dz) \quad F_{x+\Delta x} = \left(P_x + \frac{dP}{dx}dx\right)(dy \cdot dz)$$

The net pressure forcein the x-direction is then

$$F_{x+\Delta x} - F_x = \left(P_x + \frac{dP}{dx}dx\right)(dy \cdot dz) - P_x(dy \cdot dz) \tag{4.5}$$

Dividing both sides by the volume $dxdydz$ gives

$$\frac{F_{x+\Delta x} - F_x}{dx \cdot dy \cdot dz} = \overrightarrow{f_x} = \frac{dP}{dx} \tag{4.6}$$

This means that the net pressure force per unit volume is equal to the pressure gradient which suggests that the pressure gradient creates a net force to produce fluid flow; and fluid flows from a higher pressure value to a lower pressure value. In Fig. 4.5 the fluid element is moving from left to right where the pressure acting

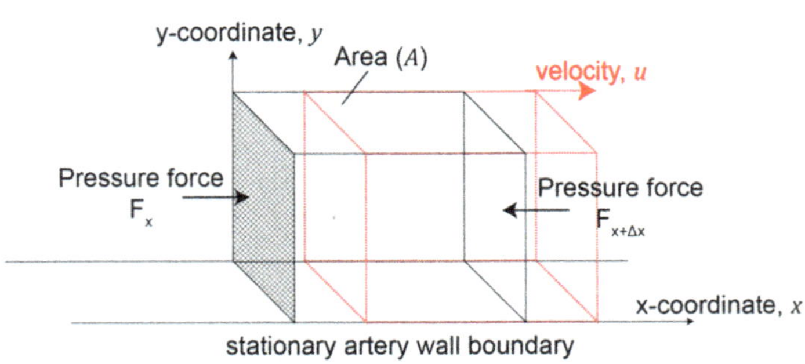

Fig. 4.5 Net pressure force in the x-direction which drives the flow from *left* to *right*. In this case the pressure is decreasing from *left* to *right*, and is a case of a negative pressure gradient

by F_x is greater than the pressure acting by $F_{x+\Delta x}$. The pressure gradient for this case is a negative value (e.g. $-dp/dx$) since it is decreasing from left to right. This is given the term "favourable pressure gradient" in fluid dynamics. Conversely if pressure is increasing from its initial position on the left to its latter position on the right, then the pressure gradient has a positive value (e.g. $+dp/dx$) and is given the term "adverse pressure gradient". If the fluid is static, then the pressure on opposing faces is the same and the pressure gradient is zero.

4.5.3 Laminar and Turbulent Flow

Let us imagine that fluid is made up of individual clusters of fluid particles moving, sliding, and mixing together. If the fluid moves sufficiently slow without much mixing between fluid particles, then the flow is considered laminar. If there is intense and chaotic mixing between the fluid particles then the flow is considered turbulent. To what extent of mixing do we consider whether a fluid is laminar or turbulent? For flow in pipes the dimensionless Reynolds number, Re is used to help establish this flow regime. The Reynolds number is a ratio of the inertial force, produced by the flow in motion, to the viscosity of the fluid.

$$\mathrm{Re} = \frac{\text{inertia force}}{\text{viscous force}} = \frac{\rho U D_h}{\mu}$$

where ρ is density, D_h is the hydraulic diameter, U is the fluid velocity, and μ is the viscosity. Physically the inertia force can be viewed as the ability of the fluid to move and deform freely while the viscosity acts to hold the fluid together, resisting any deformation. The inertia force is contributed by larger blood vessels, fluid density, or high flow velocity (ρ, U, D_h) which leads to increased fluid movement and high energy flows that are turbulent. A greater viscosity will hold the fluid together preventing significant deformation, leading to laminar flow behaviour. If we assume arteries as circular cylinders, having diameter of D_h, then the flow regime is indicated by the following Reynolds number ranges

- $\mathrm{Re} < 2000$ the flow is laminar
- $2000 < \mathrm{Re} < 4000$ the flow is transitional
- $\mathrm{Re} > 4000$ the flow is turbulent

The Re number range that classify whether a flow is laminar or turbulent also depends on any disturbances and roughness that are present in the vessel and flow. For example the presence of rough vessel walls or a stenosis can disrupt the orderly and regular flow pattern of an otherwise laminar flow, and trigger the flow into turbulence at lower Re numbers than expected.

If we were to inject dye into the middle of a flow stream in a pipe, and observe its path, the dye streak for each flow regime would look like those in Fig. 4.6. The

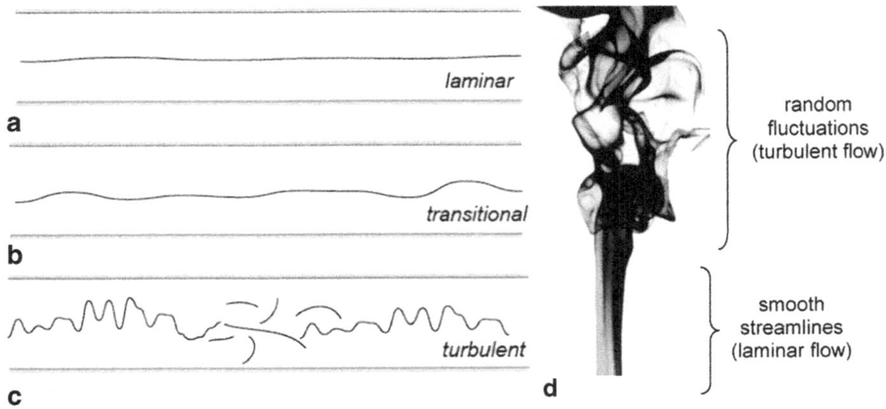

Fig. 4.6 Streaklines for laminar, transitional and turbulent flows

dye streak for a laminar flow is characterised by a smooth relatively straight path as the fluid particles pass over each other without little mixing. The effect of the fluid viscosity dominates the flow 'holding' the fluid together in its layered formation. The dye streak for a turbulent flow is characterised by the random fluctuations, irregular and chaotic motion, and local regions of vortices. We may even observe a breakdown of the dye streak because of such intense mixing.

The inertia of the fluid flow (i.e. its momentum) dominates, enabling transport of fluid particles across many layers and regions. The dye streak is an irregular path with no observable pattern. The presence of turbulence increases energy diffusion, mixing, dissipation, and heat transfer. Many flows found in practice are turbulent in nature such as smoke rising from a cigarette which starts of as a laminar flow and transitions to turbulence as it progresses.

4.6 Introduction to Internal Pipe Flow

Blood flow through the circulatory system occurs primarily in vessels that essentially form a network of pipes. It is for this reason that an introduction to internal pipe flows is important.

4.6.1 Developing and Fully Developed Regions

If we consider fluid flowing with a constant velocity, then its profile is uniform throughout. As the fluid enters a pipe its velocity profile exhibits a change starting from the wall to the centre of the pipe (Fig. 4.7). Fluid particles at the wall inherit a zero velocity due to the stationary wall, referred to as the "no-slip" condition. Because of the fluid viscosity, wall adjacent fluid particles are significantly influenced

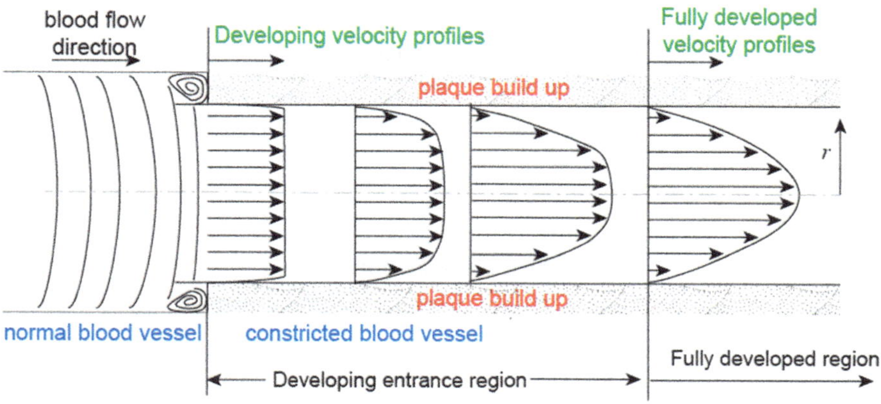

Fig. 4.7 Fluid velocity profile development in a pipe for a laminar flow. The developing region shows a change in the velocity profile shape but in the fully developed region, the shape becomes constant. The parabolic profile shape is characteristic of a laminar flow. For turbulent flows the profile would be a *blunt*, *flatter* shape

by the zero velocity at the wall. This influence continues to consecutive adjacent fluid particles but with a diminishing value. The fluid continues downstream caused by the pressure force pushing the fluid through, while the fluid viscosity tries to hold the fluid together.

These conditions give rise to a velocity profile having a parabolic shape from the wall to the centre of the pipe. The velocity profile development finally reaches a fully developed stage where the profile remains constant as it travels further downstream. The length it takes to become fully developed is referred to as the entrance length. Kays and Crawford (1993) reported the entrance lengths for laminar flow in pipes as

$$L_{e,lam} \cong 0.05\text{Re}D \tag{4.7}$$

For turbulent flows in pipes Bhatti and Shah (1987) reported the entrance lengths as

$$L_{e,turb} \cong 1.359\text{Re}^{1/4} D \tag{4.8}$$

Entrance regions are important if its distance is comparable to the total pipe length distance. This is because the pressure drop is higher in the entrance regions of a pipe and this influences the overall average friction factor for the entire pipe. However if the entire pipe length is several times the length of the entrance length, then the developing entrance length can be ignored and the entire pipe can be assumed to be fully developed.

Earlier we saw that a shear stress arises when there is a velocity gradient in a viscous fluid. In pipe flows where the velocity profile is not uniform, a shear stress

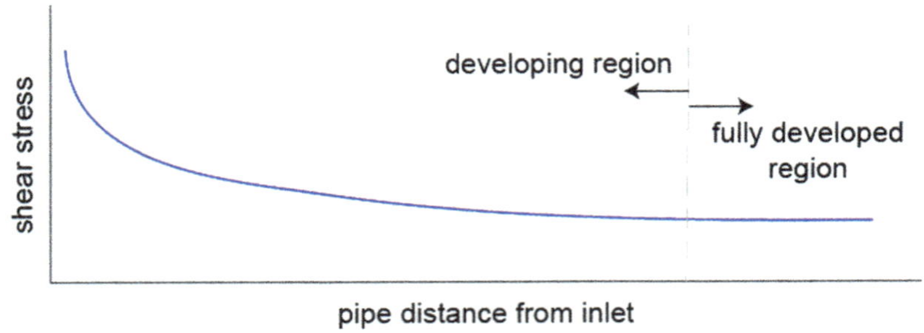

Fig. 4.8 Variation of wall shear stress in a pipe in the developing and fully developed region

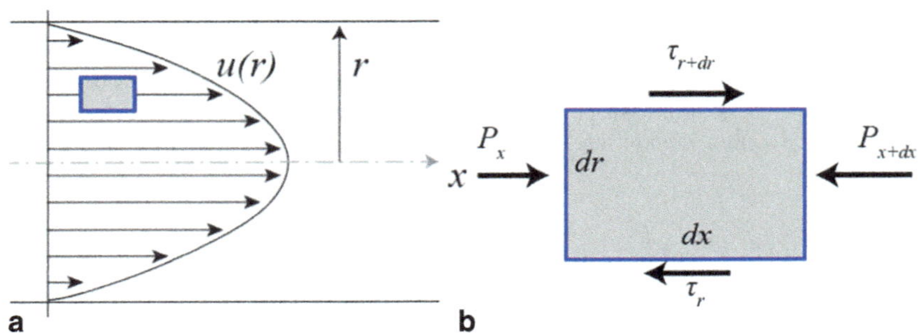

Fig. 4.9 **a** Schematic of fully developed laminar velocity profile with, **b** differential fluid element

acts on the pipe walls τ_w dependent on the velocity gradient at the wall surface. At the pipe entrance there is a sharp velocity gradient, du/dy because the velocity profile with a uniform velocity is abruptly brought to zero at the wall surface, thereby producing a greater shear stress. This reduces as the velocity profile develops and the gradient becomes less sharp. In the fully developed region the gradient becomes constant and thus the shear stress is also constant(Fig. 4.8).

4.6.2 Laminar and Turbulent Velocity Profiles in a Pipe

We can obtain the velocity profile in the fully developed region for a steady laminar incompressible flow, by solving the force balance applied to the fluid. In this region, the fluid particles move along constant streamlines and the velocity profile in the radial direction $u(r)$ does not change in the axial flow direction (Fig. 4.9a).

Earlier we showed that a fluid element is subjected to shear, and pressure forces that act in pairs on opposing surfaces of the fluid element. Figure 4.9b shows a small differential fluid element with thickness dr and length dx and the applied forces. Since the flow is steady there is no change in the net force in the axial direction and therefore the sum of forces is zero,

$$(P_x A_x - P_{x+\Delta x} A_x) + (\tau_r A_r - \tau_{r+\Delta r} A_r) = 0$$

$$(P_x 2\pi r\, dr - P_{x+\Delta x} 2\pi r\, dr) + (\tau_r 2\pi r\, dx - \tau_{r+\Delta r} 2\pi r\, dx) = 0$$

Cancelling out 2π and grouping the x-terms and r-terms gives

$$r\frac{(P_x - P_{x+\Delta x})}{dx} + \frac{(\tau_r r - \tau_{r+\Delta r} r)}{dr} = 0$$

Taking the limits as dx and $dr \to 0$

$$r\frac{dP}{dx} + \frac{d(\tau r)}{dr} = 0$$

Substituting Newton's law of viscosity formula ($\tau = -\mu\, du/dy$) we get

$$\frac{dP}{dx} = \frac{\mu}{r}\frac{d}{dr}\left(r\frac{du}{dr}\right)$$

which is rewritten as

$$\frac{dP}{dx} = \mu\frac{d^2u}{dr^2} + \frac{\mu}{r}\frac{du}{dr} \tag{4.9}$$

This is the general equation for a steady laminar incompressible pipe flow in the fully developed region. Its general solution is

$$u(r) = \frac{1}{4\mu}\frac{dP}{dx}r^2 + A\,ln(r) + B \tag{4.10}$$

To obtain the velocity profile we define the boundary conditions which are:

I. At the centre of the pipe $r=0$, $u=U_{max}$ which means A must be equal to zero for the equation to balance.
II. Combining the above condition with the wall boundary condition of $r=R$ (R is the distance from pipe centre to pipe wall), $u=0$ means that

$$u(r) = U_{max}\left(1 - \frac{r^2}{R^2}\right) \tag{4.11}$$

where

$$u(0) = U_{max} = -\frac{dP}{dx}\left(\frac{R^2}{4\mu}\right) \tag{4.12}$$

The velocity profile is parabolic in shape with a maximum at the centreline where $r=0$. At the wall where $r=R$, then the velocity becomes zero. If we integrate Eq. 4.11 from $r=0$ to $r=\mathrm{R}$ we obtain the average velocity which gives

$$u(r) = 2U_{\text{avg}}\left(1-\frac{r^2}{R^2}\right) \tag{4.13}$$

This means that

$$U_{\text{avg}} = \frac{1}{2}U_{max} = -\frac{1}{2}\frac{dP}{dx}\left(\frac{R^2}{4\mu}\right) \tag{4.14}$$

This means that the average velocity in the fully developed region of a laminar pipe flow is half that of its maximum velocity.

When the Reynolds number is greater than 4000 in pipes, turbulent flow behaviour is present. Fully developed turbulent velocity profiles are flatter than laminar flow profiles, and not uniform nor a "plug" flow. The power-law empirical velocity profile expression provides a simple method to describe the profiles,

$$\frac{u(r)}{U_{\text{max}}} = \left(1-\frac{r}{R}\right)^{1/n} \tag{4.15}$$

where the exponent n is a constant value dependent on the Reynolds number shown in Fig. 4.10, but can also be calculated based on the average and maximum velocity (which occurs in the pipe centre) of the flow,

$$\frac{U_{\text{ave}}}{U_{\text{max}}} = \frac{2n^2}{(n+1)(2n+1)} \tag{4.16}$$

The laminar profiles develop into a parabolic shape because the fluid inertia cannot overcome the viscous forces sufficiently causing a gradual decrease in velocity away from the zero velocity at the wall. The turbulent profile has much greater flow inertia and this energy allows greater mixing between each of the fluid particles. This results in a more even distribution of velocity throughout the profile. This mixing is referred to as diffusion of fluid inertia or its momentum (Fig. 4.11).

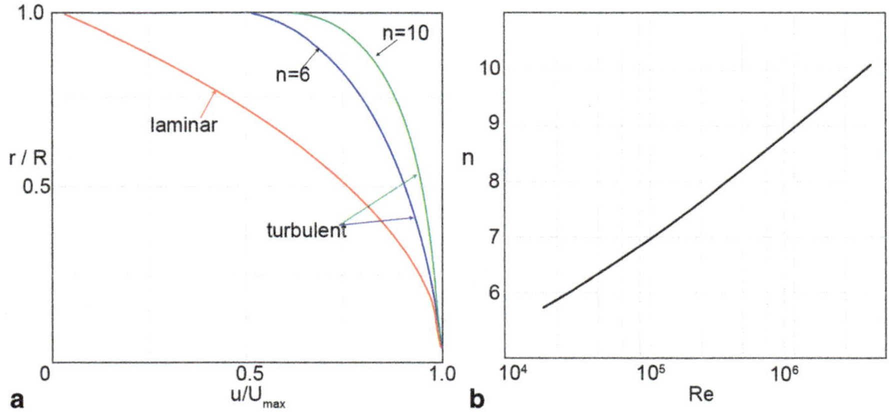

Fig. 4.10 **a** Turbulent and laminar velocity power-law empirical velocity profile defined by Eq. 4.15. **b** Exponent *n* as a function of the Reynolds number defined by Eq. 4.16

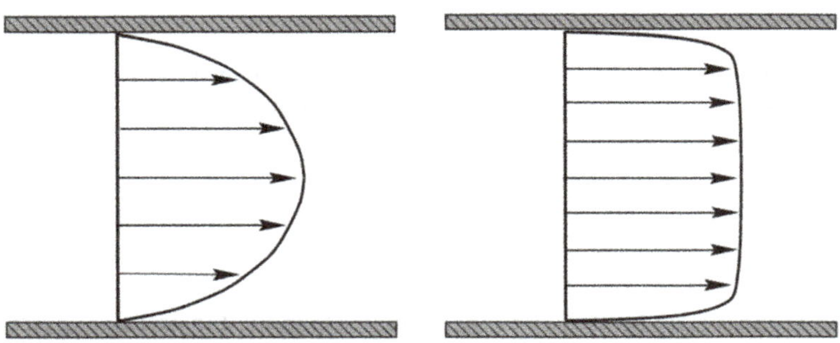

Fig. 4.11 Fluid particle layers moving in a laminar and turbulent flow. The diffusion (or mixing) of the fluid particles in a laminar flow is limited between adjacent layers because of the ordered flow structure. For turbulent flow, there is intense mixing and fluid particles throughout mix which creates a more even velocity distribution

4.6.3 Poiseulle's Law

The amount of blood that flows through the vessels depends on the difference in pressure at the two ends P_1 and P_2, the radius r and the length L of the pipe, and the viscosity μ of the blood. Poiseuille's Law relates these parameters under steady laminar flow condition, for a Newtonian fluid in a rigid pipe. If turbulence is present, then the predicted flow rate does not comply since in general this value is smaller than in laminar flows.

In many instances of the circulatory system, the vessels are rarely straight and rigid to allow a parabolic profile. The aorta is a large curved vessel which exhibits a Re number of around 1300 and an entrance length of 200 cm, much longer than the entire aorta. This means that the flow profile can never reach a fully developed

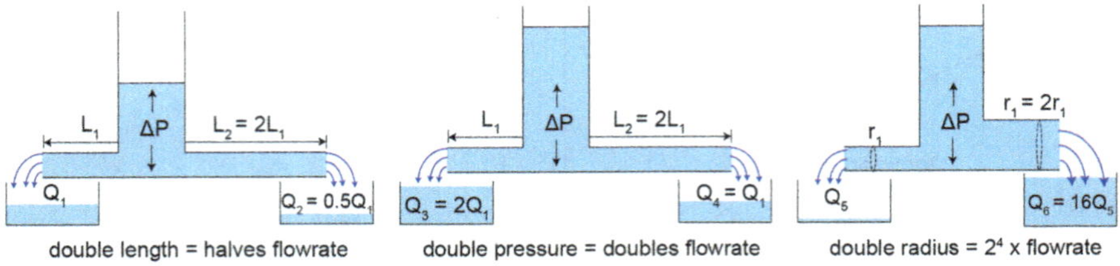

Fig. 4.12 Poiseuilles' Law that relates flow rate with pipe length, pressure, and radius

parabolic shape. Despite this, Poiseulle's Law provides a conceptual estimate to relate pressure drop to flow rate.

To obtain the correlation we substituting $\dot{Q} = U_{\text{avg}} A_c$ into Eq. 4.12 and the definition of a cross-sectional area of a pipe $A_c = \pi R^2$ we get

$$\dot{Q} = -\frac{dP}{dx}\left(\frac{R^2}{8\mu}\right)\pi R^2 = \frac{\pi R^4}{8\mu}\left(-\frac{dP}{dx}\right) \tag{4.17}$$

Poiseuille's Law relates a steady laminar flow rate, Q through a rigid pipe, with the difference in pressure at the two ends P_1 and P_2, the radius r and the length L of the pipe (represented by dx in the equation), and the viscosity μ of the blood (Fig. 4.12). The flow rate is linearly proportional to the pressure, inversely proportional to the length of the pipe and viscosity, and increases to the power of 4 to the radius of the pipe.

The flow rate is defined as the average velocity of the fluid passing through a cross-sectional area. An increase in the pipe length increases the overall resistance so reduces flow through the pipe by half. Doubling the pressure increases the amount of energy that drives the fluid motion, leading to a doubling of the flow rate. The biggest significant factor on flow rate is the change in radius. Doubling the pipe radius produces a 16-times increase in flow rate.

The significance of blood vessel radius reduction can be seen by considering a small amount of arterial occlusion and its effect on the required pressure to maintain the same blood flow rate. Under a fixed pressure, an occlusion of 20 % reduces the flow rate to 41 % of its original rate, and this is even worse when the occlusion is 50 % where the flow rate is reduced to 6.3 %. To restore the normal flow rate the corresponding percentage increase in pressure required are 244 % and 1600 % (Fig. 4.13). Such high pressure changes are unlikely to be achieved, and instead body has the ability to increase the volume flowrate by vasodilation of the small vessels called arterioles. These smaller vessels provide most of the resistance to flow, and play a critical role in delivering oxygen and nutrients to the body via the capillaries that are next to them. A vasodilation of the arteries by 19 % will double the blood flow.

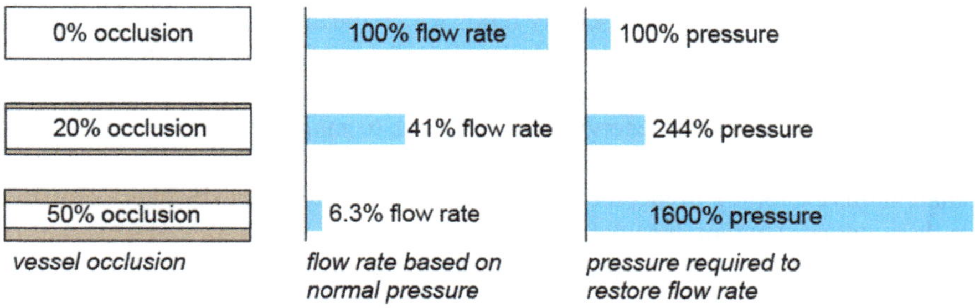

Fig. 4.13 Blood flow examples showing the significance of occluded arteries on flow rate and pressure

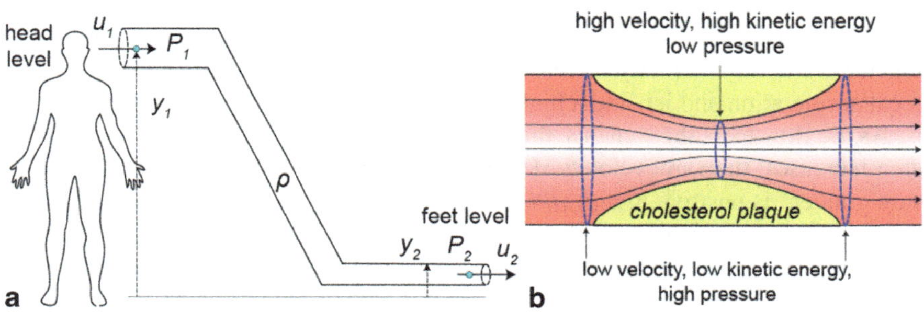

Fig. 4.14 **a** A moving fluid exhibits kinetic energy $(1/2\rho u^2)$, potential energy $(\rho g y)$, and press (P). **b** Bernoulli's equation in a *horizontal line* shows the interchanging energy from kinetic to pressure as the fluid passes through a stenosis

4.6.4 Bernoulli's Equation

The Bernoulli equation is derived from the energy conservation law, relating blood pressure with flow velocity (Fig. 4.14). This is given as

$$\underbrace{P_1}_{\text{pressure energy}} + \underbrace{\frac{1}{2}\rho u_1^2}_{\text{kinetic energy}} + \underbrace{\rho g y_1}_{\text{potential energy}} = P_2 + \frac{1}{2}\rho u_2^2 + \rho g y_2 \tag{4.18}$$

where P is pressure, u is velocity, g is gravity, and y is height.

In Poiseulle's Law, we saw that pressure is a driving force for blood flow. However this is one contribution to the total energy that drives flow between two points. Bernoulli's equation states that the total energy of the flow is given by the pressure force exerted on the flow plus its kinetic energy produced from its mass and velocity, plus the potential energy produced by gravitational effects. Bernoulli's equation however, does not include a viscosity term, which means it neglects any frictional losses, (e.g. neglects the influence of viscosity and the fluid is considered an ideal fluid). Referring to Eq. 4.18, we see that as the contribution from the pressure energy increases, the kinetic energy must decrease to maintain energy conservation (here we have neglected the influence of potential energy since changes in h are small in this example). In fact the kinetic and pressure energy can be inter-converted so that total energy remains unchanged.

The Bernoulli equation predicts flow velocity and explains the presence of local high-velocity jets found in arterial disease states that are characterized by stenosis. Consider steady flow in a stenosed artery with a fluid that has constant density and negligible friction loss. In the regions before and after the stenosis, the flow streamlines are parallel which and the pressure is constant, neglecting effects of gravity over the small difference in height, h. As the flow enters the stenosis, the cross-sectional area decreases forcing streamlines to converge and become closer together. The flow velocity increases, and the pressure decreases. This produces a local acceleration and jet, and a high wall shear stress in the stenosed area.

As the flow exits the stenosis, the streamlines now diverge as the cross-sectional area increases; the flow velocity decreases, and the pressure increases. This pressure gradient is positive in that $dP/dx > 0$ and is known as an adverse pressure gradient. Close to the wall where the flow is slowest, an adverse pressure gradient can slow the velocity as far as to zero and even a negative (reverse) flow. When this occurs the flow is said to be separated from the surface which significantly modifies the pressure distribution along the wall surface.

Let us quantify the relationship between pressure, and velocity during the flow through a stenosis. Mass conservation tells us that as the artery diameter decreases the velocity increases to maintain a constant flow rate by an inverse proportion to the square power, $u \propto 1/d^2$. This means if the artery is occluded by half due to cholesterol build-up and atherosclerosis, the velocity is increased by four times. Since the kinetic energy is proportional by the square power to the velocity, $k \propto u^2$ there is a 16 fold increase in the kinetic energy. As a result, this energy can be inter converted with the pressure energy (when we neglect the potential energy) which means that pressure decreases by 16 fold. After the stenosis, the kinetic energy returns to its original value when we consider no viscous effects. Bernoulli's equation assumes a laminar flow and that the distance between the two points is short enough so that viscous losses can be neglected. In reality viscous effects, and the likelihood of turbulence, will reduce the post-stenosis total energy.

The effects of the potential energy become relevant when we consider the blood flow throughout the body from feet up to the head (Fig. 4.14a). If we consider a constant kinetic energy, in that the blood is flowing at similar velocity, then the change in pressure is related to the potential energy by

$$P_{\text{head}} - P_{\text{feet}} = \rho g (y_{\text{head}} - y_{\text{feet}}) \tag{4.19}$$

If were lying down, with head and feet and equal level, then the pressure between the two points are now the same and there is zero potential energy difference. As we stand up from a lying position our body has to adjust, and if this adjustment is not quick enough, then we experience dizziness as a result a sudden pressure drop that reduces blood flow to the brain. Furthermore there is an increase in fluid build-up in the legs because of the larger pressure difference needed for fluids to travel upwards.

4.6.5 Pressure Drop Estimates

Flow through a pipe experiences a pressure difference between an upstream and downstream location due to viscous effects. This pressure loss for all types of fully developed internal flows is given as

$$\Delta P = f\frac{L}{D}\frac{\rho U_{\text{ave}}^2}{2} \tag{4.20}$$

where L is the pipe length, D is the diameter, and f is the friction factor. For laminar flow, the friction factor is a function of the Reynolds number only and is independent of the roughness of the pipe surface

$$f = \frac{64}{\text{Re}} \tag{4.21}$$

For turbulent flows we saw that the velocity profile is blunt in the centre of the pipe with a sharper velocity gradient at the wall. This effect has two important consequences: (i) shear stress increases and therefore the resistance to flow greatly increases, (ii) and wall roughness must be accounted for in determining the friction factor since the thickness of this roughness becomes comparable with the sharp velocity gradient. The friction factor can be determined from the well-known Moody chart or by the Colebrook equation (Colebrook 1939)

$$\frac{1}{\sqrt{f}} = -2.0\log\left(\frac{\varepsilon / D}{3.7} + \frac{2.51}{\text{Re}\sqrt{\text{f}}}\right) \tag{4.22}$$

where ε/D is the relative roughness. The Colebrook equation is a representation of the Moody chart which is plots the friction factor for pipe flow as a function of Re number and the relative roughness ε/D.

4.7 Fluid Dynamics of Blood Flow Examples

The features of fluid dynamics discussed in the previous sections that are relevant in haemodynamics are presented here through examples. Bifurcations of arteries, stenosis, curved geometries, and aneurysms are all features that can be found in the arterial system. In this section, we present a simplified artery system that incorporates four kinds of arterial vasculature to demonstrate the fluid dynamic characteristics associated with the existence of flow disturbance (Fig. 4.15).

4.7.1 Carotid Artery Bifurcation

In this section, the tuning-fork-shaped idealized carotid bifurcation model is used, which is reconstructed from 12 averaged human carotid anatomical data (Ding et al. 2001) shown in Fig. 4.16a. The fluid domain is a structured mesh with hexahedral elements. Near wall grid refinement is imposed to provide better resolution for near wall quantities, such as wall shear stress.

A constant velocity of 0.4 m/s representing peak systolic flow at the CCA (common carotid artery) is used which produces a Reynolds number of Re = 800 at the entrance. A flow rate partition of 70 % through the ICA (internal carotid artery) and 30 % through the ECA (external carotid artery) is prescribed for the bifurcating flow. This accounts for a physiological time-varying flow partition between the two branches during one cardiac cycle (Morbiducci et al. 2010).

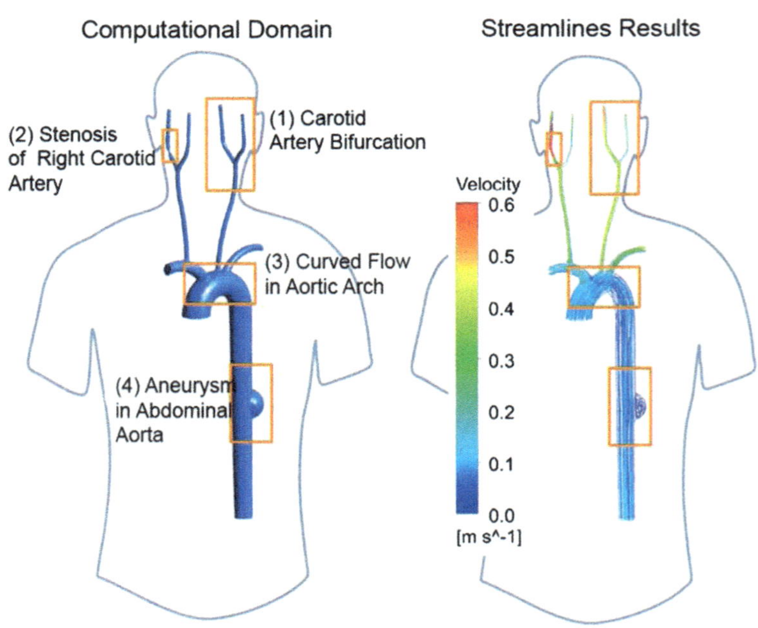

Fig. 4.15 Schematic diagram of computational fluid dynamics of arterial system

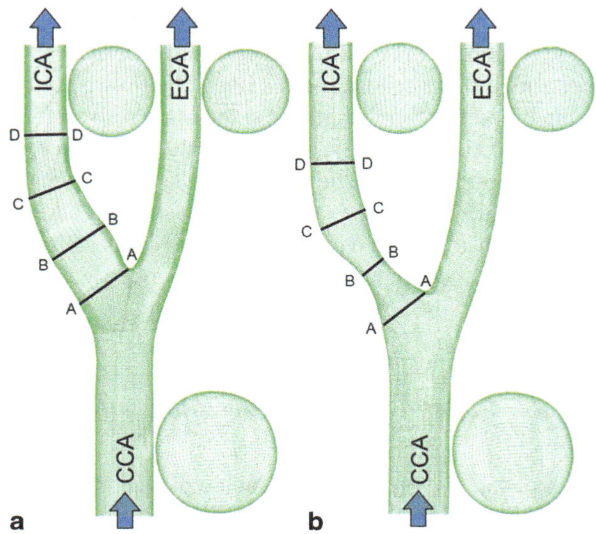

Fig. 4.16 Idealized carotid bifurcation models **a** Healthy carotid bifurcation model; **b** Diseased carotid bifurcation model with internal carotid artery (ICA) stenosis. Common carotid artery (*CCA)*; internal carotid artery (*ICA*); external carotid artery (*ECA*)

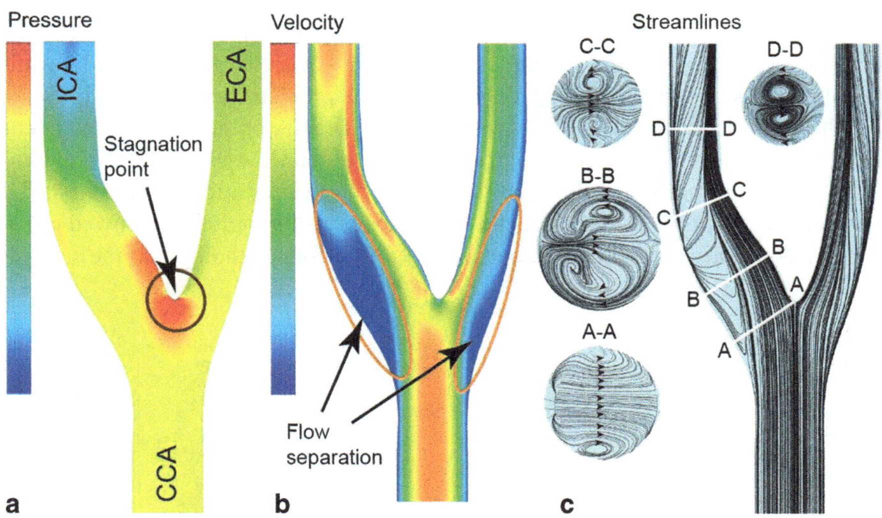

Fig. 4.17 CFD simulation results of the healthy carotid bifurcation model showing **a** the stagnation point, **b** flow separation, and **c** secondary flow in the form of vortices

Pressure, velocity, and streamline distributions at the median plane of a healthy model are shown in Fig. 4.17. Bifurcation flows are characterised by a stagnation point at the bifurcation apex as the blood splits into the two paths with the resistance through each artery influencing the pressure drop. The pressure difference between the upstream CCA and ICA is greater than CCA and ECA (Fig. 4.17a). The reason for this phenomenon can be explained by a pressure drop estimate given in Eq. 4.20

which relates pressure drop with flow velocity and geometry. The flow through the ICA is more than twice the amount through the ECA branch, producing a greater pressure difference to drive this flow. The flow patterns in Fig. 4.17b and c show the main blood flow concentrating at the inner wall regions rather than the outer wall of the bifurcation due to the vessel curvature. The presence of high velocity gradients within the inner wall region of the ICA causes a high shear field along the inner wall from the bifurcation apex to the carotid sinus. In addition, the abrupt cross-sectional area expansion at the vicinity of carotid sinus causes flow separation along with a pair of spiral secondary recirculating vortices located symmetrically about the median plane of the bifurcation (plane B-B).

4.7.2 Carotid Artery Bifurcation with Stenosis

To replicate the presence of a severe plaque, the healthy carotid artery used in the previous section is modified to include a stenosis in the ICA by reducing the artery sinus diameter just after the bifurcation apex (Fig. 4.16b). Further details of this example can be found in Dong et al. (2013). The stenosis reduces the cross-sectional area by 63 % from the ICA branch entrance (plane A-A), to the throat of the stenosis (plane B-B). To examine the plaque build-up influence on intravascular haemodynamics, a steady numerical simulation is performed over the stenosed model while keeping the boundary conditions identical with the previous healthy model.

The stenosis increases the flow resistance of the ICA branch, which requires a greater pressure difference needed to conduct the blood flow through this branch—which is more than twice the requirement for the healthy model (Fig. 4.18a). A characteristic jet flow is evident as the blood squeezes through the narrowed cross-section. A negative pressure gradient $dP/dy < 0$ is found leading into the stenosis

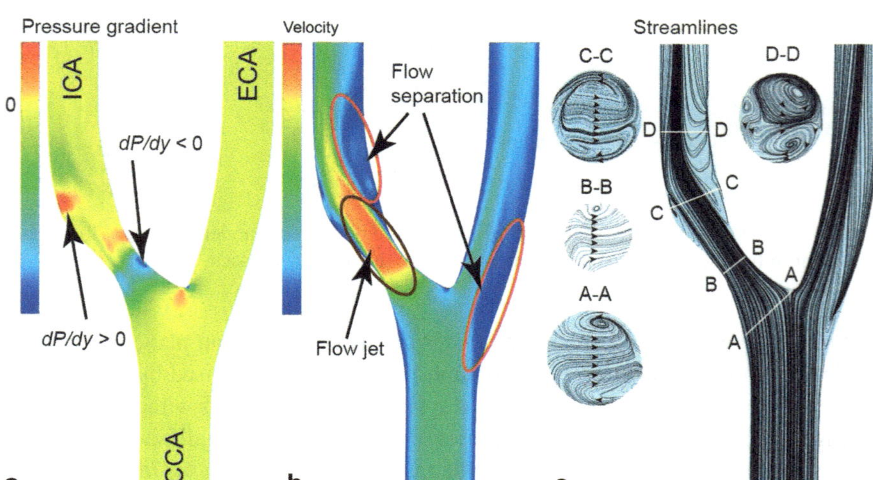

Fig. 4.18 CFD simulation results of the stenosed carotid bifurcation model

which produces acceleration in the blood flow. After the stenosis, an adverse pressure gradient $dP / dy > 0$ is formed and the blood flow velocity is reduced. The change in velocities also coincides with the cross-section area reduction, and expansion as the blood passes through the stenosis.

A jet-like flow forms and impinges on the outer wall of the carotid sinus. Meanwhile, the velocity at the throat of the stenosis is doubled compared with the mean free stream velocity value at CCA branch (Fig. 4.18b).

Secondary flow streamlines show a more disturbed pattern, and in the post-stenosis region for three spiral secondary flow vortex cores exist at the cross section plane D-D. This is induced by the laminar-turbulent transitional flow at the post-stenosis region, which is linked with the presence of severe stenosis lesion at the ICA branch. This flow disturbance has been linked to increased thromboembolic activities (Tan et al. 2008; Kefayati et al. 2014).

4.7.3 Curved Flow in Aortic Arch

The aorta is the largest artery in the human body, originating from the left ventricle of the heart and extending down to the abdomen, where it bifurcates into two smaller branches. The aorta distributes oxygenated blood to all parts of the body. Therefore, blood flow simulation in aorta plays important roles in better understanding of fluid dynamics in healthy and pathological situations. A patient-specific aorta model with abdominal aneurysm is reconstructed from CT images, and two regions of interest, aorta arch and abdominal aneurysm, are highlighted in Fig. 4.19.

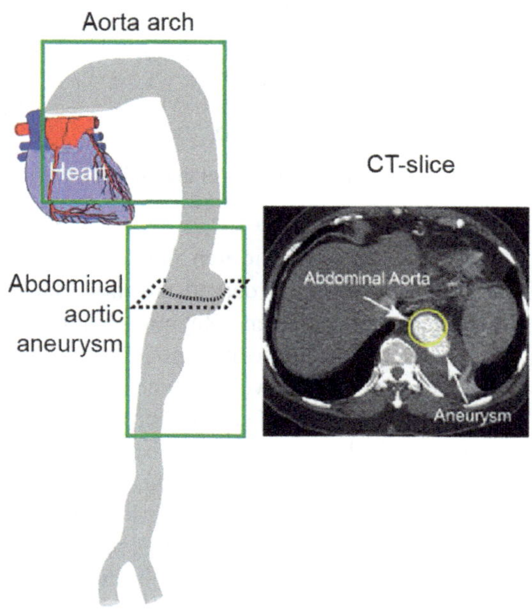

Fig. 4.19 Reconstructed aorta model based on CT images

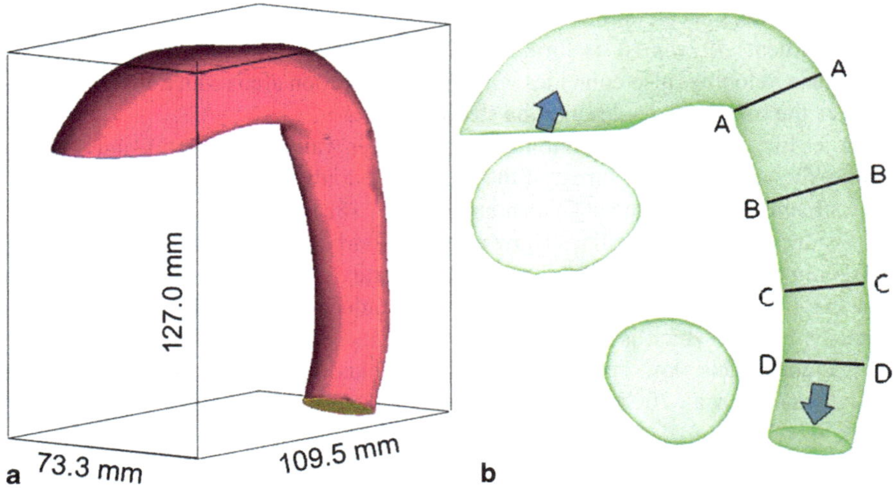

Fig. 4.20 Geometry and structured mesh result of the studied aorta arch

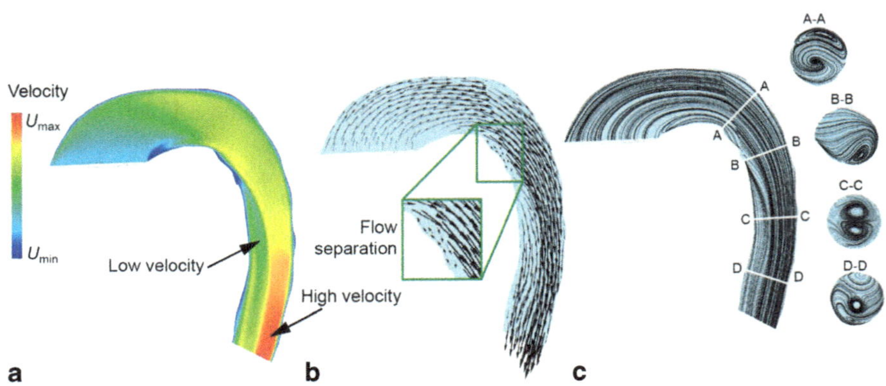

Fig. 4.21 CFD simulation results of the studied aorta arch model

The reconstructed aortic arch and the computational mesh are shown in Fig. 4.20. The geometry is highly curved and non-planar, with the vessel bend angle approximately 90°. The same Reynolds number Re = 800 is prescribed at the inlet (u = 0.08 m/s) of the aortic arch model, and a steady CFD simulation is performed.

The velocity contour, velocity vectors, and flow streamlines in the median plane are shown in Fig. 4.21. Due to a strongly curved geometry, the main blood flow moves towards the outer wall of the aortic arch, while low velocities are present in the inner curvature of the vessel (Fig. 4.21a). The low velocity region enhances particle residence times of blood cells which promotes the initiation of atherosclerosis. The velocity vector plot (Fig. 4.21b), reaffirms the flow directed towards the outer wall. We observe flow separation occurring at the proximal descending thoracic aorta, highlighted in the figure.

Secondary flow arises because of a lateral (centripetal) acceleration, which causes a radial pressure gradient that drives slower moving fluid near the vessel wall towards the center, while faster moving fluid in the vessel core is swept outwards. The strength of the secondary flow is characterised by the Dean number *De*. For a unifromly curved vessel with slight curvature, the Dean number is defined as:

$$De = \frac{2a\bar{U}}{\nu}\left(\frac{a}{R}\right)^{1/2} = 2\delta^{1/2} Re \tag{4.23}$$

where δ is the ratio of the vessel radius (a) to the radius of curvature (R), $\bar{U}$ is the mean axial velocity in the vessel. Therefore, the Dean number is the ratio of the square root of the product of the inertial and centrifugal forces to the viscous forces, and equals to the Reynolds number modified by the vessel curvature.

Although the arotic arch geometry used here is much more complex than a uniformly bent pipe, some conculsions regarding the secondary flow motion can be drawn using the above concept. At the distal descending thoracic arota (Fig. 4.21c), plane C-C, the Dean number is relatively large and the blood flow experiences a strong radial acceleration, which pushed the symmetirc pair of counter-rotating vortices towards the outer wall of the bend. At plane D-D, as a comparison, the vessel curvature is reduced. Its corresponding Dean numer becomes small. Consequently, a single vortex is obsevred close to the vessel center and the maximum circumferential velocity is attained in the center of the vessel.

4.7.4 Aneursym in Abdominal Aorta

Using the same anatomical model, the haemodynamics of the abdominal aortic aneurysm is investigated here. Abdominal aortic aneurysm (AAA) is an irreversible, localized growth, typically in the infra-renal region of the aorta. The causes and details of an aneurysm were discussed in Chap. 2, but essentially an aneurysm is a balloon-like swelling extending out from an artery. Artery walls can withstand normal blood pressure. However certain medical problems or trauma and injury can weaken its walls, leading to a growth of an aneurysm.

Fluid dynamics has been found to play an important role in understanding and predicting the initiation, growth, and rupture of aneurysms. Computational modelling allows for the investigation of flow patterns and drag forces on stent-grafts, which may influence stent-graft migration after its implantation. In this section, an AAA model with a diameter of 36 mm (Fig. 4.22a) is used to investigate the basic intra-aneurysmal flow characteristics.

The fluid domain is first meshed with hexahedral cells, and a near wall refinement is imposed at the wall to improve the grid resolution (Fig. 4.22b). Using the same CFD modelling strategy, the inlet of the interest fluid domain is set as a velocity inlet, and the velocity value is determined according to the same Reynolds

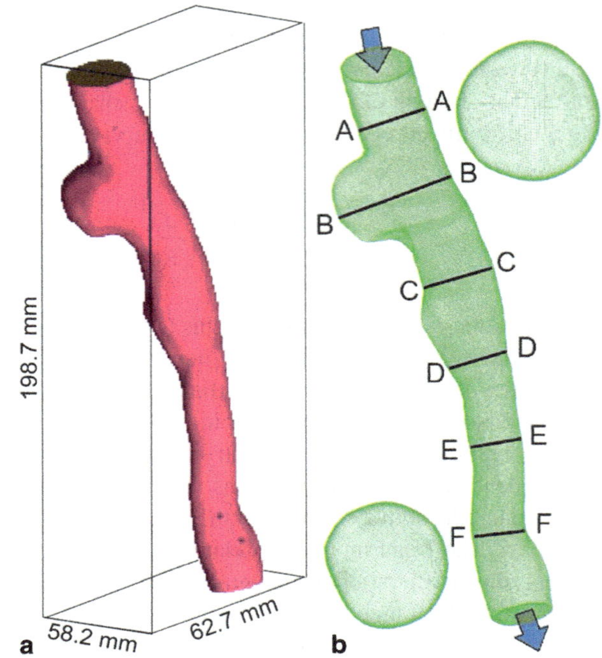

Fig. 4.22 Geometry and structured mesh result of the studied abdominal aortic aneurysm model

number 800. Numerical simulation results in terms of pressure at the wall, velocity, vector and streamlines at the median plane of the aorta are provided in Fig. 4.23.

Based on the pressure distribution contour (Fig. 4.23a), the pressure drop increases when the flow passes downstream of the abdominal aorta. The reason for

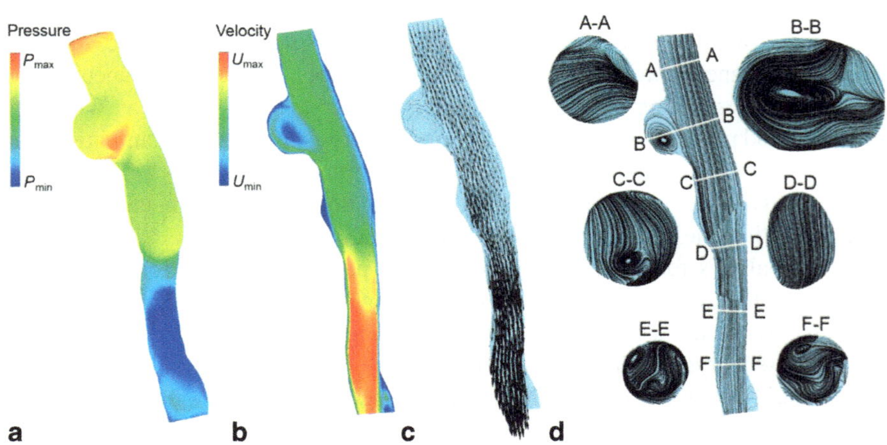

Fig. 4.23 CFD simulation results showing **a** pressure contour, **b** velocity contour, **c** velocity vector, and **d** streamlines of the studied abdominal aortic aneurysm model

this phenomenon can be explained from the velocity and vectors shown in Fig. 4.23b and c. As the blood flows over the location of the aneurysm, part of it impinges at the bottom aneurysmal wall, and its flow direction is reversed along the outer wall, forming a clockwise-rotating vortex inside the aneurysm. The aortic artery narrows downstream of abdominal aorta, hence the blood flow velocity is accelerated and pressure drop consumption increases. The streamlines demonstrate that the aneurysm experiences strong secondary flow due to the flow recirculation, and as a consequence consumes more flow energy and increases the patient's cardiac workload.

4.8 Summary

Fundamentals of haemodynamics were introduced in this chapter. The main properties of blood that govern its flow behavior are viscosity and density which wield great influence on our health and well-being. As such clinical implications can be deduced from measurements of these two properties. During blood flow the blood experiences shear and pressure forces caused by the variation in flow velocity from the arterial wall to the artery centre. When it travels at high momentum and/or in large arteries relative to its viscosity, then its flow behaviour is characterised by turbulence. At low momentum and/or small arteries the flow is typically laminar.

Since blood flow is enclosed by the arterial walls, it can be classed as a type of internal pipe flow. The influence of the artery walls give rise to the different boundary layer formation as it develops through the pipe. It was shown that Poiseulle's Law and Bernoulli's Equation provides useful estimates of pressure drop across the artery network.

The chapter culminates with specific examples in the carotid bifurcation, aortic arch, and in aneurysms which highlighted fluid dynamics characteristics. These included flow a stagnation point, flow separation, and secondary flow in the form of Dean vortices in bifurcating arteries, and curved aortic arch. The presence of an aneurysm caused localised recirculation and contributed to flow resistance.

The fundamental haemodynamic theory from this chapter has a strong connection with fluid dynamics. Therefore this chapter provides a solid foundation for understanding the derivations of the governing equations of fluid and structural dynamics that is presented in Chap. 5.

4.9 Review Questions

1. Describe the constituents of blood.
2. What is the haematocrit of blood?
3. Describe the main factors that influence the viscosity of blood.
4. What happens to blood viscosity as a blood vessel decreases in diameter?
5. Under what circumstances could blood be assumed to behave as a Newtonian fluid?
6. How do the changes in blood viscosity and its hematocrit levels affect an athlete's sports performance?
7. What is the primary cause for shear forces in blood flow?
8. How does the pressure force acting on blood differ from shear forces?
9. What does the Reynolds number describe in terms of blood flowing through an artery vessel?
10. What factors would contribute to a laminar blood flow? What would cause the flow to change to a turbulent flow?
11. What does a developing region refer to for internal pipe flows, and what is the entrance length?
12. As the Reynolds number increases is the entrance length longer or shorter? Explain why?
13. Describe the difference between a laminar, and a turbulent velocity profile.
14. What does Poiseuilles's Law describe? If a vessel's diameter is reduced by half, how much of an influence does it have on the pressure?
15. What does Bernoulli's equation describe? How does the equation help describe the flow through a constricted vessel such as a stenosis?
16. What are Dean vortices?

Chapter 5
Computational Fluid Structure Interaction

5.1 Introduction

In a Fluid-Structure-Interaction model, (FSI) pressure forces caused by the fluid flow is applied onto a structure. With sufficient loading, the structure deforms and creates a new boundary for the fluid flow analysis. The interaction between fluid and solid take place at a shared interface or boundary. Force loadings and structural deformation results are passed from one analysis to the other and hence is given the name fluid-structure interaction.

This chapter introduces the fluid dynamics and structural mechanics set of equations. These sets of equations are applied to the separate computational domains that represent either the fluid or structural component. The domains are connected together via a common boundary where boundary conditions and constitutive models are applied.

This chapter focusses on obtaining a physical understanding of the applied equations avoiding details or complicated proofs of the equations. Furthermore the energy conservation equations have been left out as it is not essential to the basic understanding of FSI modeling. For a more complete description of these equations, the interested reader can refer to several books on: continuum mechanics (Lai et al. 2009; Reddy 2013); structural dynamics (Craig and Kurdila 2006; Paz and Leigh 2004) and; computational fluid dynamics (Tu et al. 2012; Versteeg and Malalasekera 2007).

5.2 Introduction to Fluid Dynamics

The fluid dynamics equations are mathematical statements of the *conservation laws* in physics. It describes the transport of mass, momentum, and energy through a physical domain. These equations are:

- the conservation of mass;

J. Tu et al., *Computational Hemodynamics—Theory, Modelling and Applications*,
Biological and Medical Physics, Biomedical Engineering,
DOI 10.1007/978-94-017-9594-4_5

- the conservation of momentum (Newton's second law, the rate of change of momentum equals the sum of forces acting on the fluid) and;
- the conservation of energy (first law of thermodynamics, the rate of change of energy equals the sum of rate of heat addition to and the rate of work done on the fluid).

5.2.1 Mass Conservation

The mass flow rate through a surface of a control volume is

$$\dot{m} = -\oint_s \rho \boldsymbol{u} \cdot \mathbf{n}\, ds \tag{5.1}$$

where s is the surface of a control volume, and $\boldsymbol{u} \cdot \mathbf{n}$ is the velocity component acting perpendicular to the surface area, A. The sign of the normal component of velocity determines whether the fluid flows in or out of the control volume, and a negative sign is added to denote that a net outflow of mass produces a positive value (Fig. 5.1).

For a one-dimensional steady flow the mass flow through a surface is given as

$$\dot{m} = -\oint_s \rho \boldsymbol{u} \cdot \mathbf{n}\, ds = \rho u A \tag{5.2}$$

The amount of mass entering a control volume surface is equal to the mass leaving the control volume surface and this is the principle of mass conservation, summarised as (Fig. 5.2)

$$\sum \dot{m}_{in} - \sum \dot{m}_{out} = 0 \;\rightarrow\; \sum (\rho u_{in} A_{in}) - \sum (\rho u_{out} A_{out}) = 0 \tag{5.3}$$

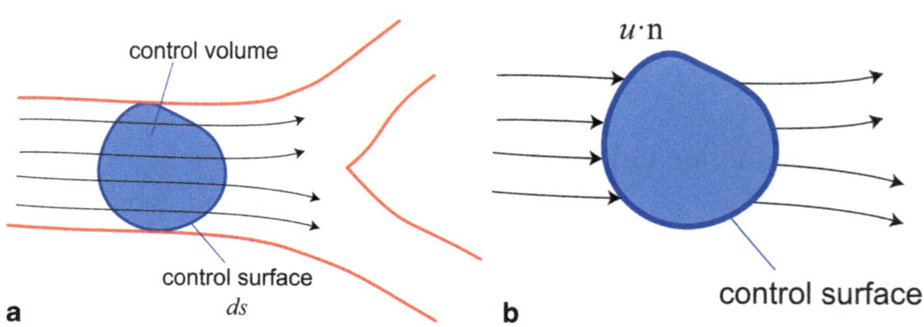

Fig. 5.1 **a** Mass flow rate passing through an arbitrary control surface. Flow streamlines that enter the control volume are accounted by its vector component normal to the surface face denoted by $\boldsymbol{u} \cdot \mathbf{n}$

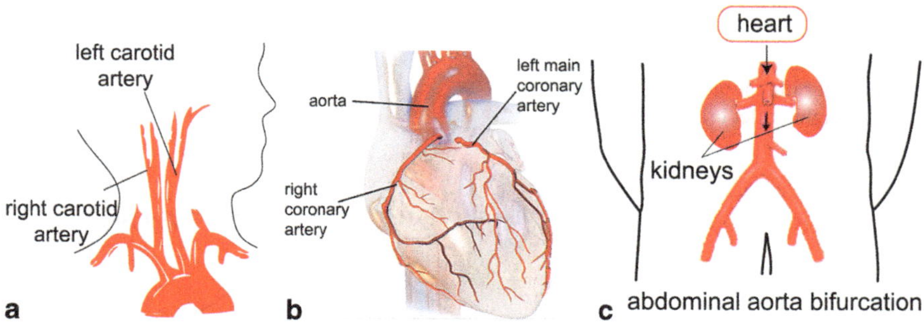

Fig. 5.2 Main artery bifurcations in the body which include **a** carotid arteries **b** coronary arteries, and **c** abdominal aorta artery

5.2.1.1 Artery Bifurcation Example

Throughout the circulatory network, many parts of the arteries bifurcate into subsequent smaller branches. The major bifurcations include the carotid artery, coronary artery, and abdominal aorta bifurcation (Fig. 5.2). There are two common carotid arteries, the left and right, which sub divide in the neck to form the external and internal carotid arteries to deliver oxygenated blood to the head. Both the left and right main coronary arteries bifurcate into smaller branches which in turn bifurcate again into even smaller branches to surround the heart with a network of branches. The abdominal aorta is the largest artery in the abdominal cavity and is a continuation the aorta in the heart. It bifurcates at the fourth lumbar vertebrae into the left and right common iliac arteries.

In these geometries the main flow is divided into two separate streams, and therefore its mass flow rate is divided, and in the case of the main coronary artery, subsequent bifurcation continues. Subdivision of the flow rate is dependent on the resistance to flow and this is determined by the branch diameters and its alignment with the main branch. Nevertheless the mass flow must be conserved throughout the entire branch network. Figure 5.3 demonstrates the mass conservation through an idealised bifurcation network. If we consider the mass flow rate at the inlet as $\dot{m}_1$, then the sum of all the mass flow rates through the outlets must equal to $\dot{m}_1$,

$$\dot{m}_1 = \dot{m}_2 + \dot{m}_3 = \dot{m}_4 + \dot{m}_5 + \dot{m}_6 + \dot{m}_7 + \dot{m}_8 + \dot{m}_9 + \dot{m}_{10} + \dot{m}_{11}$$

5.2.1.2 One-Dimensional Vessel Flow Example

In Chap. 4 we saw that from the inlet, the velocity profile changes from a constant velocity profile $u_1 = U$ at the pipe entrance to a parabolic profile in the fully developed region (Fig. 5.4).

The mass flow rates at the inlet and the fully developed region must equal, for mass conservation. The velocity profiles are different but the cross-sectional areas

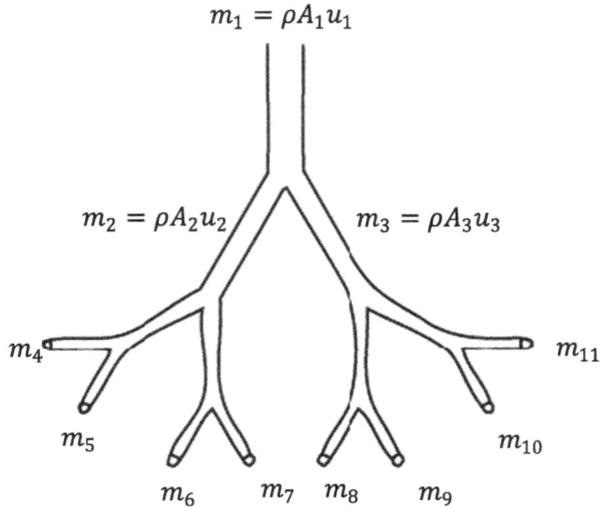

Fig. 5.3 Schematic of an idealised bifurcation network having one main inlet and subsequent bifurcations to produce eight outlet branches

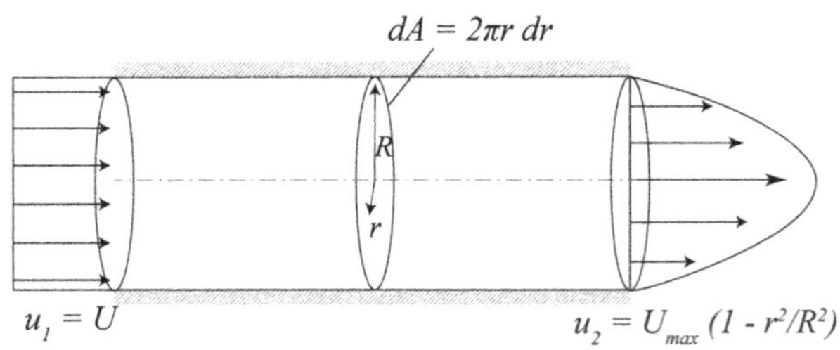

Fig. 5.4 Schematic of velocity profiles at the inlet, u_1 where the flow hasn't developed, and at the fully developed region u_2

are the same. From this we demonstrate that the uniform averaged velocity is twice the maximum velocity, as.

$$\int_0^R \rho u_1 dA = \int_0^R \rho u_2 dA$$

$$\int_0^R \rho U \cdot 2\pi r\, dr = \int_0^R \rho U_{max}\left(1 - \frac{r^2}{R^2}\right) \cdot 2\pi r\, dr$$

$$\rho \pi R^2 U = \rho 2\pi U_{max}\left(\frac{R^2}{2} - \frac{R^2}{4}\right)$$

$$U = \frac{U_{max}}{2}$$

We note that the area is constant and that while the velocity profiles are different, they provide the same amount of mass flow.

5.2.1.3 General Mass Conservation—Continuity Equation

The previous examples were solved using the steady one-dimensional form of the mass conservation equation. In general cases where the flow varies in three dimensions and also unsteady in time, a more general equation is needed. Following the same principle of conservation, the rate of change of mass inside a control volume is equal to the net mass of fluid flowing in or out of the control volume across its boundary surface. Taking outflow of mass as positive we get

$$\underbrace{\frac{\partial}{\partial t}\int_V \rho dV}_{\text{rate of change of mass}} = \underbrace{-\oint_s \rho \boldsymbol{u}\cdot\mathbf{n}\, ds}_{\text{net inflow of mass}} \tag{5.4}$$

This equation is also called the continuity equation. Using Gauss's theorem Eq. (5.4) is rearranged to give the mass conservation in differential equation form as

$$\frac{\partial \rho}{\partial t} + \nabla \cdot \rho \boldsymbol{u}_i = 0 \tag{5.5}$$

Expanding this we get

$$\frac{\partial \rho}{\partial t} + \frac{\partial(\rho u)}{\partial x} + \frac{\partial(\rho v)}{\partial y} + \frac{\partial(\rho w)}{\partial z} = 0 \tag{5.6}$$

5.2.1.4 Steady Converging Pipe Example

Consider an incompressible 2D steady fluid flow through a converging pipe shown in Fig. 5.5.

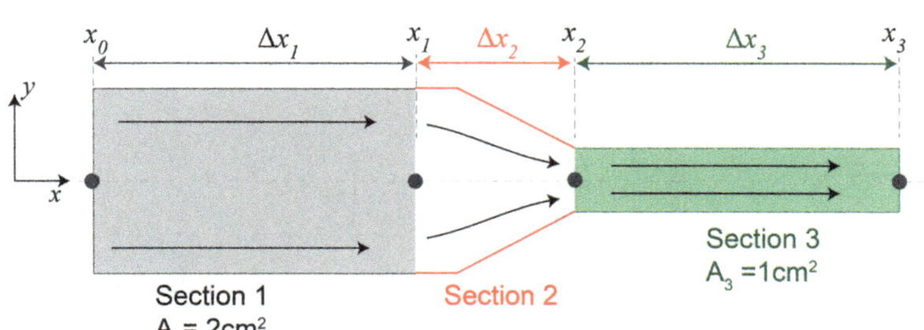

Fig. 5.5 Schematic of fluid flow through a converging pipe. Section 1 of the pipe has the largest cross-sectional area. Section 2 is the converging section of the pipe. Section 3 has the smallest cross-sectional area

As the flow is incompressible, steady and in two dimensions, the density term, the unsteady term and the *z*-coordinate term cancel out to give

$$\frac{\partial u}{\partial x} + \frac{\partial v}{\partial y} = 0 \tag{5.7}$$

For Sects. 1 and 3 there is a constant velocity in the *x*-direction and no variation in the vertical velocity component in the *y*-direction $\partial v / \partial y = 0$. The constant *x*-velocity component *u* means that $\partial u / \partial x = 0$ as well and upon integration the problem simplifies to $m = \rho A_1 u_1 = \rho A_3 u_3$. The cross-sectional area for Sect. 1 is twice that of Sect. 3, which means the velocity in Sect. 3 is twice that of Sect. 1.

In Sect. 2 we see that the cross-sectional area reduces in size and the velocity variation occurs in two dimensions. If we consider the *y*-velocity component above the pipe centreline, it is flowing downwards when $y>0$, while it points upwards when $y<0$ thus giving a negative velocity gradient in the *y*-direction,

$$\frac{\partial v}{\partial y} < 0$$

this means that

$$\frac{\partial u}{\partial x} > 0$$

i.e. that the velocity in the *x*-direction is increasing to satisfy mass conservation in Eq. (5.7).

5.2.2 Momentum Conservation

The conservation of momentum is derived from *Newton's second law of motion*, $\sum F = ma$, where $\sum F$ is the sum of all forces acting on a control volume. Two types of forces are considered, namely *body forces* and *surface forces*. We present the momentum equation directly in its differential form. For an incompressible flow, the force per unit volume is,

$$\rho \frac{D\mathbf{U}}{Dt} = F_{\text{body}} + F_{\text{surface}} \tag{5.8}$$

where the acceleration is defined as $D\mathbf{U}/Dt$. This represents the rate of change of the velocity vector of a moving fluid particle.

Body forces act over the entire volume which includes gravity, centrifugal, Coriolis and electromagnetic forces that act at a distance to the control volume. These effects are usually incorporated by introducing them into the momentum equation as an additional term to the contribution of the surface forces. In many cases the only body force that is considered is the gravitational force, and therefore

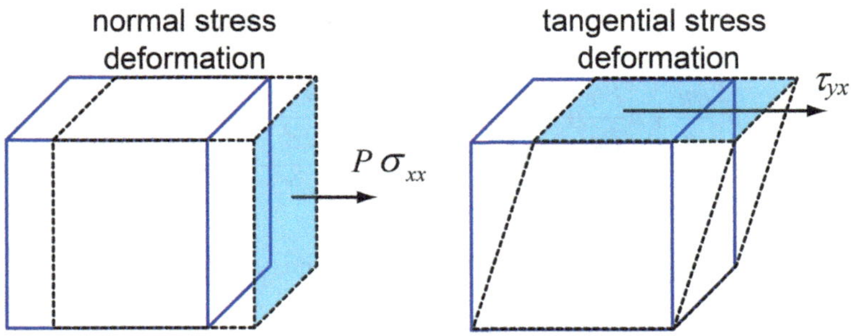

Fig. 5.6 Deformed fluid element due to the action of the surface forces, in the form of normal and tangential stresses

$$F_{\text{body}} = \rho \mathbf{g} \tag{5.9}$$

Surface forces are forces that act on the surface of the fluid element causing it to deform (Fig. 5.6).

This includes the normal stress σ_{xx}, which are a combination of pressure p exerted by the surrounding fluid and normal viscous stress components τ_{xx} that both act perpendicular to fluid element, and tangential stresses τ_{yx} and t_{zx} that act on the surfaces of the fluid element. The sum of these surface forces on a control volume in the x-direction can be written in terms of the pressure gradient and the viscous forces (known as the diffusion term)

$$F_{\text{surface}} = \underbrace{-\frac{\partial P}{\partial x}}_{\text{pressure}} + \underbrace{\nabla \cdot \sigma_{ij}}_{\text{diffusion}} \tag{5.10}$$

where σ_{ij} is the stress tensor that accounts for the viscous forces acting on the fluid element. Combining Eqs. 5.8–5.10 gives

$$\underbrace{\rho \frac{D\mathbf{U}}{Dt}}_{\text{inertia}} = \underbrace{\partial P}_{\text{pressure}} + \underbrace{\nabla \cdot \sigma_{ij}}_{\text{diffusion}} + \underbrace{\rho g_i}_{\text{gravity}} \tag{5.11}$$

Equation (5.11) is the fluid momentum equation in the form of Cauchy's equation momentum and can be used as a base for determining deformation in structures as well (see Sect. 5.3). This can be expanded out to see its individual terms for the x- y- and z- components of momentum. The term $D\mathbf{U}/Dt$ is a material derivative and is defined as the local and advection inertial force through its acceleration as

$$\frac{D\mathbf{U}}{Dt} = \underbrace{\frac{\partial u}{\partial t}}_{\text{local}} + \underbrace{\nabla \cdot u_i u_j}_{\text{advection}} \tag{5.12}$$

In two-dimensions the x-momentum equation in differential form is

$$\rho\left(\underbrace{\frac{\partial u}{\partial t}}_{\text{local}}+\underbrace{u\frac{\partial u}{\partial x}+v\frac{\partial u}{\partial y}}_{\text{advection}}\right)=\underbrace{-\frac{\partial P}{\partial x}}_{\text{pressure}}+\underbrace{\mu\left(\frac{\partial^2 u}{\partial x^2}+\frac{\partial^2 u}{\partial y^2}\right)}_{\text{diffusion}}+\underbrace{\rho g_x}_{\text{gravity}} \tag{5.13}$$

Similarly the momentum equation in the y-direction is

$$\rho\left(\underbrace{\frac{\partial v}{\partial t}}_{\text{local}}+\underbrace{u\frac{\partial v}{\partial x}+v\frac{\partial v}{\partial y}}_{\text{advection}}\right)=\underbrace{-\frac{\partial P}{\partial y}}_{\text{pressure}}+\underbrace{\mu\left(\frac{\partial^2 v}{\partial x^2}+\frac{\partial^2 v}{\partial y^2}\right)}_{\text{diffusion}}+\underbrace{\rho g_y}_{\text{gravity}} \tag{5.14}$$

5.2.2.1 Interpretation of the Local and Advection Term

Let us consider the x-momentum equation in Eq. (5.13) and its application in a stenosed artery in Fig. 5.7. Point A is located before the stenosis; Point B is located in the constricted passageway; while Point C is located after the stenosis. Under a steady state flow assumption the velocity over time at Point A is constant, and is represented with a horizontal line. This means that the local acceleration is $\partial u / \partial t = 0$. During the systolic phase of a heart beat the velocity variation over time is represented by the fluctuating profile given in Fig. 5.7b. The local acceleration, $\partial u / \partial t$ is the derivative of this curve which describes the motion of the fluid changing locally at a fixed point, and varying with time

The second term of the inertial force is the advection term which describes the fluid acceleration spatially. If we follow a fluid particle as it passes through points A, B, and C under a steady flow so that velocity itself is not fluctuating with time, the velocity component u now varies spatially mainly caused by the change in geometry. The velocity is accelerating between the locations of A, B, and C i.e. the velocity gradient in the term $u\,\partial u / \partial x$ of Eq. (5.13) is increasing. Similarly, after the stenosis, from point B to C, the velocity is decreasing and the spatial *acceleration*

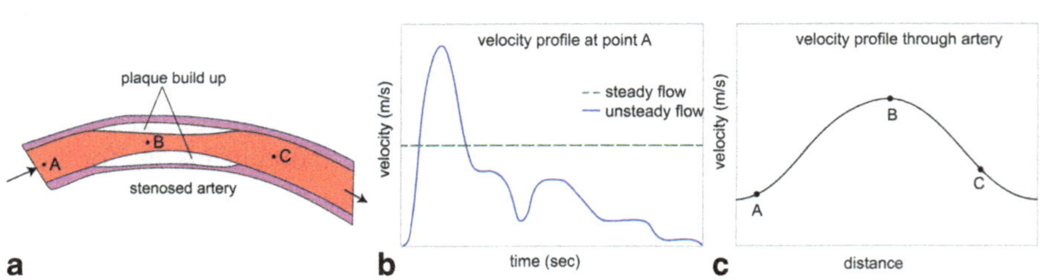

Fig. 5.7 **a** Blood flow motion through a stenosed artery with three locations for tracing its velocity. **b** Velocity profile monitored at a fixed location at point A for a steady and unsteady flow. **c** The motion of a fluid particle passing through points A, B, C

gradient is negative, i.e. a deceleration. We describe the fluid motion sweeping past points in space by the advection term in the momentum equations. If the geometry between points A, B, and C is fixed and does not vary, and in the absence of any other forces (e.g., gravity, thermal etc.) then the advection becomes zero.

5.2.2.2 Interpreting the Pressure Term

During fluid flow, the pressure term is a relative measure of the local intensity of material in moving. It is a stress force that acts normal to the control surface of the fluid element. When these forces are summed, the net pressure in each coordinate determines the fluid motion. Thus it is not the pressure itself which causes a net pressure but rather the pressure gradient, (e.g., $-\partial P / \partial x$) across the fluid element. A positive pressure gradient, i.e. increasing pressure, slows the fluid down, while a negative pressure gradient i.e. decreasing pressure, accelerates it.

To illustrate this, let us consider a steady flow through a bifurcation where we assume that at the bifurcation, the geometry has a circular shape shown in Fig. 5.8.

From fluid dynamics we can use the known flow variation along the approaching stagnation streamline (A–B) which is

$$u(x) = U_\infty \left(1 - \frac{R^2}{x^2}\right) \tag{5.15}$$

If we neglect the viscosity term in the momentum equation so that we can focus on the pressure term, then we have an inviscid flow. The diffusion terms are therefore zero. Along the stagnation streamline, the vertical velocity component v is zero. The x-momentum equation becomes

$$\rho\left(\underbrace{\frac{\partial u}{\partial t}}_{\text{local}} + \underbrace{u\frac{\partial u}{\partial x}}_{\text{convection}}\right) = \underbrace{-\frac{\partial P}{\partial x}}_{\text{pressure}} \tag{5.16}$$

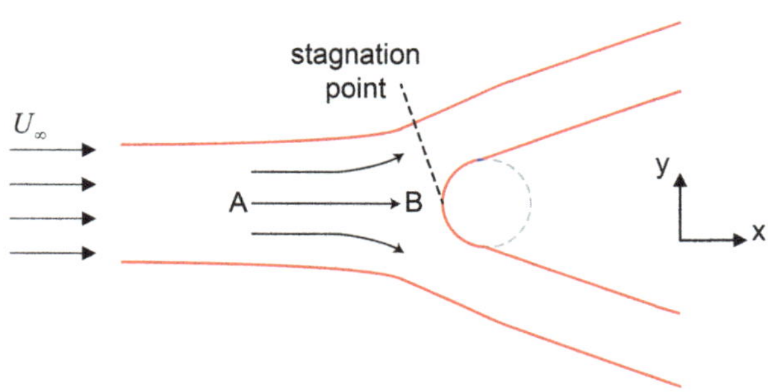

Fig. 5.8 Schematic of a steady flow through a bifurcation

and we see that total acceleration (i.e. sum of the local *acceleration* and *convection*) of the fluid is driven by the pressure gradient in the x-direction. The negative sign denotes that a positive pressure produces a decrease in acceleration, and vice-versa.

For a steady flow, we get

$$\rho\left(\underbrace{u\frac{\partial u}{\partial x}}_{\text{convection}}\right) = -\underbrace{\frac{\partial P}{\partial x}}_{\text{pressure}} \tag{5.17}$$

and when this is integrated along the streamline, produces Bernoulli's equation which we have seen from Chap. 4. For example,

$$u\frac{\partial u}{\partial x} = -\frac{1}{\rho}\frac{\partial P}{\partial x} \Rightarrow \frac{P(x)}{\rho} + \frac{u^2(x)}{2} = \frac{P_\infty}{\rho} + \frac{U_\infty^2}{2}$$

where P_∞ is the upstream pressure, and U_∞ is the freestream velocity. If we set the upstream velocity to $U_\infty = 1$ m/s and a radius of $R = 1$ m the above equation becomes

$$P(x) - P_\infty = \frac{\rho}{2}(U_\infty^2 - u^2(x)) = \frac{\rho}{2}\left[1 - \left(1 - \frac{1}{x^2}\right)^2\right]$$

Along the streamline, the velocity profile $u(x)$ and the acceleration is plotted in Fig. 5.9. The velocity drops very rapidly as the fluid approaches the bifurcation. At the bifurcation midpoint, the velocity is zero (stagnation point) and the surface pressure is a maximum. The pressure difference $p(x) - p_{atm}$ demonstrates that the pressure increases as the fluid approaches the stagnation point. With the density ρ set to 1 kg/m^3 it reaches a maximum value of 0.5, i.e. $p_{stag} - p_{atm} = (1/2)\rho U_\infty^2$ as $u(x) \rightarrow 0$ near the stagnation point.

The total acceleration profile depicted in Fig. 5.10 also shows strong deceleration of the fluid as it approaches the cylinder. The maximum deceleration occurs at $x = -1.29$ m with a magnitude of -0.372 m/s^2.

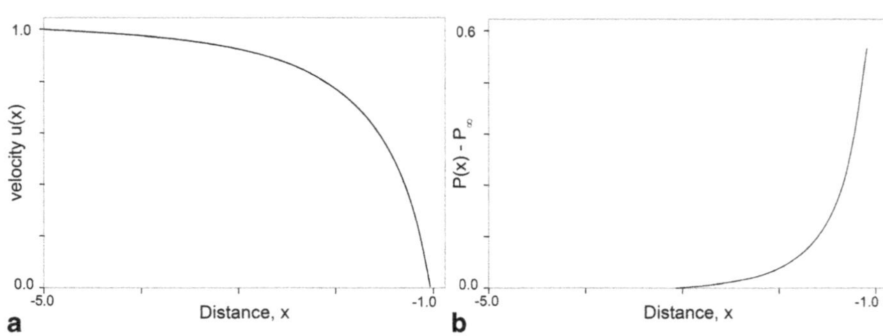

Fig. 5.9 **a** Velocity profile $u(x)$ along the stagnation streamline. **b** Pressure difference profile $P(x) - P_\infty$ along the stagnation streamline

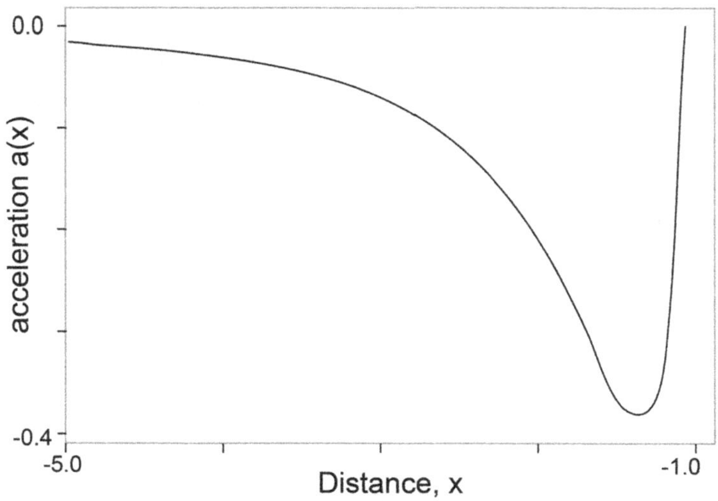

Fig. 5.10 Acceleration profile a(s) along the stagnation streamline

5.2.2.3 Interpreting the Diffusion Term

Diffusion is a physical process regarding flow properties that are transported by the random motion of gas molecules. Fluid momentum diffusion is caused by the molecular friction between fluid particles that move at different velocities. Its influence on the flow field is primarily dependent on its diffusion coefficient, μ which is the blood viscosity. The physical behaviour of viscosity was discussed in Chap. 4, where its main characteristic is to resist fluid deformation thereby inhibiting fluid flow. In contrast the convection term promotes fluid deformation and momentum. These two contrasting terms compete with each other to conserve momentum transfer.

In fluid dynamics, the concept of *dynamic similarity* is frequently adopted. This involves normalizing the mathematical equations to yield the non-dimensional governing equations. The non-dimensional form of the momentum equation is achieved by dividing all the dependent and independent flow variables by relevant constant quantities. For lengths the variable can be divided by a characteristic length D (e.g. diameter of blood vessel), all velocities by a reference velocity u_{in} (e.g. inlet velocity), and pressure by ρu_{in}^2 (twice the dynamic pressure),

$$x^* = \frac{x}{D}, \quad u^* = \frac{u}{u_{in}}, \quad v^* = \frac{v}{u_{in}}, \quad P^* = \frac{P}{\rho u_{in}} \tag{5.18}$$

where the asterisks denote the non-dimensional variables. Introducing these variables into the x-momentum equation produces

$$u^* \frac{\partial u^*}{\partial x^*} + v^* \frac{\partial u^*}{\partial y^*} = -\frac{\partial P^*}{\partial x^*} + \frac{1}{\mathrm{Re}} \left(\frac{\partial^2 u^*}{\partial x^{*2}} + \frac{\partial^2 u^*}{\partial y^{*2}} \right) \tag{5.19}$$

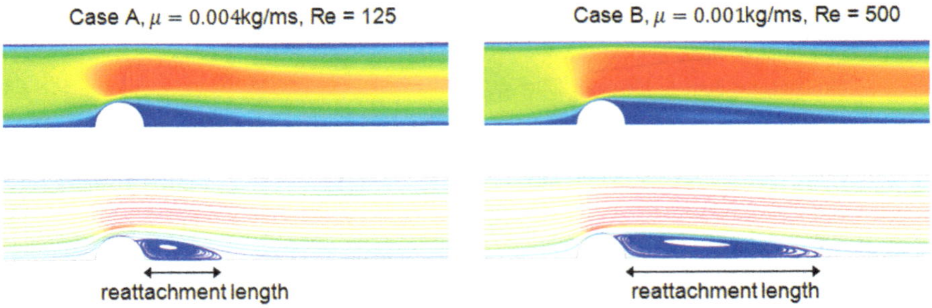

Fig. 5.11 Velocity contour and flow streamline plots for a flow through a stenosed blood vessel

where $\mathrm{Re} = \rho u D/\mu$ is the Reynolds number that was discussed in Chap. 4. A major advantage of non-dimensionalizing is the significant reduction of parameters, now involving just one, the Reynolds number. This confirms that the diffusion term competes with the convection term. This means that for various combinations of fluid velocities, dynamic viscosities, and vessel diameter, the same fluid flow effect is obtained if the Re is the same.

Let us consider an artery vessel that exhibits a semi-stenosis shown in Fig. 5.11. In case A, the viscosity is $\mu = 0.004\mathrm{kg/ms}$, four times as great as Case $\mathrm{B}=0.001\mathrm{kg/ms}$. Setting the density as $\rho = 1000\mathrm{kg/m^3}$, $D=0.01\mathrm{m}$, and the inlet velocity at $u = 0.05\mathrm{m/s}$, the corresponding Re numbers are: Case A = 125; Case B = 500.

Figure 5.11 shows the velocity contours and streamlines as the fluid passes over the stenosis. The flow accelerates, and separates as it passes over the stenosis, due to the reduced cross-sectional area. A region of recirculation is found immediately after the stenosis where the flow separates. After a distance downstream, the flow reattaches. The increased diffusion (larger μ) in Case A increases the resistance to flow thereby causing a reduction in the reattachment length.

If we now change the velocity for Case B from $u = 0.05\mathrm{m/s}$ to $u = 0.0125\mathrm{m/s}$, we obtain a Re = 125. The velocity reduces the overall momentum and thereby allows diffusion to have a relatively greater influence than before. The relative influence of the momentum and diffusion is then exactly the same as for Case A. This means that the velocity contour and streamline plot showing the reattachment length for the new case will be the same as that for Case A having a Re = 125.

By normalising the momentum equation, we have shown that the diffusion term competes with the convection term to influence fluid momentum through the use of the non-dimensional Re number. Similar flow characteristics can be achieved for different flow scenarios if the variables ρ, u, D, and μ produce the same Re number.

5.2.3 Introduction to Turbulence

In Chap. 4 we discussed the influence of turbulence on the flow character, and its comparison against laminar flow. In the laminar regime, flows are completely described by the continuity and momentum equations. Small disturbances within the fluid streamlines of a laminar flow can eventually lead to a chaotic and random state of motion—a condition of turbulence. These disturbances may originate from the free stream of the fluid motion, or induced by the surface roughness where they are amplified in the direction of the flow, in which case turbulence will occur. The onset of turbulence depends on the ratio of the inertia to viscous force, which is indicated by the Reynolds number, discussed in the previous section. At low Reynolds number, inertia forces are smaller than the viscous forces. The naturally occurring disturbances are dissipated away and the flow remains laminar. At high Reynolds number, the inertia forces are sufficiently large to amplify the disturbances, and a transition to turbulence occurs. The velocity and all other flow properties vary in a random and chaotic manner.

Turbulence is associated with the existence of *random fluctuations* in the fluid. If we consider flow through a blood vessel in Fig. 5.12, then at any moment in time its motion is random and unpredictable. If we measured its velocity at Point X over time, then the velocity variation would exhibit random fluctuations.

We observe that over time there is an averaged value for the velocity $\bar{u}$, with some deviation, defined as the fluctuating component $u'(t)$ at any moment in time. This means that the instantaneous velocity is decomposed as the sum of the average and its instantaneous fluctuation: $u(t) = \bar{u} + u'(t)$. In general, to characterize a turbulent flow the mean values of other flow properties ($\bar{u}$, $\bar{v}$, $\bar{w}$, $\bar{P}$ etc.) and its fluctuating component (u', v', w', p' etc.) is used.

The random fluctuations from turbulent flow should not be confused with time dependent oscillatory fluctuations. There is a common misconception that if a flow fluctuates then it is turbulent. To illustrate this, we consider a transient laminar flow over a cylinder (Fig. 5.13). Flow over a cylinder exhibits vortices that shed away (known as vortex shedding) from the cylinder in an oscillatory and periodic motion. Taking Point × in the flow field and tracking its velocity over time, we get a velocity profile shown in Fig. 5.13d that fluctuates periodically over time. This suggests that the velocity fluctuates but is orderly and is predictable—unlike turbulence which is random.

Turbulence models involve additional equations or modifications to the governing equations (continuity, momentum, and energy) to account for the turbulent fluctuations in the flow field by finding a solution for the Reynolds stresses in Eq. (5.23). They have been derived and improved over time by many researchers, based on experimental measurements, boundary layer theory, wall bounded flows and simple free shear flows. There are a number of turbulence models that range in complexity which is summarised later in this section.

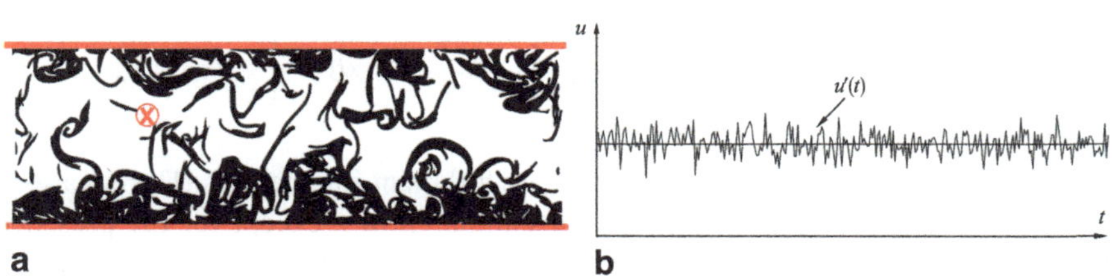

Fig. 5.12 **a** Schematic of instantaneous flow fluctuations in a blood vessel. **b** Velocity measurement taken at Point X over time displaying a averaged value with random deviations from the averaged value at any moment in time

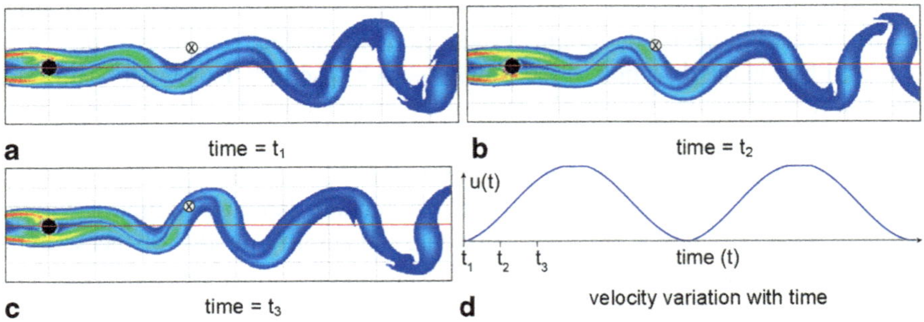

Fig. 5.13 Laminar flow over a cylinder showing the wake region that oscillates periodically over time

The purpose of turbulence modelling is to account for the averaged flow equations with the inclusion of a fluctuating component. The continuity and momentum equations are known as the Navier-Stokes equations (named after famous mathematicians) and its averaged form is called the Reynolds Averaged Navier-Stokes (RANS). This produces a set of time averaged equations with the turbulent features encapsulated by a new term called the Reynolds stress. It is this term that research effort and development in turbulence modelling aims to describe and resolve.

The RANS form of the continuity and momentum equations in 2D are

$$\frac{\partial \bar{u}}{\partial x} + \frac{\partial \bar{v}}{\partial y} = 0 \tag{5.20}$$

$$\begin{aligned}\frac{\partial \bar{u}}{\partial t} + \frac{\partial(\bar{u}\,\bar{u})}{\partial x} + \frac{\partial(\bar{v}\,\bar{u})}{\partial y} = -\frac{1}{\rho}\frac{\partial \bar{p}}{\partial x} + \frac{\partial}{\partial x}\left(\nu\frac{\partial \bar{u}}{\partial x}\right) + \frac{\partial}{\partial y}\left(\nu\frac{\partial \bar{u}}{\partial y}\right) + \frac{\partial}{\partial x}\left[\nu\frac{\partial \bar{u}}{\partial x}\right] \\ + \frac{\partial}{\partial y}\left[\nu\frac{\partial \bar{v}}{\partial x}\right] - \left[\frac{\partial\left(\overline{u'u'}\right)}{\partial x} + \frac{\partial\left(\overline{u'v'}\right)}{\partial y}\right]\end{aligned} \tag{5.21}$$

$$\begin{aligned}\frac{\partial \bar{v}}{\partial t} + \frac{\partial(\bar{u}\,\bar{v})}{\partial x} + \frac{\partial(\bar{v}\,\bar{v})}{\partial y} = -\frac{1}{\rho}\frac{\partial \bar{p}}{\partial y} + \frac{\partial}{\partial x}\left(\nu\frac{\partial \bar{v}}{\partial x}\right) + \frac{\partial}{\partial y}\left(\nu\frac{\partial \bar{v}}{\partial y}\right) + \frac{\partial}{\partial x}\left[\nu\frac{\partial \bar{u}}{\partial y}\right] \\ + \frac{\partial}{\partial y}\left[\nu\frac{\partial \bar{v}}{\partial y}\right] - \left[\frac{\partial\left(\overline{u'v'}\right)}{\partial x} + \frac{\partial\left(\overline{v'v'}\right)}{\partial y}\right]\end{aligned} \tag{5.22}$$

where $\bar{u}$, $\bar{v}$, $\bar{p}$ are mean values and u', v', p' are turbulent fluctuations. The equations above are similar to those formulated for laminar flows, except for the presence of $\overline{u'u'}$, $\overline{u'v'}$, and $\overline{v'v'}$ terms. As a result, we have three additional unknowns (in three dimensions, we will have nine additional unknowns), known as the Reynolds stresses, in the time-averaged momentum equations. How the Reynolds stresses is resolved is the key feature of all RANS-based turbulence models.

A first order approach is to relate the Reynolds stresses to the mean rates of deformation, (known as the Boussinesq assumption), similar to the way a laminar shear stress is related to the velocity gradient. For example:

Laminar shear stress: Turbulent shear stress:

$$\tau = \mu \frac{du}{dy} \qquad \tau_{turb} \approx \mu_t \frac{d\overline{u}}{dx} \approx -\rho\overline{u'u'}$$

$$\tau_{turb} \approx \mu_t \frac{d\overline{v}}{dy} \approx -\rho\overline{v'v'} \tag{5.23}$$

$$\tau_{turb} \approx \mu_t \left(\frac{d\overline{v}}{dy} + \frac{d\overline{v}}{dy} \right) \approx -\rho\overline{u'v'}$$

Equation (5.23) relates the turbulent momentum transport to be proportional to the mean gradients of velocity. The term μ_t is sometimes called the eddy viscosity (or turbulent viscosity) because the mixing effect is based on the motion of small recirculating eddies or vortices that transport the fluid particles for momentum exchange. The total shear stress is then

$$\tau = (\mu + \mu_t)\frac{d\overline{u}}{dy} \tag{5.24}$$

The eddy viscosity is not a fluid property but rather is a conceptual term that describes the Reynolds stresses with mean strain rates (i.e. velocity gradients). It is dependent on the flow field and determining its value is the objective of RANS turbulence models. This involves the inclusion of additional equations based on the convection and diffusion of the flow parameter in question. In the first order method for relating Reynolds stresses, the additional equations can range from zero to two which gives rise to its turbulence model classification. This includes:

- Zero equation model: mixing length model.
- One equation model: Spalart-Almaras.
- Two equation models: k-ε models, k-ω models

A second order approach is to directly use the continuity and momentum equations for the second order moments, which are the Reynolds stresses and turbulent fluxes instead of the Boussinesq assumption. The motivation is to overcome the limitations of first order models which assume isotropic turbulence and additional strains. This results in more additional equations and introduces more unknown variables. Second order turbulence models include the Reynolds stress models.

5.2.3.1 LES and DNS Approach

During the averaging of the governing equations, some physics may be lost. In the LES model (Large Eddy Simulation) a filtering function is applied to the equations whereby large turbulent structures, in the form of eddies/vortices are resolved directly (i.e. no modelling) while smaller eddies are modelled. Large scale turbulence depends strongly on the flow and boundary conditions, thus is flow-dependent and is difficult to model. On the other hand small scale turbulence is nearly homogenous and isotropic, making it independent of the flow and easier to model. LES is a time-dependent and 3D model which uses the mesh size as the filtering scale. This places a high demand on computational resources, since the mesh and time step sizes need to be fine enough to resolve down to the smallest turbulence scales, which becomes increasingly severe in near-wall regions where the scales are much smaller than those in the bulk flow region. LES is an advanced approach to RANS-based models which can handle transitional flows. It has the ability to represent coherent turbulent structures, and large scale turbulence, while modelling the small-scale effects.

The most complete approach is DNS (Direct Numerical Solution) which in fact is not a modelling approach at all. Rather it directly solves the governing equations without any modelling by using sufficiently fine mesh and time steps to capture all scales of turbulence. This places an enormous demand on computational resources. For example a typical flow domain having a cross-sectional area of 0.1 m by 0.1 m with a high Reynolds number turbulent flow might contain eddies as small as 10 or 100 μm in size. This means that to resolve the flow at all length scales, a computational mesh of 10^9 to 10^{12} grid points are needed. Furthermore, the fastest events that can take place have a frequency of the order of 10 kHz, which would require a time step of about 100 μs. At this stage and even in the near future, DNS of turbulent flows at high Reynolds numbers are not feasible. Only low to moderate Reynolds numbers under very simple geometries can be handled with the fastest computers today. This means that for practical engineering applications, DNS is an unrealistic and inefficient approach since the smallest scales of turbulence may not be necessary. Figure 5.14 shows a comparison of the different modelling approaches in terms of how much of the flow is modelled or resolved, and the required computational demand for each approach.

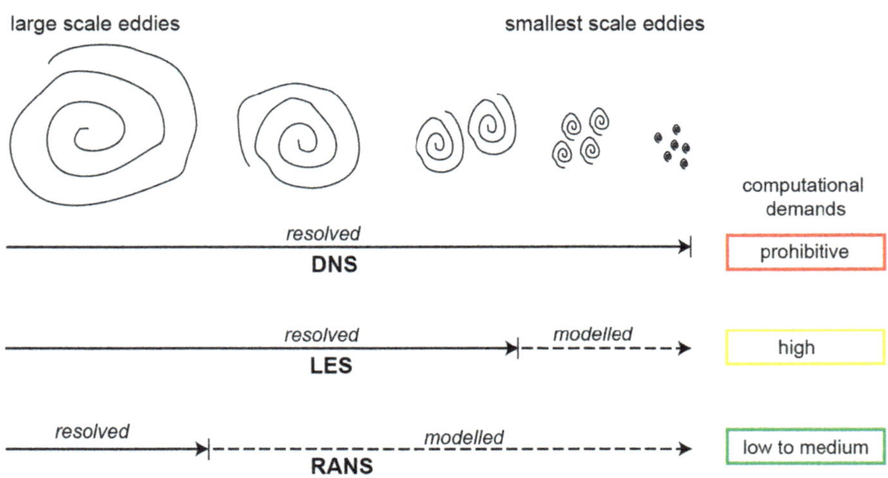

Fig. 5.14 Comparison of turbulence models and the amount of flow physics that is resolved or modelled

Table 5.1 Summary of common RANS turbulence models

Turbulence model	Notes	Advantages	Disadvantages
mixing length	No additional equations, but relies on a mixing length theory to find the Reynolds stress	Fast calculation times. Good predictions for simple flows where experimental correlations for the mixing length exist	Cannot describe flow separation or recirculation where the turbulent length scale varies
Spalart-Allmaras	One additional equation for the turbulent viscosity	Good for attached wall-bounded flows, and flows with mild separation and recirculation	Bad for large separation, free shear flows, and decaying turbulence
k-ε models	Two additional equations one for k and ε each. Most widely used model along with the k-ω. Assumes fluid flow is fully turbulent. The model leads to equal normal stresses, and isotropic turbulence	Stable calculations and reasonable predictions for many flows. Most general turbulence of all RANS models	Poor predictions for swirling and rotating flows, strong separation, severe pressure gradient. Lack of sensitivity to adverse pressure gradients
k-ω models	Two additional equations one for k and ω each. Its numerical behaviour is similar to the k-ε and suffers from similar disadvantages such as the isotropic turbulence assumption. Allows for a more accurate turbulent profile near the wall but requires a fine mesh to resolve the thin turbulent boundary layer. Is generally superior to k-ε for wall-bounded, free shear. and low Reynolds number flows, but separation is typically predicted to be excessive and early		
Reynolds Stress Models	Seven additional equations, one each for the six independent Reynolds stresses, and one turbulent dissipation. Unlike other RANS model, the isotropic turbulence assumption is avoided	Accurately predicts more complex flows, accounting for streamline curvature, swirl, rotation, high strain rates, and separation	More equations contain more unknown terms that need to be modeled. Requires more computational time due to additional equations

We summarise the more common turbulence models in Table 5.1 highlighting the advantages, disadvantages, and notes.

5.3 Introduction to Solid Mechanics

5.3.1 Elasticity

A material is elastic if it deforms under an applied force, and returns to its original position when the force is removed. Conceptually we consider an elastic body subjected to a force, F sufficient to initiate deformation (Fig. 5.15). As a result of

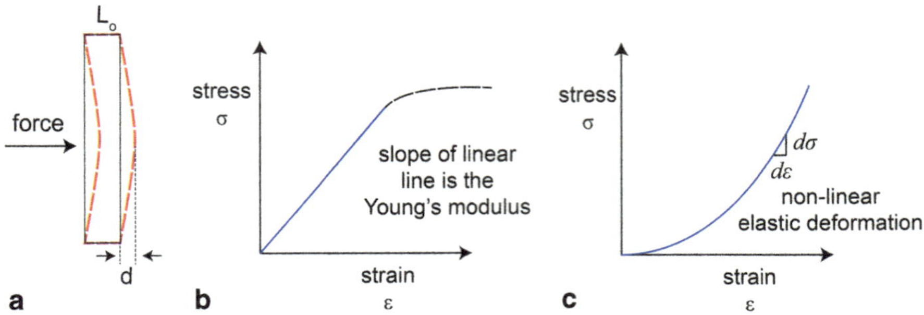

Fig. 5.15 A body such as a wall structure subjected to a force load. **a** The initial length is L_o and the deflection caused by the force load is d. For elastic materials **b** the stress-strain relationship is governed by Hooke's Law producing a linear line that is the Young's modulus. **c** For non-linear elastic deformation the Young's modulus is found by taking the incremental gradient

the force, the body deforms and the amount of deformation, d as a ratio of its initial shape, e.g. length, L_o is defined as the strain, ε. The amount of force applied per unit area, A is the stress loading, δ. During linear elastic deformation the relationship between stress and strain, known as Hooke's Law is shown in Fig. 5.15b where the slope of the linear line is the material's Young's modulus, E. These relationships, stress, strain, and Hooke's Law are given below:

$$\varepsilon = \frac{d}{L_o} \qquad \sigma = \frac{F}{A} \qquad E = \frac{\sigma}{\varepsilon} \tag{5.25}$$

Young's modulus, E describes how *stiff* a material is; a higher E produces a steeper slope, which means that a greater force loading is needed to deform the material.

Arteries exhibit non-linear elastic behaviour, where the material does not follow Hooke's Law (Fig. 5.15). The relationship is non-linear showing that as the stress in an artery increases, the material becomes stiffer and resists strain. In such cases Young's modulus is defined as the slope of the curve at a given stress-strain point, which is termed the incremental Young's modulus, defined as

$$E_{inc} = \frac{d\sigma}{d\varepsilon} \tag{5.26}$$

If we consider the strain in all three directions, then Eq. (5.25) which refers to deformation in one direction can be rewritten in its separate coordinates as

$$\varepsilon_x = \frac{\sigma_x}{E} \qquad \varepsilon_y = \frac{\sigma_y}{E} \qquad \varepsilon_z = \frac{\sigma_z}{E} \tag{5.27}$$

During material deformation, axial strain occurs in the direction of the force load. In addition there is deformation in the other two directions, laterally and perpendicular to the force direction (Fig. 5.16).

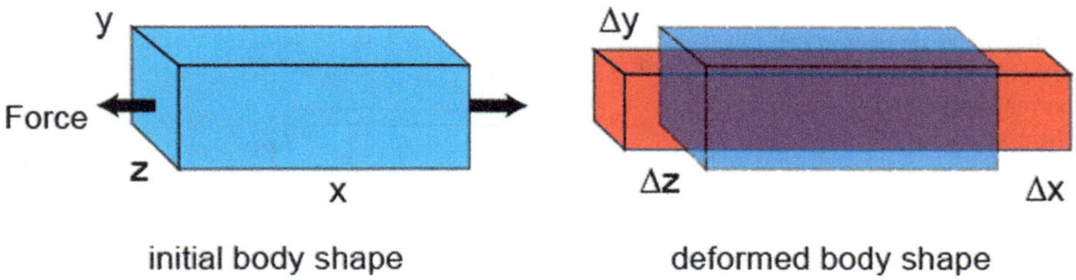

Fig. 5.16 Deformation in all three directions of an elastic material subjected to a force in tension along the x-axis. The initial body shape expands in the x-direction but contracts in the y- and z-directions

Deformation in all three directions caused by an axial force load is referred to as the Poisson effect and for isotropic materials, the Poisson's ratio, ν is

$$v = \frac{\text{lateral strain}}{\text{axial strain}} = \frac{-\varepsilon_y}{\varepsilon_x} = \frac{-\varepsilon_z}{\varepsilon_x} \tag{5.28}$$

which measures the change in shape in the lateral direction in relation to the change in shape in the axial direction.

5.3.2 *Plane Stress*

Stresses occurring on arterial walls can be assumed as a plane stress, where we assume the walls are subjected to parallel blood shear forces. By using this assumption the stress analysis is simplified considerably into a 2D analysis (Fig. 5.17).

The stress-strain relations are given here without its derivation (a full description can be found in Hutton 2000) as,

$$\begin{Bmatrix} \sigma_x \\ \sigma_y \\ \tau_{xy} \end{Bmatrix} = \frac{E}{1-v^2} \begin{bmatrix} 1 & v & 0 \\ v & 1 & 0 \\ 0 & 0 & \frac{1-v}{2} \end{bmatrix} \begin{Bmatrix} \varepsilon_x \\ \varepsilon_y \\ \gamma_{xy} \end{Bmatrix} \tag{5.29}$$

which can be rewritten as

$$\{\sigma\} = [\mathbf{D}]\{\boldsymbol{\varepsilon}\}$$

where σ is a column matrix of stress components, **D** is the elastic material property matrix for plane stress, and $\boldsymbol{\varepsilon}$ is the column matrix of strain components. The strain energy per unit volume, u_e, becomes

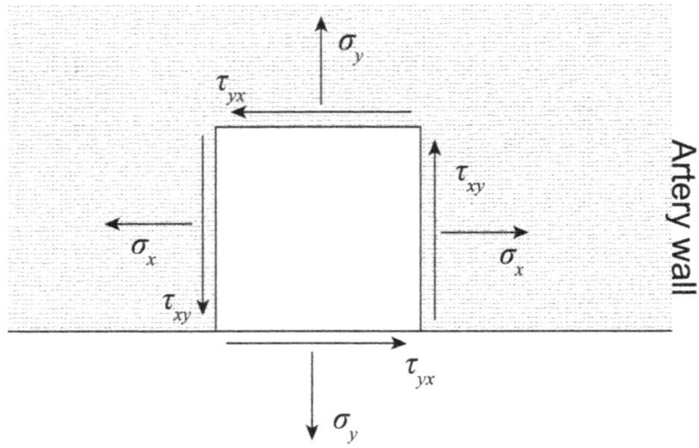

Fig. 5.17 Plane stresses acting on an element in the arterial wall, where. σ is the normal stress, and τ is the shear stress

$$u_e = \int_0^{\varepsilon} \sigma d\varepsilon = \frac{1}{2}(\sigma_x \varepsilon_x + \sigma_y \varepsilon_y + \tau_{xy}\gamma_{xy}) \tag{5.30}$$

which in matrix notation is

$$u_e = \frac{1}{2}\{\varepsilon\}^T \{\sigma\} = \frac{1}{2}\{\varepsilon\}^T [D]\{\varepsilon\} \tag{5.31}$$

The total strain energy of a body subjected to plane stress is then

$$U_e = \frac{1}{2}\int\int\int \{\varepsilon\}^T [D]\{\varepsilon\}\mathrm{dV} \tag{5.32}$$

where V is total volume of the body.

5.3.3 Structural Dynamics Equations

Structural analysis mainly concerns the behaviour of a physical structure when subjected to a force. We present an introduction to structural dynamics by exploring a vessel wall analogous to a simple single degree of freedom (i.e. one displacement direction) of a spring-mass-damper system. Vessel walls, which are elastic in behaviour, are subjected to hydrodynamic forces from the blood that it transports causing deformation in the wall structure. If the applied force deforms the vessel wall sufficiently to overcome the structure's natural frequency, an oscillatory motion will occur. In response to an applied force, the vessel walls produce three opposing forces:

- an inertia force, which is proportional to acceleration, $F_i = m\ddot{\mathbf{d}}$
- a damping force, which is proportional to velocity, $F_c = c\dot{\mathbf{d}}$
- a stiffness force, which is proportional to displacement, $F_k = k\mathbf{d}$

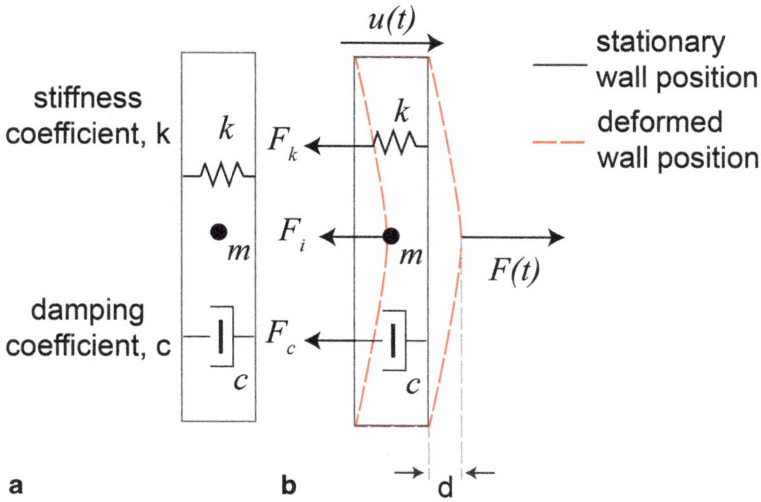

Fig. 5.18 Schematic of **a** static vessel wall and **b** deformed vessel wall due to a dynamic external force F(t). The vessel wall structure exhibits material properties: stiffness coefficient, a function of its Young's modulus; and damping coefficient, a function of friction and viscous effects

The coefficient m is the mass, c is the damping coefficient, and k is the stiffness coefficient based on the elastic material properties of the vessel walls, and **d** is the displacement. The elasticity describes its tendency to deform non-permanently. Putting these forces together we get the general equation of motion (Fig. 5.18).

$$m\ddot{\mathbf{d}} + c\dot{\mathbf{d}} + k\mathbf{d} = \mathrm{F(t)} \tag{5.33}$$

The *inertia force* is based on d'Almeberts Principle which states that a mass develops an inertial force, F_i proportional to its acceleration in an opposing direction. Figure 5.19 shows this principle where if we denote **d** as the displacement, then the velocity is equal to its derivative with respect to time given as $\dot{\mathbf{d}}$, and acceleration is the second derivative with respect to time given as $\ddot{\mathbf{d}}$.

The *damping force*, F_d is the product of the velocity and damping coefficient, c. This force attenuates motion and structural deformation. In fluid-structure-interaction models an applied momentary force will induce a displacement leading to an excitation in the structure causing oscillation. The damping force attenuates this oscillation, reducing the amplitude with time (Fig. 5.19c). In blood vessels damping is caused by friction/viscosity, and thermal effects, which dissipates the energy stored in the oscillation.

The *stiffness force* represents the restoring body force when it has deformed, allowing the body to restore to its original position. The magnitude of this force is proportional to the deformation **d,** and the material property k, which is the stiffness coefficient. Various elastic moduli may be used to describe the stiffness, but typically for an FSI problem, the Young's modulus is sufficient.

In a fluid-structure-interaction problem the structural equations are described in Lagrangian coordinates, typically applied to a finite element mesh which differs to

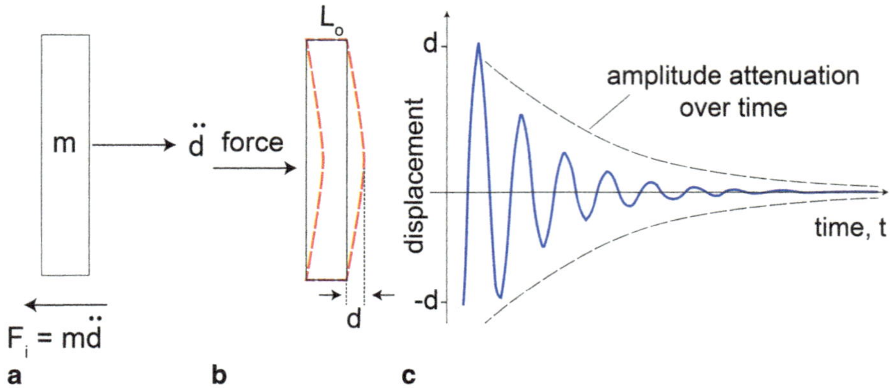

Fig. 5.19 **a** Deformation of a vessel wall caused by an applied force develops an inertial force in response to an applied load. **b** Upon release of the force the wall will deform towards its original shape but may overshoot its original position giving rise to an oscillatory deformation. **c** The damping force attenuates the oscillation with time

the finite volume approach that is often used for fluid flow equations (although fluid flow can also be cast in finite element method). Equation (5.33) is rewritten with the stiffness force replaced by a stress term, per unit volume to give

$$\rho_s \ddot{\mathbf{d}} + c\dot{\mathbf{d}} + \nabla \cdot \sigma_{ij} = \mathbf{f} \tag{5.34}$$

ρ_s is the density of the solid, and σ_{ij} is the stress tensor with the divergence representing internal elastic forces. The external force, **f** represents forces induced by blood flow that includes fluid pressure and shear.

The relationship between the stress tensor and deformation (i.e. strain) represents the constitutive behaviour of the vessel wall. This may be linear or non-linear depending on the material. In general the relationship is of the form $\sigma_{ij} = D\varepsilon_{ij}$ where D is the constitutive elastic stiffness matrix (typically a function of the material elasticity, Young's modulus, and Poisson ratio), and ε_{ij} is the strain tensor. In-depth descriptions of these relationships are given in Xia and Lin (2008) and Zienkiewicz and Taylor (2000).

5.3.4 Elastic Properties of Arteries

Arteries carry blood from the heart, which exhibits high pressure, while veins conduct blood from the capillaries back to the heart which have lower pressure. These pressure forces act on the vessel walls causing elastic deformation. During systolic phase, healthy arteries expand, and absorb the peak systolic pressure. During diastole the blood pressure reduces and the vessel walls return to their original form to maintain the flow. When arteries become diseased, such as arteriosclerosis, they become stiffer, and lose the ability to deform freely, increasing pressure at peak systolic phase.

Artery walls are made up of four types of tissues, (i) endothelial cells, (ii) connective elastic and collagenous fibres, (iii) smooth muscle, and (iv) irregular connective elastic and collagenous fibres. These are found within three layers: the tunica intima (inner), the tunica media (middle), and the tunica adventitia (outer). Figure 5.20 shows a schematic of the artery wall structure and composition of tissues within each of the three layers. It also shows the relative tissue composition within each artery type. The tunica intima consists of endothelial cells; the tunica media is the middle layer and is composed of smooth muscle and elastic fibres; and the tunica adventitia contains connective collagenous fibres that anchor the blood vessel to nearby organs, giving it stability.

The amount and composition of different tissues in the artery varies based on the artery size and this determines its elasticity. More elastin in the arteries allows them to expand and contract. Accordingly the arteries are either predominatly muscular or elastic. The aorta and larger arteries contain more elastin which stretches as the blood is forced into the artery. The deformed artery holds this energy (like the stiffness constant, k in Eq. (5.33)) and during diastole releases this energy, recoiling back to its original shape during diastole to propel the blood forward. The ability to absorb the peak systolic pressure dampens sudden loads on the artery wall. Arterioles contain less elastin and more smooth muscle which make the walls more rigid and stiff.

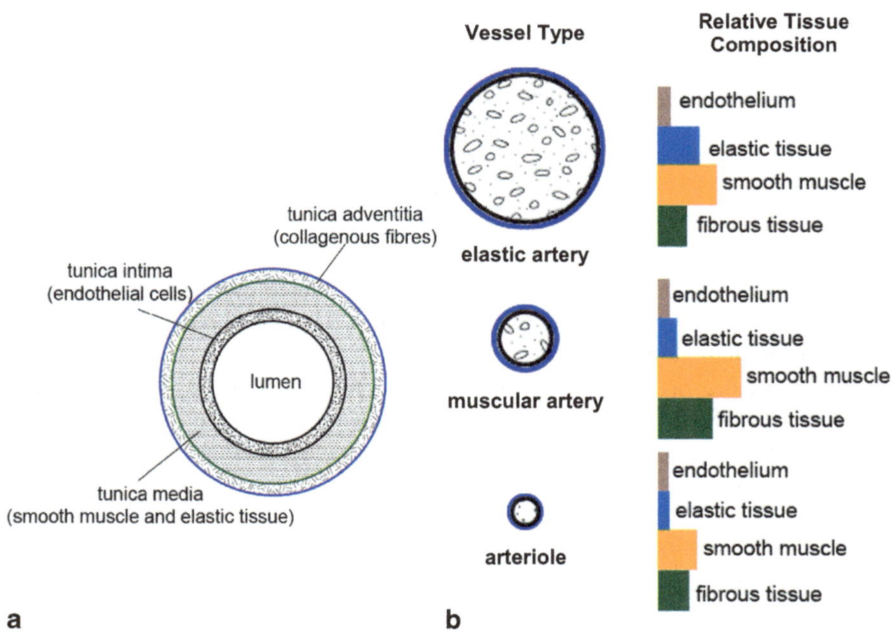

Fig. 5.20 Arterial wall has a three-layered anatomy made up of tunics—tunica intima, tunica media, and tunica adventitia. The intima consists of endothelial cells, the media consists of smooth muscle and elastic tissue, and the adventitia consists of collagenous fibres. Larger arteries are typically elastic containing a larger relative composition of elastin, while smaller arteries are predominantly made up of smooth muscle

5.4 Computational Methods

The conservation equations for fluid flow (i.e. mass, momentum) and structural domain (i.e. momentum) are partial differential equations (PDEs). Computational methods are used to transform these equations into discrete algebraic equations which are then applied to mesh points within the fluid and structural domain. Discretisation of the fluid equations are most commonly use the finite difference or finite volume approach, while for the structural equations, it is the finite-element approach.

5.4.1 Finite Difference Method

In the finite difference method, the partial derivatives are approximated by algebraic equations at grid nodal points through the Taylor series expansion. The Taylor series is defined as:

$$f(x) \approx f(a) + \frac{1}{1!}\frac{\partial f\ (a)}{\partial x}(x-a) + \frac{1}{2!}\frac{\partial f^{\ 2}(a)}{\partial x^2}(x-a)^2 + \frac{1}{3!}\frac{\partial f^{\ 3}(a)}{\partial x^3}(x-a)^3 + \cdots \tag{5.35}$$

and is rewritten in sigma notation as

$$f(x) \approx \sum_{n=0}^{\infty} \frac{f^{(n)}(a)}{n!}(x-a)^n \tag{5.36}$$

This is applied to a one-dimensional uniformly distributed Cartesian grid shown in Fig. 5.21, where location a is replaced by the index i. In two-dimensions the indices are (i, j) and in three-dimensions (i, j, k).

Following the Taylor series, we substitute $f(a)$ as ϕ_i and $f(x)$ as ϕ_{i+1} where ϕ is a generic flow field variable. This produces the variable at point $(i+1)$ expanded about the point (i). Rearranging the equation, the approximation for the partial derivative becomes

$$\left(\frac{\partial \phi}{\partial x}\right)_i = \frac{\phi_{i+1} - \phi_i}{\Delta x} + \underbrace{O(\Delta x)}_{Truncation\ error} \quad \text{where} \quad O(\Delta x) = \left(\frac{\partial^2 \phi}{\partial x^2}\right)_i \frac{\Delta x}{2} + \left(\frac{\partial^3 \phi}{\partial x^3}\right)_i \frac{\Delta x^2}{6} + \cdots \tag{5.37}$$

The term $O(\Delta x)$ signifies the truncation error related to the finite difference approximation. It measures the accuracy of the approximation and determines the rate

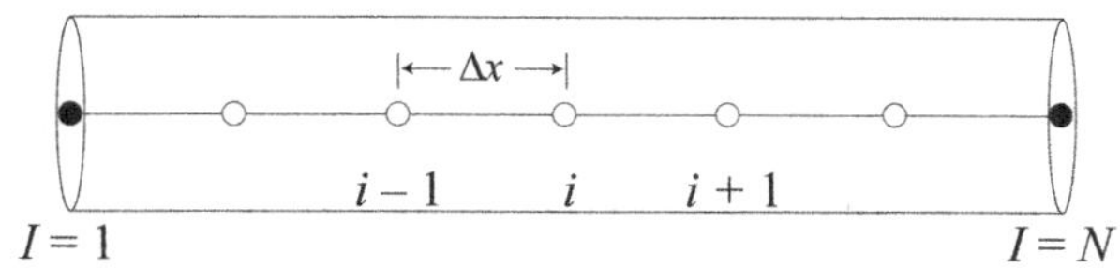

Fig. 5.21 A representation of a one-dimensional and two-dimensional uniformly distributed Cartesian grid for the finite difference method (full symbols denote boundary nodes and open symbols denote computational nodes)

at which the error decreases based on the lowest-order term in the truncated terms. Since the equation describes the function from point (i) to a position in front of it at ($i+1$), the finite difference formulation is called a *forward difference* and is influenced by information to the right or in front of the origin.

Similarly the Taylor series expansion for the variable at point ($i-1$) with respect to point (i) gives the *backward difference* approximation as

$$\left(\frac{\partial \phi}{\partial x}\right)_i = \frac{\phi_i - \phi_{i-1}}{\Delta x} + \underbrace{O(\Delta x)}_{\textit{Truncation error}}$$
$$\textit{where} \quad O(\Delta x) = \left(\frac{\partial^2 \phi}{\partial x^2}\right)_i \frac{\Delta x}{2} - \left(\frac{\partial^3 \phi}{\partial x^3}\right)_i \frac{\Delta x^2}{6} + \cdots \tag{5.38}$$

Combining the forward and backward difference equations we get the *central difference* approximation

$$\left(\frac{\partial \phi}{\partial x}\right)_i = \frac{\phi_{i+1} - \phi_{i-1}}{2\Delta x} + \underbrace{O(\Delta x^2)}_{\textit{Truncation error}} \quad \textit{where} \quad O(\Delta x^2) = \left(\frac{\partial^3 \phi}{\partial x^3}\right)_i \frac{\Delta x^2}{3} + \cdots \tag{5.39}$$

For second order derivatives, such as the diffusion term in the momentum equation, we sum the Taylor series expansions from Eqs. (5.37) and (5.38), which gives

$$\left(\frac{\partial^2 \phi}{\partial x^2}\right)_i = \frac{\phi_{i+1} - 2\phi_i + \phi_{i-1}}{(\Delta x)^2} + \underbrace{O(\Delta x^2)}_{\textit{Truncation error}} \quad \textit{Central difference} \tag{5.40}$$

This equation represents the central finite difference for the second order derivative with respect to x evaluated at the point (i).

Visual Representation Figure 5.22 shows the grid points that are involved and their contribution in terms of being added or subtracted towards the finite difference approximation for the first and second order derivatives.

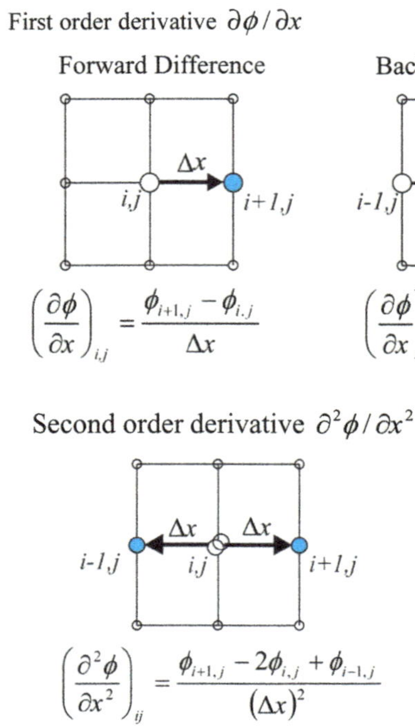

$$\left(\frac{\partial\phi}{\partial x}\right)_{i,j}=\frac{\phi_{i+1,j}-\phi_{i,j}}{\Delta x}\qquad\left(\frac{\partial\phi}{\partial x}\right)_{i,j}=\frac{\phi_{i,j}-\phi_{i-1,j}}{\Delta x}\qquad\left(\frac{\partial\phi}{\partial x}\right)_{ij}=\frac{\phi_{i+1,j}-\phi_{i-1,j}}{2\Delta x}$$

$$\left(\frac{\partial^2\phi}{\partial x^2}\right)_{ij}=\frac{\phi_{i+1,j}-2\phi_{i,j}+\phi_{i-1,j}}{(\Delta x)^2}$$

Fig. 5.22 Finite difference approximations for the first order derivatives $\partial\phi/\partial x$ and $\partial\phi/\partial y$, and second order derivatives $\partial^2\phi/\partial x^2$. White circles represent a negative value, while shaded circles represent a positive value, at a specified point each of which contribute towards the finite difference approximation

Example: 1D Convection-Diffusion Momentum Flow Through a Pipe
We consider momentum equation from Eq. (5.13) in the form of a 1D convection-diffusion equation as

$$\underbrace{\frac{\partial u}{\partial t}}_{\text{local}}+\underbrace{A\frac{\partial u}{\partial x}}_{\text{convection}}=+\underbrace{D\left(\frac{\partial^2 u}{\partial x^2}\right)}_{\text{diffusion}} \tag{5.41}$$

where A is treated as a constant velocity, and D is μ/ρ, the kinematic viscosity. The velocity u is a function of time, t and distance, x. The local acceleration term is discretised with a forward difference, the convection term with a central differencing, and the diffusion term with a second difference summarised as

$$\frac{\partial u}{\partial t}=\frac{u_i^{n+1}-u_i^n}{\Delta t}\qquad\frac{\partial u}{\partial x}=\frac{u_{i+1}^n-u_{i-1}^n}{2\Delta x}\qquad\frac{\partial^2 u}{\partial x^2}=\frac{u_{i+1}^n-u_i^n+u_{i-1}^n}{\Delta x^2} \tag{5.42}$$

Substituting Eq. (5.42) into (5.41) and rearranging for the new value forward in time we get

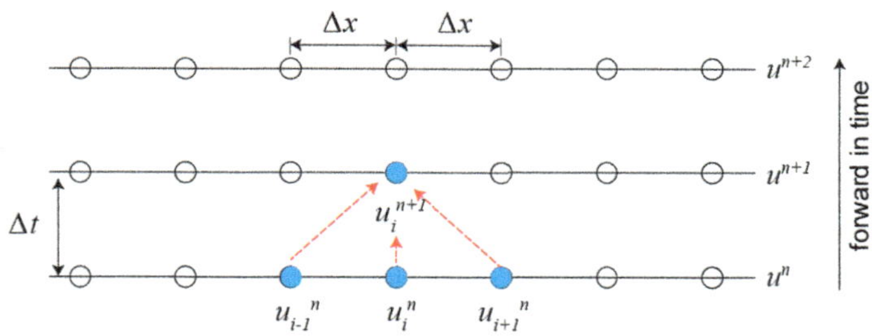

Fig. 5.23 Numerical approximation of the convection-diffusion equation onto discrete nodes. The values on the new time step $n+1$, are found from explicitly from values at the current time step n

$$u_i^{n+1} = u_i^n - A\frac{\Delta t}{2\Delta x}(u_{i+1}^n - u_{i-1}^n) + D\frac{\Delta t}{\Delta x^2}\left(u_{i+1}^n - 2u_i^n + u_{i-1}^n\right) \tag{5.43}$$

This equation is looped over each node in a grid mesh. Schematically this is shown in Fig. 5.23, where the value of each node at the next time step $n+1$ is given explicitly in terms of the values at the current time step n. Constant values for the problem include:

$$A = 1 \qquad \Delta t = 0.005 \qquad \Delta x = 0.05$$

$$D = 0.1 \qquad \text{n} = 200 \qquad \text{total } x = 1.0$$

The solution is obtained by calculating Eq. (5.43) over every node. Since the solution solves for nodes at $n+1$ based on nodes at time n, we need to provide an initial condition for all nodes at time $t = 0$ to start the calculations. We set the flow domain with an initial sine wave velocity profile shown in Fig. 5.24 defined by

$$u(x,t=0) = -\sin(2\pi x) + 1 \tag{5.44}$$

The solution over $n=150$ time steps are shown in Fig. 5.25 where the solution is a decaying travelling wave. The convection term transports the initial wave profile, while the diffusion term dissipates it. We see that for Case B the initial velocity wave dissipates rapidly due to the increased diffusion ($D=0.3$ for Case B compared with

Fig. 5.24 Initial velocity profile applied onto the discrete nodes in the 1D domain

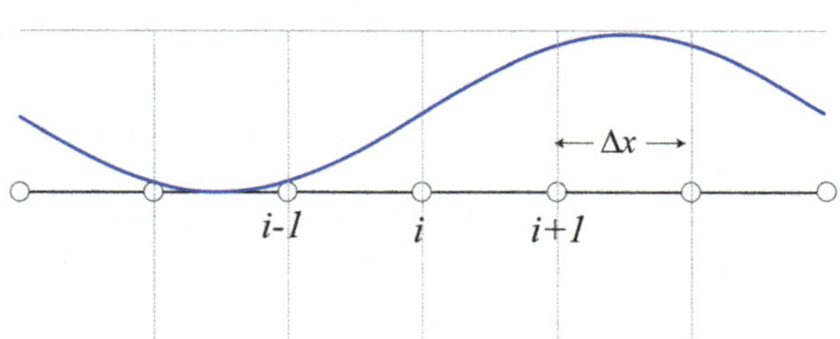

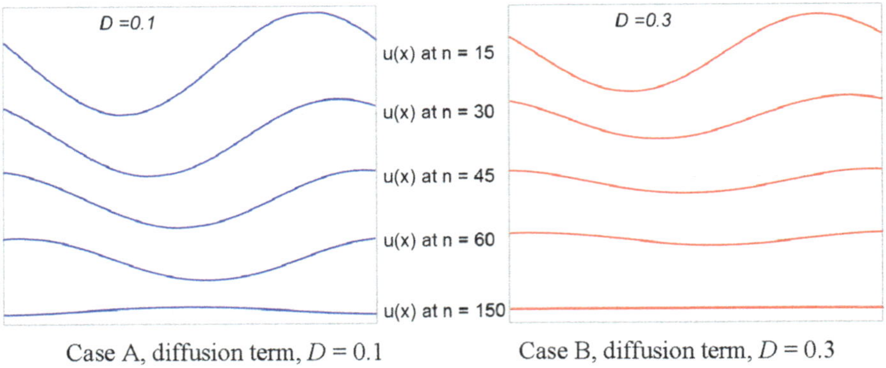

Fig. 5.25 Velocity profile in the one-dimensional domain produced from the convection-diffusion equation over time. The time dependent solution displays a decaying travelling wave. Case A uses a diffusion term with value $D=0.1$, and Case B uses a diffusion three times as great, with $D=0.3$

$D = 0.1$ for Case A). The relative strength of convection from $A \cdot \partial u / \partial x$ and diffusion from $D \cdot \partial^2 u / \partial x^2$ is related by the dimensionless Peclet number, $P_e = AL / D$, where L is a characteristic length, usually taken as the blood vessel diameter. This number is analogous to the Reynolds number that is used for the non-linear form, and is an important number used to define the stability of a solution which is discussed in later sections in this chapter.

By enforcing constant values for the coefficients A, and D, we obtain the linear form of the convection–diffusion equation which differs to its non-linear form of the momentum equation. This means that the flow and diffusion don't interact.

5.4.2 Finite Volume Method

The finite volume approach is based on conservation principles we saw earlier in developing the momentum equations. If we consider a single control volume (Fig. 5.26) and apply the fundamental principle that mass is conserved, then

That is, '*the flux of a variable (i.e. net rate of a mass of a variable that crosses the control surface) is equal to the net change in quantity of the variable inside the control volume*'. This conservation principle is the cornerstone of the finite volume method. Since the control volumes can take arbitrary shapes, it allows more flexibility in representing the grid by either structured or unstructured mesh, as it is not limited by the cell shape. A disadvantage of finite volume is that it is susceptible to false/numerical diffusion when simple numerics and poor choice of mesh design is applied.

The physical domain is divided into discrete control volumes and nodes are placed between control volume boundaries. The analysed node is given the variable P, and its neighbour nodes are E and W denoting its east and west neighbour (Fig. 5.27).

Upon discretisation, the resulting equation for nodal point P is generalised as

$$a_P \phi_P = \sum a_{nb} \phi_{nb} + b \tag{5.45}$$

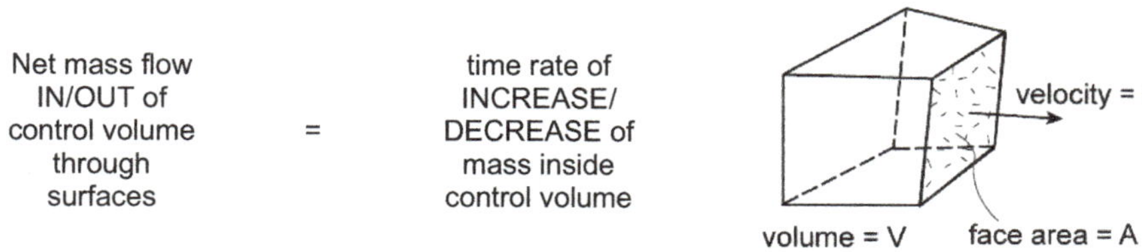

Fig. 5.26 Conservation principle applied to a single control volume

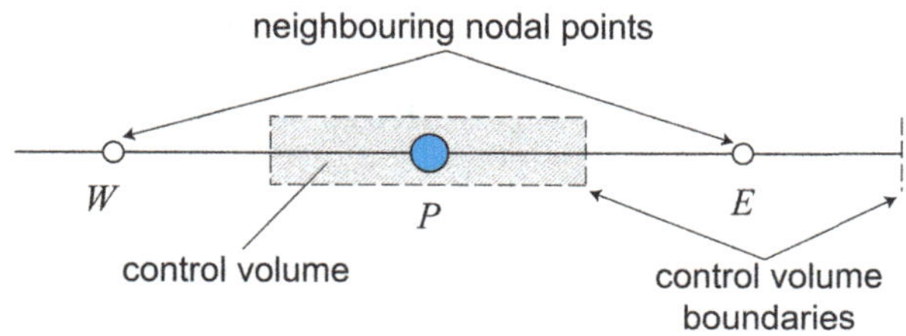

Fig. 5.27 Control volume schematic for the finite volume approach

where a_{nb} is the neighbouring node coefficient, ϕ_{nb} is the flow variable at the neighbouring node, and b is a constant. There are a number of discretisation schemes available but for it to produce physically realistic solution, it must satisfy rules of: conservativeness, boundedness, and transportiveness.

Conservativeness refers to flux consistency at the control volume faces so that the flux leaving a control volume is equal to the flux entering the adjacent control volume through the same face. Boundedness requires the discretized equations to have coefficients satisfying

$$\frac{\sum |a_{nb}|}{|a_p - S_p|} \tag{5.46}$$

where S_p comes from source terms. This constraint can be satisfied if a_p is as large as possible, and S_p is negative. All coefficients should be of the same sign, preferably positive so that an increase in ϕ at one node should result in an increase in ϕ at neighbouring nodes. For transportiveness, the Peclet number is used which we saw earlier is the relative strengths of convection and diffusion

$$\mathrm{Pe} = \frac{\text{convection}}{\text{diffusion}} = \frac{u}{v}L \tag{5.47}$$

where u is the fluid velocity, v is the kinematic velocity, and L is a characteristic length. The Peclet number, Pe is related with the directionality of influence and this is determined by the discretisation scheme. This is presented schematically in Fig. 5.28 where a value of Pe=0 implies pure diffusion, and the contour of influence are concentric circles for constant values of ϕ since diffusion tends to spread out evenly in all directions. As Pe increases, the influence curve becomes more

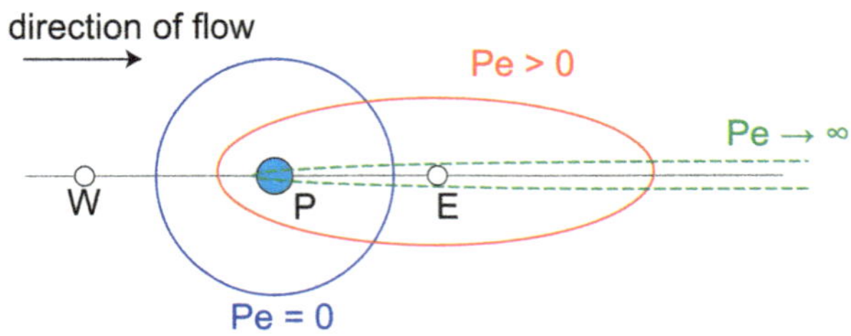

Fig. 5.28 Influencing regions from the flow field on nodal points in the discretised domain

elliptic and the value of ϕ at the E node is influenced more by the upstream node. When $\mathrm{Pe} \rightarrow \infty$ the flow is in pure convection, and the influence contour is stretched where the value of ϕ at E is affected only by upstream conditions.

5.4.3 One-Dimensional Steady State Convection-Diffusion in Finite Volume

The convection terms move a scalar variable, ϕ in the flow direction by the velocity component while diffusion distributes the variable in all directions by its diffusion coefficient, Γ. If we ignore source terms and focus on steady convection and diffusion of the variable ϕ in one dimension, its equation becomes

$$\frac{d(u\phi)}{dx} = \frac{d}{dx}\left[\Gamma \frac{d\phi}{dx}\right] \tag{5.48}$$

This convection diffusion equation, is representative of the momentum equation if $\phi = u$ and $\Gamma = \nu$. The 1D mesh for nodal point P, and its neighbouring nodes is given in Fig. 5.29.

Integration of Eq. (5.48) over the control volume gives

$$\underbrace{(u\phi)_e - (u\phi)_w}_{\text{convective flux}} = \underbrace{\left(\Gamma_e \frac{d\phi}{dx}\right)_e - \left(\Gamma_w \frac{d\phi}{dx}\right)_w}_{\text{diffusive flux}}$$

where the two diffusive fluxes are approximated with a linear interpolation as

$$\left(\Gamma_e \frac{d\phi}{dx}\right)_e = \Gamma_e \frac{\phi_E - \phi_P}{\Delta x_{PE}} = D_e(\phi_E - \phi_P); \quad \left(\Gamma_w \frac{d\phi}{dx}\right)_w = \Gamma_w \frac{\phi_P - \phi_W}{\Delta x_{WP}} = D_w(\phi_P - \phi_W)$$

where $D_e = \Gamma_e / \Delta x_{PE}$ and $D_w = \Gamma_w / \Delta x_{WP}$

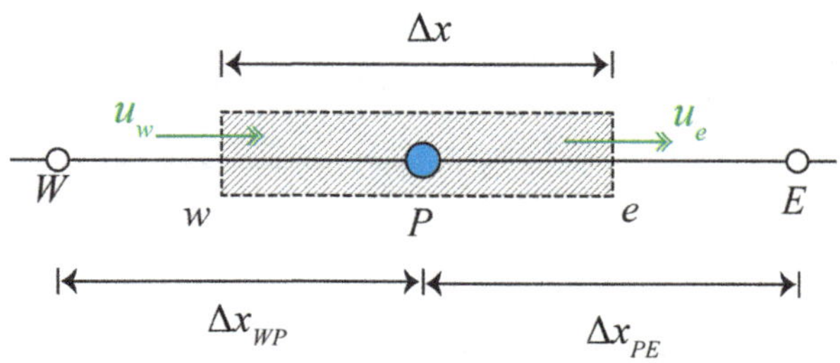

Fig. 5.29 A one-dimensional mesh used to solve the convection-diffusion equation. The mesh is uniform, but the finite volume method is not restricted to uniform meshes

Central Differencing The values of ϕ_e and ϕ_w in the convective flux needs to be evaluated, since the values of the scalar are stored at the nodes *P, E*, and *W*. Using central differencing on a uniform grid, we obtain

$$\phi_e = (\phi_P + \phi_E)/2 \qquad \phi_w = (\phi_W + \phi_P)/2 \tag{5.49}$$

Putting the terms together, and rearranging, the discretised equation becomes

$$\frac{u_e}{2}(\phi_P + \phi_E) - \frac{u_w}{2}(\phi_W + \phi_P) = D_e(\phi_E - \phi_P) - D_w(\phi_P - \phi_W) \tag{5.50}$$

We rearrange the terms to identify the coefficients of ϕ_E and ϕ_W and ϕ_P as

$$\left[\left(D_e - \frac{u_e}{2}\right) + \left(D_w + \frac{u_w}{2}\right) + (u_e - u_w\right)]\phi_P = \left(D_e - \frac{u_e}{2}\right)\phi_E + \left(D_w + \frac{u_w}{2}\right)\phi_W \tag{5.51}$$

so that

$$a_P\phi_P = a_E\phi_E + a_W\phi_W \tag{5.52}$$

where $a_P = a_E + a_W + (u_e - u_w)$; $\qquad a_E = D_e - \frac{u_e}{2}$; $\qquad a_W = D_w + \frac{u_w}{2}$;

The central differencing discretisation is second order accurate however it is unable to exhibit any bias in the flow direction by the transportiveness property. The scheme fails in strongly convective flows as it is unable to identify the flow direction. To overcome this, variations to the interpolation scheme are made. Two commonly used schemes are the Upwind Differencing and QUICK which are introduced here (Fig. 5.30).

Upwind Differencing In upwind differencing, the value of a cell face is equal to the value at the upstream node. If we limit our example to a flow direction from left to right, then $\phi_e = \phi_P$ and $\phi_w = \phi_W$

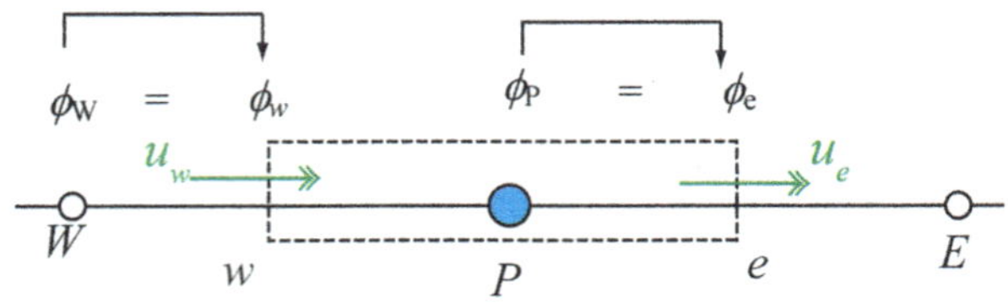

Fig. 5.30 Control volume representation for Upwind Differencing discretisation of nodal point P

The resulting discretised equation becomes

$$u_e\phi_P - u_w\phi_W = D_e(\phi_E - \phi_P) - D_w(\phi_P - \phi_W)$$

and rearranging the terms to identify the coefficients of ϕ_E and ϕ_W and ϕ_P as before

$$a_P\phi_P = a_E\phi_E + a_W\phi_W \tag{5.53}$$

where $a_P = a_E + a_W + (u_e - u_w)$; $\quad a_E = D_e$; $\quad a_W = D_w + u_w$;

Effectively, the convection term is discretised by applying a backward difference scheme which is first order accurate. If a flow direction is from right to left then a forward difference scheme is used (Fig. 5.31).

Quick The Quadratic Upstream Interpolation for Convective Kinetics, (QUICK) scheme uses a quadratic approximation across two variable points at the upstream and one at the downstream. The unequal weighting influence of this particular scheme hinges on knowledge biased towards the upstream flow conditions. For a flow direction from left to right the nodal points are:

$$\phi_w = -\frac{1}{8}\phi_{WW} + \frac{3}{8}\phi_P + \frac{6}{8}\phi_W \quad \text{and} \quad \phi_e = -\frac{1}{8}\phi_W + \frac{6}{8}\phi_P + \frac{3}{8}\phi_E$$

The resulting discretised equation becomes

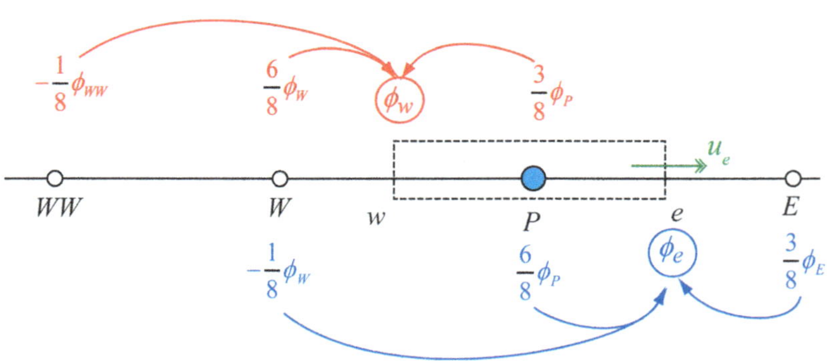

Fig. 5.31 Control volume representation for the QUICK discretisation of nodal point P

$$u_e\left(-\frac{1}{8}\phi_w+\frac{6}{8}\phi_P+\frac{3}{8}\phi_E\right)-u_w\left(-\frac{1}{8}\phi_{WW}+\frac{3}{8}\phi_P+\frac{6}{8}\phi_W\right) \\ = D_e(\phi_E-\phi_P)-D_w(\phi_P-\phi_W) \tag{5.54}$$

and rearranging the terms to identify the coefficients of ϕ_E and ϕ_W and ϕ_P as before

$$a_P\phi_P = a_E\phi_E + a_W\phi_W \tag{5.55}$$

where $a_P = a_E + a_W + a_{WW} + (u_e - u_w)$; $a_E = D_e - \frac{3}{8}u_e$; $a_W = D_w + \frac{6}{8}u_w + \frac{1}{8}u_e$; $b = -\frac{u_w}{8}$

Example: Comparisons of Discretisation Schemes In central differencing for a uniform mesh, $a_E = D_e - \frac{u_e}{2}$ which means that the coefficient becomes negative if $2D_e < u_e$ leading to a diverging and unstable solution. The Peclet number is used as a guide to determine stability of a problem. If we consider the characteristic length scale L_c as the mesh cell length Δx, then the quantity is referred to as the cell Peclet number. Across face *e*, $Pe_c = u_e\Delta x / D_e$ and that as long as $Pe_c \le 2$ then we are guaranteed positive coefficients and the solution will converge. Similarly in the QUICK scheme we obtain $a_E = D_e - \frac{3}{8}u_e$ which means the stability criterion is $Pe_c \le 8/3$. The Peclet number is a function of the grid size, and this implies a minimum spatial grid size to ensure that Pe is small enough to maintain the stability criterion. In Upper Differencing, the value of ϕ on a cell face is determined by the flow direction and the neighbouring coefficients are guaranteed to have positive values.

To demonstrate the dependence of Pe and the discretisation scheme, on the solution, we can compare the numerical solution with the exact analytical solution. The analytical solution for a 1D steady convection-diffusion problem is

$$\frac{d(u\phi)}{dx} - \frac{d}{dx}\left[\Gamma\frac{d\phi}{dx}\right] = 0 \rightarrow \frac{\phi-\phi_0}{\phi_L-\phi_0} = \frac{\exp(ux/\Gamma)-1}{\exp(uL/\Gamma)-1} \tag{5.56}$$

In this example, the length of the domain is 1 m divided into five control volumes. The boundary conditions are $\phi_0=0$ and ϕ_L=The velocity, *u* is 1.6 m/s and diffusion coefficient, Γ is 0.1. The resulting Pe number is Pe=3.2, producing an unstable solution for the Central Differencing. Figure 5.32 shows the 'undershoots' and 'overshoots' occurring in the solution for Central Differencing. A similar oscillating pattern is found for the QUICK scheme as well (although not shown in the figure) since it has a restriction of Pe<8/3. The Upper Diffrencing has no restrictions on the Pe number and produces the trendline of the analytical solution. However the accuracy of the Upper Difference solution drops off as *x* approaches 1. For both

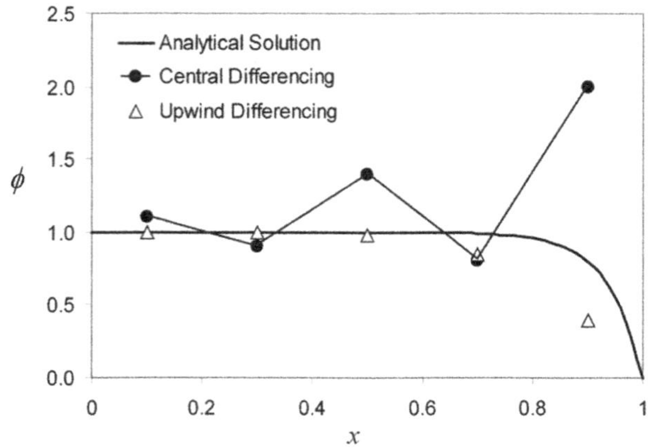

Fig. 5.32 Flow variable solution using Central Differencing and Upper Differencing schemes. Un-physical oscillations are produced with the Central Differencing scheme whereas the Upper Differencing shows a more realistic solution

discretisation schemes, improved results can be obtained if the grid spacing/mesh resolution is refined (i.e. increase the number of control volumes).

False Diffusion Despite the Upper Differencing scheme being unconditionally stable, it is essentially a backward differencing formula of first order accuracy. It produces an artificial or false diffusion that is non-physical resulting in an incorrect distribution of a flow variable ϕ through the flow domain for small Peclet numbers. This occurs when the flow direction is not aligned with the grid lines and hence a nonzero gradient exists in the direction perpendicular to the flow.

To demonstrate this, the transport equation for a pure 2D convection flow is used, $\frac{d(u\phi)}{dx}+\frac{d(v\phi)}{dy}=0$ (no diffusion term). When the flow moves horizontally through the domain in-line with the rectangular grid (Fig. 5.33), the Upper Difference solution provides a good representation and no false diffusion occurs. However, if the flow is now oriented at 45° to the grid lines, then false diffusion occurs.

In the absence of physical diffusion, the transported variable should exhibit a uniform temperature of 0 below the dashed line, and 100 above the dashed line. If we take the temperature values along the horizontal line half way up the domain labelled as *a-a'* in Fig. 5.33b, the results show the Upper Difference scheme (using a 5×5 control volume discretised domain), cannot capture the step profile of the exact solution. Instead the variable, ϕ exhibits false diffusion behaviour spreading across the domain rather than producing a step profile. Improvements to the result can be obtained if the mesh is refined to a 10×10 control volume mesh (Fig. 5.34).

To illustrate why this occurs we consider the Upper Differenced discretised 1D convection-diffusion equation where the diffusion term is set to zero:

$$\frac{d(u\phi)}{dx}=0 \Rightarrow u\left(\frac{\phi_P-\phi_W}{\Delta x}\right)=0 \tag{5.57}$$

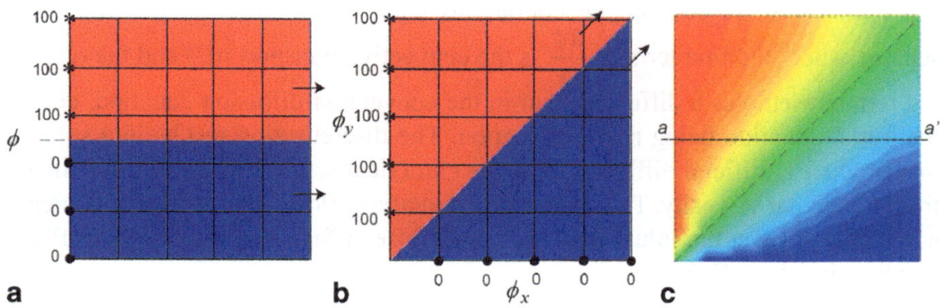

Fig. 5.33 Transport of the scalar variable ϕ by pure convection. Line *a-a'* is a horizontal line half way up the domain. **a** The variable is convected through the domain aligned with the computational grid lines. **b** The variable is convected through the domain at an angle of 45°to the computational grid lines. **c** false diffusion of temperature in computational domain

assuming $u_e = u_w$. The convective fluxes are defined at the control volume faces as $\phi_e = \phi_P$ and $\phi_w = \phi_W$. Expanding ϕ_W through a Taylor series we get

$$\phi_W = \phi_P - \Delta x \frac{d\phi}{dx} + \frac{\Delta x^2}{2!}\frac{d^2\phi}{dx^2} - \ldots$$

and rearranging

$$\frac{\phi_P - \phi_W}{\Delta x} = \frac{d\phi}{dx} - \frac{\Delta x}{2}\frac{d^2\phi}{dx^2} + O(\Delta x^2) \tag{5.58}$$

If we substitute Eq. (5.58) into (5.57) we get

$$u\frac{d\phi}{dx} - \frac{u\Delta x}{2}\frac{d}{dx}\left[\frac{d\phi}{dx}\right] + \mathrm{O}(\Delta x^2) = 0 \tag{5.59}$$

Fig. 5.34 Solutions for the convection transport of ϕ. The upper differencing scheme using a 10×10 control volume mesh shows that the profile is closer to the exact solution

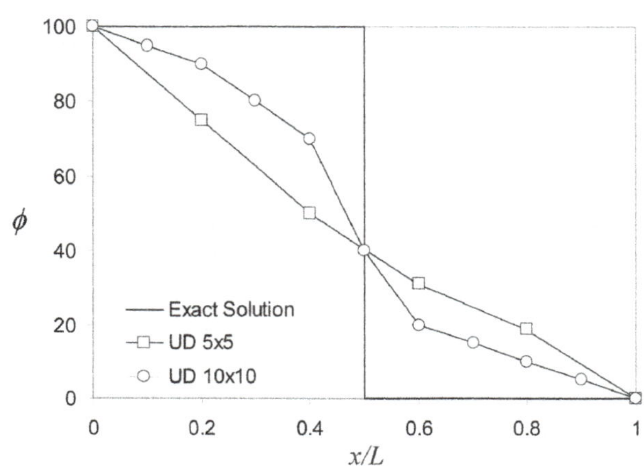

Equation (5.59) represents the continuous equation of the Upper Differencing scheme where the first term, $u\frac{d\phi}{dx}$ is the convective transport of ϕ and the remaining terms resembles a diffusion term in the convection-diffusion equation, despite the initial equation being pure convection. The discretisation results in a solution that produces artificial diffusion from the numerical scheme and not real diffusion from molecular viscosity. This is most problematic in flows that exhibit small Peclet numbers. However the solution improves and false diffusion reduces for large Peclet numbers in Upper Differencing.

The Central Differencing also suffers from false diffusion but to a lesser extent at low Pe numbers. At higher Pe numbers Central Differencing produces unrealistic results while using the QUICK scheme produces undershoots and overshoots. To overcome the false diffusion problem, the following can be performed: refining the mesh; aligning grid lines more in the direction of flow; and including more neighbour nodes in the discretisaiton.

5.4.4 Finite Element Method (FEM)

Although the finite element method can be used to discretise the fluid flow equations, its use for structural analysis is more widespread. This method considers the solution region comprised of smaller, interconnected elements. Each element approximates the governing equations by applying interpolation functions that describe its behaviour between nodal points. The continuum problem which has an infinite number of unknowns is reduced to one with a finite number of unknowns at specified locations called *nodes*.

Each element produces a matrix relation called the stiffness matrix, k that relates the applied force to its displacement. The local stiffness matrix corresponds to an individual element while the *global* stiffness matrix defines the stiffness of the entire domain from assembling of all local stiffness matrices. In this introduction we present the mathematical steps leading to the local stiffness matrix.

The main steps involved with FEM are summarized below:

I. Discretisation—A physical domain is discretized by dividing the region into non-overlapping elements or sub-regions defined by nodes, and linked together through a shape function. The elements are formed by joining its nodal points which are also shared with neighbouring connected elements.

II. Interpolation—Shape functions are used to interpolate field variables across the element from node to node, and are typically denoted by the letter N which appear as coefficients in the interpolation polynomial. A shape function is written for each individual node and has the property that its magnitude is 1 at that node and 0 for all other nodes in that element

III. Assembling the equations—The nodal values of the unknown field variable are related to other parameters to establish a matrix equation. The element equations are then assembled together to form a global equation system for the

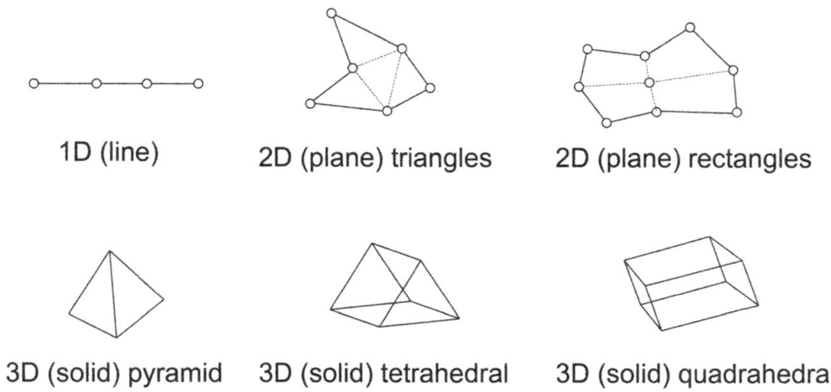

Fig. 5.35 Types of finite elements. 1D elements are lines that can represent springs, beams, pipes. 2D elements can be triangles that represent plates, shells

whole computational domain. Boundary conditions are then imposed to solve the equations.

The discretized finite elements can be one, two or three dimensions, and also special elements with zero dimensionality (lumped springs or point masses). Lower dimensional elements can be expanded to build a model in 2D or 3D space. Nodes are usually located at the corners or end points of the elements (Fig. 5.35). Most elements used in practice have fairly simple geometries. In one-dimension, elements are straight lines or curved segments, in two dimensions they are triangular or quadrilateral, and in three dimensions they can be tetrahedra, pentahedra (wedges or prisms), and hexahedra (cuboids or "bricks").

Unlike the finite volume method, a finite element is not a differential element of size $dx \times dy$ but rather is an element at which the value of a field variable is explicitly calculated. This quantity is approximated over each element using polynomial interpolation to produce a very large set of simultaneous equations. In this section, we present simplified examples to allow the reader to appreciate the discretisation process in FEM. For a more complete coverage of the FEM method notable books by Hutton (2005), Zienkiewicz and Taylor (2000), and Stasa (1985) are recommended.

One-Dimensional Finite Element Example The simplest finite element is a linear one-dimensional element shown in Fig. 5.36 which is a bar that obeys Hooke's law, with forces applied at its ends. This model is a two-force member where the forces exerted on the ends are equal in magnitude, and opposite in sense. The element has length L, and we denote the axial displacement at any position along the length of the bar as $u(x)$. The nodes are placed at each end as nodes-1 and −2, with nodal displacements $u(x=0)=u_1$ and $u(x=L)=u_2$. The continuous displacement field variable $u(x)$ is expressed (approximately) in terms of the two nodal variables u_1 and u_2 through discretisation of the bar into a finite element given as

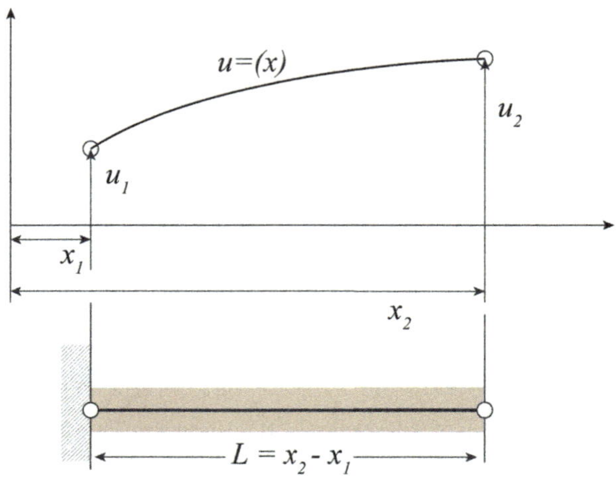

Fig. 5.36 A single 1D finite bar element with forces applied at the ends of the bar

$$u(x) = N_1(x)u_1 + N_2(x)u_2 = [\mathbf{N}]\{u\} \tag{5.60}$$

where [] is a row matrix and {} is a column matrix.
The shape functions for each node are

$$N_1 = \frac{x_2 - x}{x_2 - x_1} = 1 - \frac{x}{L} \qquad N_2 = \frac{x - x_1}{x_2 - x_1} = \frac{x}{L} \tag{5.61}$$

which provides a linear interpolation from node-1 to node-2, since it produces $N_1 = 1$ for $x = x_1$, and $N_1 = 0$ for $x = x_2$, and similarly for node-2.

The continuous displacement function represented by the discretisation becomes

$$u(x) = [\mathbf{N}]\{u\} = \left(1 - \frac{x}{L}\right)u_1 + \left(\frac{x}{L}\right)u_2 \tag{5.62}$$

where [ ] is a row matrix of the interpolation functions, and {} is a column matrix of the nodal displacements. We now determine the relation between the nodal displacements and applied forces to obtain the stiffness matrix for the bar element. The fundamental equation for deflection δ of an elastic bar having modulus of elasticity E, length L and uniform cross-sectional area A when subjected to axial load P is given by $\delta = PL/AE$ and the equivalent spring constant is

$$k = \frac{P}{\delta} = \frac{AE}{L} \tag{5.63}$$

To compute the nodal displacements from some given loading condition on the element we obtain the necessary equilibrium equations relating the displacements to applied forces. Firstly the strain-displacement relation is

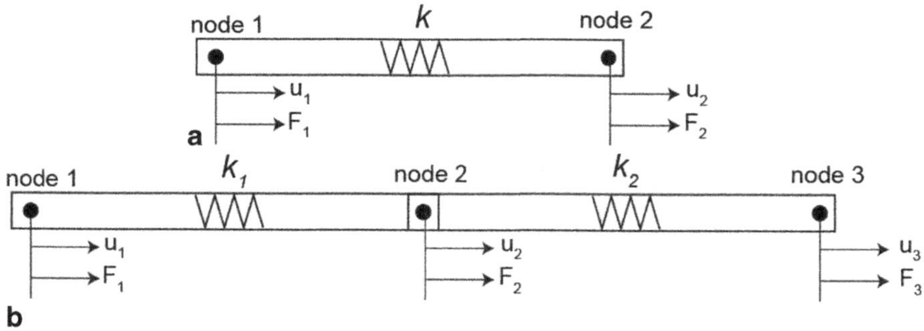

Fig. 5.37 1D finite element discretisation of a solid structure treated as a spring-mass system

$$\varepsilon_x = \frac{du}{dx} = \frac{u_2 - u_1}{L} \tag{5.64}$$

and the stress-strain relation is taken from Hooke's law, giving

$$\sigma_x = E\varepsilon_x = E\frac{u_2 - u_1}{L} \tag{5.65}$$

and finally we relate the stress to the applied load,

$$P = \sigma_x A = \frac{AE}{L}(u_2 - u_1) \tag{5.66}$$

Applying Eq. (5.66) to relate the nodal forces, f_1 and f_2 to the nodal displacements u_1 and u_2, we get

$$f_1 = -\frac{AE}{L}(u_2 - u_1) \qquad f_2 = \frac{AE}{L}(u_2 - u_1) \tag{5.67}$$

and in matrix form we get

$$\underbrace{\frac{AE}{L}\begin{bmatrix} 1 & -1 \\ -1 & 1 \end{bmatrix}}_{\text{stiffness matrix}}\begin{Bmatrix} u_1 \\ u_2 \end{Bmatrix} = \begin{Bmatrix} f_1 \\ f_2 \end{Bmatrix} \rightarrow [k_e]\begin{Bmatrix} u_1 \\ u_2 \end{Bmatrix} = \begin{Bmatrix} f_1 \\ f_2 \end{Bmatrix} \tag{5.68}$$

Worked Example Consider the displacement field variable for a spring-mass system is due to an applied force (Fig. 5.37).

For a single element the net displacement and its corresponding force is given by

$$d = u_2 - u_1 \quad F_1 = -k(u_2 - u_1) \quad F_2 = -k(u_2 - u_1)$$

and this can be expressed in matrix form as

$$\begin{Bmatrix} F_1 \\ F_2 \end{Bmatrix} = \begin{bmatrix} k & -k \\ -k & k \end{bmatrix} \begin{Bmatrix} u_1 \\ u_2 \end{Bmatrix}$$

and similarly for a two element model the equations are

$$\underbrace{\begin{Bmatrix} F_1 \\ F_2 \end{Bmatrix} = \begin{bmatrix} k_1 & -k_1 \\ -k_1 & k_1 \end{bmatrix} \begin{Bmatrix} u_1 \\ u_2 \end{Bmatrix}}_{\text{element1}} \quad \underbrace{\begin{Bmatrix} F_2 \\ F_3 \end{Bmatrix} = \begin{bmatrix} k_2 & -k_2 \\ -k_2 & k_2 \end{bmatrix} \begin{Bmatrix} u_1 \\ u_2 \end{Bmatrix}}_{\text{element2}}$$

For additional elements, the same matrix form is added which are in the form of

$$\underbrace{\{F\}}_{\substack{\text{matrix of} \\ \text{applied forces}}} = \underbrace{[K]}_{\substack{\text{stiffness matrix} \\ \text{with size [DOF} \times \text{DOF]}}} \underbrace{\{U\}}_{\substack{\text{unknown matrix of} \\ \text{nodal displacement} \\ \text{to be solved}}}$$

where $\{F\}$ is a column vector matrix of known nodal forces, $[K_e]$ is the stiffness matrix, $\{U\}$ is the column vector matrix of unknown nodal displacements. The stiffness matrix for a single linear spring element is a 2×2 matrix. This corresponds to two nodal displacements (or degrees of freedom) which are co-dependent (that is, the body is continuous and elastic). For the two element model, the matrix equation for each element is combined to produce a single matrix that contains three nodal displacements thereby producing a 3×3 stiffness matrix, (e.g. having three degrees of freedom) as,

$$\begin{Bmatrix} F_1 \\ \sum F_2 \\ F_3 \end{Bmatrix} = \begin{bmatrix} k_1 & -k_1 & \\ -k_1 & k_1 + k_2 & -k_2 \\ & -k_2 & k_2 \end{bmatrix} \begin{Bmatrix} u_1 \\ u_2 \\ u_3 \end{Bmatrix}$$

where $\sum F_2 = F_2^1 + F_2^2$ is the sum of the force components for element 1, F_2^1 and element 2, F_2^2. Once the displacements are known, the strain, and stress can be determined. For 1D structures these are

$$\varepsilon = \frac{\Delta u}{L} \quad \sigma = E \cdot \varepsilon$$

If we consider a structure that contains two elements with node 1 attached to a fixed support so that $s_1 = 0$, and the spring constants and applied forces are $k_1 = 30N/m$, $k_2 = 60N/m$, and $F_2 = F_3 = 2N$. The finite element analysis stiffness matrix becomes

$$\begin{Bmatrix} F_1 \\ 2 \\ 2 \end{Bmatrix} = \begin{bmatrix} 30 & -30 & \\ -30 & 90 & -60 \\ & -60 & 60 \end{bmatrix} \begin{Bmatrix} 0 \\ u_2 \\ u_3 \end{Bmatrix}$$

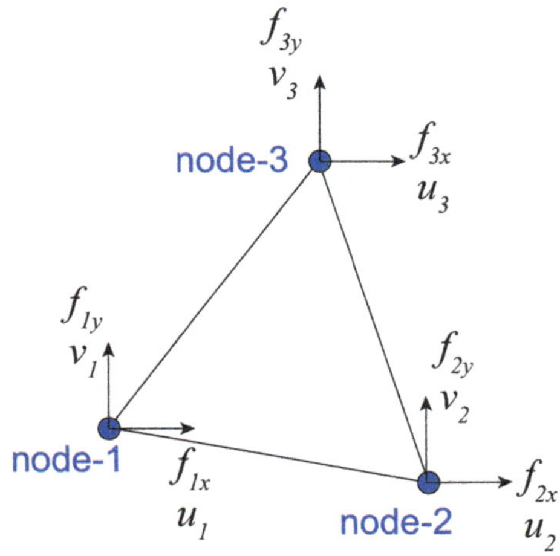

Fig. 5.38 Nodal displacement and forces occurring for a 2D triangular element

where the solution is

$$u_2 = 0.133\text{m},\ u_3 = 0.167\text{m},\ \text{and}\, F_1 = 4N$$

Two-Dimensional Finite Element Example We now consider a 2D triangular element with three nodes subjected to plane stress (no displacement in the z-coordinate) (Fig. 5.38).

The element has three nodes in the x-y coordinate, with each node having two degrees of freedom movement in x, and y directions). Displacements at some point inside a finite element $\{\mathbf{u}\}$ can be determined with the use of nodal displacements $\{\mathbf{d}\}$ and shape functions $\mathbf{N_i}$. Nodal displacements in the x-coordinate direction are u_1, u_2, and u_3, while displacements in the y direction are v_1, v_2, and v_3.

$$\begin{aligned} u(x,y) &= N_1(x,y)u_1 + N_2(x,y)u_2 + N_3(x,y)u_3 = [\mathbf{N}]\{u\} \\ v(x,y) &= N_1(x,y)v_1 + N_2(x,y)v_2 + N_3(x,y)v_3 = [\mathbf{N}]\{v\} \end{aligned} \tag{5.69}$$

and in matrix form is:

$$\begin{Bmatrix} u \\ v \end{Bmatrix} = \begin{bmatrix} N_1 & 0 & N_2 & 0 & N_3 & 0 \\ 0 & N_1 & 0 & N_2 & 0 & N_3 \end{bmatrix} \begin{Bmatrix} u_1 \\ v_1 \\ u_2 \\ v_1 \\ v_2 \\ u_3 \\ v_3 \end{Bmatrix} \rightarrow \mathbf{u} = \mathbf{Nd}$$

where $N1$, $N2$, and $N3$ are the interpolation functions as defined as

$$
\begin{aligned}
N_1 &= \frac{1}{2A}\left\{(x_2 y_3 - x_3 y_2) + (y_2 - y_3)x + (x_3 - x_2)y\right\} \\
N_2 &= \frac{1}{2A}\left\{(x_3 y_1 - x_1 y_3) + (y_3 - y_1)x + (x_1 - x_3)y\right\} \\
N_3 &= \frac{1}{2A}\left\{(x_1 y_2 - x_2 y_1) + (y_1 - y_2)x + (x_2 - x_1)y\right\}
\end{aligned}
\tag{5.70}
$$

and A is the area determined by

$$
A = \frac{1}{2}\det\begin{bmatrix} 1\, x_1\, y_1 \\ 1\, x_2\, y_2 \\ 1\, x_3\, y_3 \end{bmatrix} = \frac{1}{2}(x_1 y_2 - x_1 y_3 - x_2 y_1 + x_2 y_3 + x_3 y_1 - x_3 y_2)
$$

The element strain components are the derivatives of the displacement with respect to the strain coordinate, given as

$$
\begin{aligned}
\varepsilon_x &= \frac{\partial u}{\partial x} = \frac{\partial N_1}{\partial x}u_1 + \frac{\partial N_2}{\partial x}u_2 + \frac{\partial N_3}{\partial x}u_3 \\
\varepsilon_y &= \frac{\partial v}{\partial y} = \frac{\partial N_1}{\partial y}v_1 + \frac{\partial N_2}{\partial y}v_2 + \frac{\partial N_3}{\partial y}v_3 \\
\gamma_{xy} &= \frac{\partial u}{\partial y} + \frac{\partial v}{\partial x} = \frac{\partial N_1}{\partial x}u_1 + \frac{\partial N_2}{\partial x}u_2 + \frac{\partial N_3}{\partial x}u_3 + \frac{\partial N_1}{\partial x}v_1 + \frac{\partial N_2}{\partial x}v_2 + \frac{\partial N_3}{\partial x}v_3
\end{aligned}
$$

These are the three components of strain at a point, where ε_x and ε_y are strains in the x- and y- coordinate respectively, and γ_{xy} is the shear strain.

Similarly as presented for the 1D-case the $[B]$ (strain-displacement) matrix is

$$
\begin{aligned}
[\mathbf{B}] &= \frac{1}{2A}\begin{bmatrix} y_2 - y_3 & y_3 - y_1 & y_1 - y_2 & 0 & 0 & 0 \\ 0 & 0 & 0 & x_3 - x_2 & x_1 - x_3 & x_2 - x_1 \\ x_3 - x_2 & x_1 - x_3 & x_2 - x_1 & y_2 - y_3 & y_3 - y_1 & y_1 - y_2 \end{bmatrix} \\
&= \frac{1}{2A}\begin{bmatrix} \beta_1 & \beta_2 & \beta_3 & 0 & 0 & 0 \\ 0 & 0 & 0 & \alpha_1 & \alpha_2 & \alpha_3 \\ \alpha_1 & \alpha_2 & \alpha_3 & \beta_1 & \beta_2 & \beta_3 \end{bmatrix}
\end{aligned}
$$

Using Eqn. (5.32) for the elastic strain energy of the element gives

$$
U_e = \frac{1}{2}\iiint \{\varepsilon\}^T [D]\{\varepsilon\}\,\mathrm{dV} = \frac{1}{2}\iiint \left\{\delta^{(e)}\right\}^T [B]^T [D][B]\left\{\delta^{(e)}\right\}\mathrm{dV} \tag{5.71}
$$

Since the elastic properties are constant across the finite element, Eq. (5.71) becomes

$$U_e = \frac{1}{2}(V[B]^T[D][B])\left\{\delta^{(e)}\right\} \tag{5.72}$$

Using total potential energy theorem for an element (see Hutton 2005) the resulting matrix relation becomes

$$\begin{aligned} V[B]^T[D][B]\{\delta\} &= \{f\} \\ [k]\{\delta\} &= \{f\} \end{aligned} \tag{5.73}$$

The element volume V is equal to thickness t multiplied by area, A. The stiffness matrix, $[k]$ becomes

$$[k] = \frac{Et}{4A(1-v^2)} \begin{bmatrix} \beta_1 & 0 & \alpha_1 \\ \beta_2 & 0 & \alpha_2 \\ \beta_3 & 0 & \alpha_3 \\ 0 & \alpha_1 & \beta_1 \\ 0 & \alpha_2 & \beta_2 \\ 0 & \alpha_3 & \beta_3 \end{bmatrix} \begin{bmatrix} 1 & v & 0 \\ v & 1 & 0 \\ 0 & 0 & \dfrac{1-v}{2} \end{bmatrix} \begin{bmatrix} \beta_1 & \beta_2 & \beta_3 & 0 & 0 & 0 \\ 0 & 0 & 0 & \alpha_1 & \alpha_2 & \alpha_3 \\ \alpha_1 & \alpha_2 & \alpha_3 & \beta_1 & \beta_2 & \beta_3 \end{bmatrix} \tag{5.74}$$

and $\{f\}$ is the column matrix of the applied forces

$$\{f\} = \begin{Bmatrix} f_{1x} \\ f_{2x} \\ f_{3x} \\ f_{1y} \\ f_{2y} \\ f_{3y} \end{Bmatrix}$$

5.5 Numerical Solution of Algebraic Systems

A system of linear or non-linear algebraic equations is produced from the discretisation, and this is solved by numerical methods. The complexity and size of the set of equations depends on the dimensionality and geometry of the physical problem. In this section we present two types of numerical methods: *direct methods* and *iterative methods*.

Direct methods compute the solution to a problem in a finite number of steps. Examples are the Gaussian elimination, and the TDMA (TriDiagonal Matrix Algorithm) which applies matrix algebra such as row operations to transform the matrix into a form that is easier to solve. Typically direct methods are the preferred solution method for low to medium sized systems (e.g. < 1000 equations) as it is computationally inexpensive and requires a minimum amount of storage in the core memory.

In contrast, iterative methods are based on repeated applications of an algorithm leading to its eventual convergence after a number of repetitions. A convergence criterion is specified to determine when a suitable solution has been reached. Iterative methods are typically used for very large systems. For non-linear problems, they are used out of necessity but are just as valuable for sparse linear systems. Well-known point-by-point methods such as Jacobi and Gauss-Siedel are presented in this chapter to provide the reader with some basic understanding of iterative methods. Other variants from these two iterative methods will also be described particularly those algorithms that are used in solving CHD problems.

5.5.1 Direct Solution Methods

One of the most basic methods for solving linear systems of algebraic equations is the ***Gaussian elimination***. The algorithm derives from the basis of systematic reduction of large systems of equations to smaller ones. Let us suppose that the systems of equations are written in the form:

$$A\phi = B \tag{5.75}$$

where ϕ is the unknown nodal variables. Matrix A contains non-zero coefficients of the algebraic equations as:

$$A = \begin{bmatrix} A_{11} & A_{12} & A_{13} & \cdots & A_{1n} \\ A_{21} & A_{22} & A_{23} & \cdots & A_{2n} \\ A_{31} & A_{32} & A_{33} & \cdots & A_{3n} \\ \vdots & \vdots & \vdots & \ddots & \vdots \\ A_{n1} & A_{n2} & A_{n3} & \cdots & A_{nn} \end{bmatrix}$$

while B comprises of known values of ϕ, that are given by the boundary conditions or source/sink terms. It can be observed that the diagonal coefficients of the matrix A are represented by the entries of $A_{11}, A_{22}, \ldots, A_{nn}$. The first step is to apply *Forward Elimination* which eliminates the entries below the diagonal to yield a lower triangle of zeroes. This means eliminating the elements of $A_{21}, A_{31}, A_{32}, \ldots, A_{nn-}1$ by replacing them with zeroes. We begin the elimination process by considering the first column elements of $A_{21}, A_{31}, \ldots, A_{n}1$ in the matrix A.

Multiplying the first row of the matrix by A_{21}/A_{11} and subtracting these values from the second row, all the elements in the second row are subsequently modified, which includes the terms in B on the right hand side of the equations. The other elements A_{31}, A_{41}, …., A_{n1} in the first column of matrix A are treated similarly by repeating this process down the first column (e.g. Row3– Row1 × A_{31}/A_{11}), so that all the elements below A_{11} are reduced to zero. The same procedure is then applied for the second column, (for all elements below A_{22}) and so forth until the process reaches the n-1th column. Note that at each stage we need to divide by A_{nn} and therefore it is imperative that the value is non-zero. If it is not, then row exchange with another row below that has a non-zero needs to be performed.

After this process is complete, the original matrix A becomes an *upper triangular matrix* that is given by:

$$U = \begin{bmatrix} A_{11} & A_{12} & A_{13} & \cdots & A_{1n} \\ 0 & A_{22} & A_{23} & \cdots & A_{2n} \\ & 0 & A_{33} & \cdots & A_{3n} \\ \vdots & \vdots & \vdots & \ddots & \vdots \\ 0 & 0 & 0 & \cdots & A_{nn} \end{bmatrix}$$

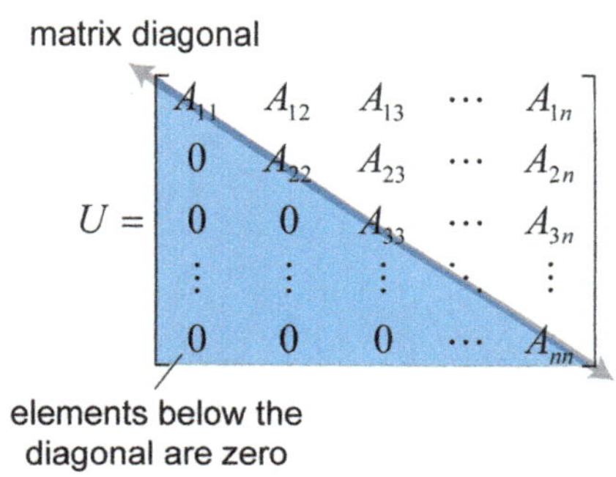

All the elements in the matrix U except the first row differ from those in the original matrix A and our systems of equations can be rewritten in the form:

$$U\phi = B$$

The upper triangular system of equations can now be solved by the *Back Substitution* process. The last row of the matrix U contains only one non-zero coefficient, A_{nn}, and its corresponding variable ϕ_n is solved by

$$\phi_n = \frac{B_n}{U_{nn}}$$

The second last row in matrix U contains only the coefficients $A_{n-1,n}$ and A_{nn} and, once ϕ_n is known, the variable ϕ_{n-1} can be solved. By proceeding up the rows of the matrix we continue substituting the known variables and ϕ_i is solved in turn. The general form of equation for ϕ_i is expressed as:

$$\phi_i = \frac{B_i - \sum_{j=i+1}^{n} A_{ij}\phi_j}{A_{ii}} \tag{5.76}$$

It is not difficult to see that the bulk of the computational effort is in the *forward elimination* process; the back substitution process requires less arithmetic operations and is much less costly. Gaussian elimination can be expensive especially for a full matrix containing a large number of unknown variables to be solved but it is as good as any other methods that are currently available.

The matrix obtained in the example above is in the form of a tri-diagonal matrix which is a special case of matrices that occurs frequently. A tri-diagonal matrix has nonzero elements only on the diagonal plus or minus one column such as:

$$A = \begin{bmatrix} A_{11} & A_{12} & 0 & \cdots & 0 \\ A_{21} & A_{22} & A_{23} & \cdots & 0 \\ 0 & A_{32} & A_{33} & \cdots & 0 \\ \vdots & \vdots & \vdots & \ddots & \vdots \\ 0 & 0 & 0 & \cdots & A_{nn} \end{bmatrix}$$

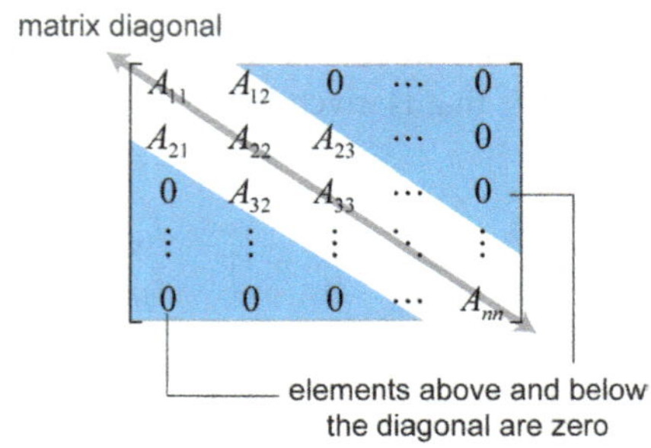

In this form it is advantageous to consider variants of Gaussian elimination such as the *TriDiagonal Matrix Algorithm* (**TDMA**), also known as the Thomas algorithm. Let us consider a general tri-diagonal form of a system of algebraic equations as:

$$\begin{bmatrix} A_{11} & A_{12} & 0 & 0 & 0 & 0 & 0 \\ A_{21} & A_{22} & A_{23} & 0 & 0 & 0 & 0 \\ 0 & \cdots & \cdots & \cdots & 0 & 0 & 0 \\ 0 & 0 & A_{ii-1} & A_{ii} & A_{ii+1} & 0 & 0 \\ 0 & 0 & 0 & \cdots & \cdots & \cdots & 0 \\ 0 & 0 & 0 & 0 & A_{nn-2} & A_{n-1n-1} & A_{n-1n} \\ 0 & 0 & 0 & 0 & 0 & A_{nn-1} & A_{nn} \end{bmatrix} \begin{bmatrix} \phi_1 \\ \phi_1 \\ \cdots \\ \phi_i \\ \cdots \\ \phi_{n-1} \\ \phi_n \end{bmatrix} = \begin{bmatrix} B_1 \\ B_2 \\ \cdots \\ B_i \\ \cdots \\ B_{n-1} \\ B_n \end{bmatrix}$$

The TDMA like the Gaussian elimination solves the system of equations above in two parts: *forward elimination* and *back substitution*. For the *forward elimination* process, the neighbouring entries are eliminated below the diagonal to yield zero entries. This means replacing the elements of A_{21}, A_{32}, A_{43},…, A_{nn-1} with zeroes. For the first row, the diagonal entry A_{11} is normalized to unity and the neighbouring entry A_{12} and the matrix B term B_1 are modified according to

$$A_{12}' = \frac{A_{12}}{A_{11}}, \quad B_1' = \frac{B_1}{A_{11}} \tag{5.77}$$

Like the Gaussian elimination, by multiplying the first row of the matrix by A_{21} and subtracting it from the second row; all the elements in the second row are subsequently modified (where A_{21} becomes zero), which also include the terms in B

on the right hand side of the equations. Applying the same procedure to the rest of the rows of the matrix, the neighbouring element entries and the matrix B terms in general form are:

$$A'_{ii+1} = \frac{A_{ii+1}}{A_{ii} - A_{ii-1}A'_{i-1i}}, \quad B'_i = \frac{B_i - A_{ii-1}B'_{i-1}}{A_{ii} - A_{ii-1}A'_{i-1i}} \tag{5.78}$$

The matrix containing the non-zero coefficients is therefore manipulated into:

$$\begin{bmatrix} 1 & A'_{12} & 0 & 0 & 0 & 0 & 0 \\ 0 & 1 & A'_{23} & 0 & 0 & 0 & 0 \\ 0 & \cdots & \cdots & \cdots & 0 & 0 & 0 \\ 0 & 0 & 0 & 1 & A_{ii+1} & 0 & 0 \\ 0 & 0 & 0 & \cdots & \cdots & \cdots & 0 \\ 0 & 0 & 0 & 0 & 0 & 1 & A_{n-1n} \\ 0 & 0 & 0 & 0 & 0 & 0 & 1 \end{bmatrix} \begin{bmatrix} \phi_1 \\ \phi_1 \\ \cdots \\ \phi_i \\ \cdots \\ \phi_{n-1} \\ \phi_n \end{bmatrix} = \begin{bmatrix} B'_1 \\ B'_2 \\ \cdots \\ B'_i \\ \cdots \\ B'_{n-1} \\ B'_n \end{bmatrix}$$

The second stage involves *back substitution*, to evaluate ϕ_{nm} and ϕ_i as

$$\varphi_n = B'_n \text{ and } \varphi_i = B'_i - \varphi_{i+1}A'_{ii+1} \tag{5.79}$$

The TDMA is more economical than the Gaussian elimination because of the absence of arithmetic operations (multiplication and divisions) in obtaining ϕ_i during back substitution.

During the forward elimination stage the term A_{nn} may consist of a zero value and to prevent ill-conditioning of a matrix, it is necessary to ensure that

$$|A_{ii}| > |A_{ii-1}| + |A_{ii+1}| \tag{5.80}$$

This means that the diagonal coefficients need to be much larger than the sum of the neighbouring coefficients. If this is not the case, matrix row swapping (otherwise known as *pivoting*) can be performed.

5.5.2 Iterative Methods

Direct methods such as Gaussian elimination can be employed to solve any system of equations. Unfortunately, in most CHD problems that usually results in a large system of non-linear equations. The cost of using this method is generally quite high as it requires a lot pre-conditioning of the matrix to prepare it for the direct method process. This leaves the option of employing iterative methods. In an iterative method, one guesses the solution, and uses the equation to systematically improve the solution until it reaches a specified level of convergence. If the number of iterations is small in achieving convergence, an iterative solver may cost less to use than a direct method.

Of the various classes of iterative methods, the simplest is the *Jacobi* method which solves the left hand side of the matrix expression, using previous values for ϕ on the right hand side. Using the system of equations, $A\phi = B$, as described in the previous section; the general form of the algebraic equation for each unknown nodal variable of ϕ is written as:

$$\sum_{j=1}^{i-1} A_{ij}\phi_j + A_{ii}\phi_i + \sum_{j=i+1}^{n} A_{ij}\phi_j = B_i \tag{5.81}$$

In Eq. (5.81), the Jacobi method requires that the nodal variables ϕ_j (non-diagonal matrix elements) are assumed at each iteration step k and the nodal variables ϕ_i are treated as unknown at iteration step $k+1$. Solving for ϕ_i, we have

$$\phi_i^{(k+1)} = \frac{1}{A_{ii}}\left(B_i - \sum_{j=1}^{i-1} A_{ij}\phi_j^{(k)} - \sum_{j=i+1}^{n} A_{ij}\phi_j^{(k)} \right) \tag{5.82}$$

The iteration process begins with an initial guess of the nodal variables ϕ_j ($k=0$). After repeated application of Eqn. (5.82) to all the n unknowns, the first iteration, $k=1$, is completed. We proceed to the next iteration step, $k=2$, by substituting the iterated values at $k=1$ into Eqn. (5.82) to estimate the new values at the next iteration step. This process is continuously repeated for as many iterations as required to converge to the desired solution.

A more immediate improvement to the Jacobi method is provided by the *Gauss-Siedel* method where the updated nodal variables $\phi_j^{(k+1)}$ are immediately used on the right-hand side of Eq. (5.81) as soon as they are available. In such a case, the previous values of $\phi_j^{(k)}$ that appear in the second term of the right-hand side of Eq. (5.82) are replaced by the current values of $\phi_j^{(k)}$, giving

$$\phi_i^{(k+1)} = \frac{B_i}{A_{ii}} - \sum_{j=1}^{i-1} \frac{A_{ij}}{A_{ii}}\phi_j^{(k+1)} - \sum_{j=i+1}^{n} \frac{A_{ij}}{A_{ii}}\phi_j^{(k)} \tag{5.83}$$

Comparing the above two iterative procedures, the Gauss-Siedel iteration is typically twice as fast as the Jacobi iteration. After repeated applications of Eqs. (5.82) and (5.83), convergence can be gauged in a number of ways. One method is to ensure that the maximum difference between each iteration, $\phi_j^{(k+1)} — \phi_j^{(k)}$ falls below some predetermined value. If the relative change is continually increasing with each iteration, then the solution is diverging.

5.5.3 Solution for a One-Dimensional Steady Diffusion Equation

TDMA Method The solution for a system of equations for a 4-cell discretisation in matrix form is given as

$$\begin{bmatrix} 3000 & -1000 & 0 & 0 \\ -1000 & 2000 & -1000 & 0 \\ 0 & -1000 & 2000 & -1000 \\ 0 & 0 & -1000 & 3000 \end{bmatrix} \begin{bmatrix} T_1 \\ T_2 \\ T_3 \\ T_4 \end{bmatrix} = \begin{bmatrix} 203000 \\ 3000 \\ 3000 \\ 803000 \end{bmatrix}$$

We first normalise the top row by making A_{11} to unity (divide Row-1 by 3000). We then multiply the first row by A_{21} and subtract it from the second row.

$$\begin{bmatrix} 1 & -1/3 & 0 & 0 \\ 0 & 1666\frac{2}{3} & 1000 & 0 \\ 0 & -1000 & 2000 & -1000 \\ 0 & 0 & -1000 & 3000 \end{bmatrix} \begin{bmatrix} T_1 \\ T_2 \\ T_3 \\ T_4 \end{bmatrix} = \begin{bmatrix} 67\frac{2}{3} \\ 70666\frac{2}{3} \\ 3000 \\ 803000 \end{bmatrix}$$

Repeating the first step, but for the second row, A_{22} is set to unity (divide Row-2 by $1666\frac{2}{3}$). We then multiply second row by A_{32} and subtract it from the third row.

$$\begin{bmatrix} 1 & -1/3 & 0 & 0 \\ 0 & 1 & -3/5 & 0 \\ 0 & 0 & 1400 & -1000 \\ 0 & 0 & -1000 & 3000 \end{bmatrix} \begin{bmatrix} T_1 \\ T_2 \\ T_3 \\ T_4 \end{bmatrix} = \begin{bmatrix} 67\frac{2}{3} \\ 42\frac{2}{5} \\ 45400 \\ 803000 \end{bmatrix}$$

Repeating for the remaining rows of the matrix, we get:

$$\begin{bmatrix} 1 & -1/3 & 0 & 0 \\ 0 & 1 & -3/5 & 0 \\ 0 & 0 & 1400 & -1000 \\ 0 & 0 & -1000 & 3000 \end{bmatrix} \begin{bmatrix} T_1 \\ T_2 \\ T_3 \\ T_4 \end{bmatrix} = \begin{bmatrix} 67\frac{2}{3} \\ 42\frac{2}{5} \\ 45400 \\ 803000 \end{bmatrix}$$

The second stage of the TDMA simply involves *back substitution*. Using Eqn. (5.79), the solution to the above system is

$$T_1 = 140.5; \quad T_2 = 218.5; \quad T_3 = 293.5; \quad T_4 = 365.5$$

For this problem, we could have obtained the solution using the Gaussian elimination instead of the Thomas (the TDMA matrix) algorithm. For such a small matrix, the additional arithmetic operations for Gaussian elimination may not be as significant compared to the Thomas algorithm. Nevertheless, this is not true when a larger number of grid or nodal points are used. This is because of the additional and more cumbersome numerical computations (multiplication and divisions) that have to be performed on the matrix entries. The algorithm degenerates and becomes inefficient once the order of the matrix becomes higher (> 10).

The Jacobi Method To illustrate the *Jacobi method*, the resulting set of algebraic equations as previously used is rewritten

$$3000T_1 - 1000T_2 + 0\times T_3 + 0\times T_4 = 203000$$

$$-1000T_1 + 2000T_2 - 1000T_3 + 0\times T_4 = 3000$$

$$0\times T_1 - 1000T_2 + 2000T_3 - 1000T_4 = 3000$$

$$0\times T_1 + 0\times T_2 - 1000T_3 + 3000T_4 = 803000$$

The above set of equations can be reorganized so that the required variable is on the left hand side of the equation.

$$T_1 = (T_2/3) + (203/3)$$

$$T_2 = (T_1/2) + (T_3/2) + 3/2$$

$$T_3 = (T_2/2) + (T_4/2) + 3/2$$

$$T_4 = (T_3/3) + (803/30)$$

By employing initial guesses: $T_1^{(0)} = T_2^{(0)} = T_3^{(0)} = T_4^{(0)} = 100$, the nodal temperatures for the first iteration are determined as:

$$T_1^{(1)} = (100/3) + (203/3) = 101.000$$

$$T_2^{(1)} = (100/2) + (100/2) + 3/2 = 101.500$$

$$T_3^{(1)} = (100/2) + (100/2) + 3/2 = 101.500$$

$$T_4^{(1)} = (100/3) + (803/3) = 301.000$$

The above first iteration values of $T_1^{(1)} = 101$, $T_2^{(1)} = 101.5$, $T_3^{(1)} = 101.5$ and $T_4^{(1)} = 317\frac{2}{3}$ are substituted back into the system of equations; the second iteration yields

$$T_1^{(2)} = (101.5/3) + (203/3) = 101.500$$

$$T_2^{(2)} = (101/2) + (101.5/2) + 3/2 = 102.750$$

$$T_3^{(2)} = (101.5/2) + (301/2) + 3/2 = 202.750$$

$$T_4^{(2)} = (101.5/3) + (803/3) = 301.500$$

After repeated applications of the iterative process up to 10 and 20 iterations, the nodal temperatures have advanced to

$$\begin{bmatrix} T_1^{(10)} \\ T_2^{(10)} \\ T_3^{(10)} \\ T_4^{(10)} \end{bmatrix} = \begin{bmatrix} 137.247 \\ 208.837 \\ 286.388 \\ 361.080 \end{bmatrix} \text{ and } \begin{bmatrix} T_1^{(20)} \\ T_2^{(20)} \\ T_3^{(20)} \\ T_4^{(20)} \end{bmatrix} = \begin{bmatrix} 140.3632 \\ 218.092 \\ 293.200 \\ 365.313 \end{bmatrix} \text{ and } \begin{bmatrix} T_1^{(40)} \\ T_2^{(40)} \\ T_3^{(40)} \\ T_4^{(40)} \end{bmatrix} = \begin{bmatrix} 140.500 \\ 218.500 \\ 293.500 \\ 365.500 \end{bmatrix}$$

From the previous Example 4.3, we obtained the exact direct solution by the TDMA algorithm. The nodal temperatures after 20 iterations converge towards the exact nodal temperature values.

The Gauss-Siedel Method: We begin as in the Jacobi method with the set of equations

$$T_1 = (T_2/3) + (203/3)$$

$$T_2 = (T_1/2) + (T_3/2) + 3/2$$

$$T_3 = (T_2/2) + (T_4/2) + 3/2$$

$$T_4 = (T_3/3) + (803/30)$$

Employing the same initial guesses, the first iteration yields

$$T_1^{(1)} = (100/3) + (203/3) = 101.000$$

$$T_2^{(1)} = (101/2) + (100/2) + 3/2 = 102.000$$

$$T_3^{(1)} = (102/2) + (100/2) + 3/2 = 102.500$$

$$T_4^{(1)} = (102.5/3) + (803/3) = 301.833$$

After performing 10 iterations, the nodal temperatures have advanced to

$$\begin{bmatrix} T_1^{(10)} \\ T_2^{(10)} \\ T_3^{(10)} \\ T_4^{(10)} \end{bmatrix} = \begin{bmatrix} 140.021 \\ 217.737 \\ 292.944 \\ 365.315 \end{bmatrix} \text{and} \begin{bmatrix} T_1^{(20)} \\ T_2^{(20)} \\ T_3^{(20)} \\ T_4^{(20)} \end{bmatrix} = \begin{bmatrix} 140.499 \\ 218.499 \\ 293.499 \\ 365.500 \end{bmatrix}$$

The temperature values obtained through the Gauss-Siedel method at this present stage are comparable to the values obtained by the Jacobi method at 20 iterations. We can infer from this example that the Gauss-Siedel iteration is twice as fast as the Jacobi iteration. Convergence is achieved quicker by the Gauss-Siedel method because of the *immediate substitution* of the temperatures to the right-hand side of the equations when they are made available. Thus far, we have not discussed the issue of terminating the iteration process for this particular problem. The degree to which convergence is achieved is set by the user. If the absolute maximum difference $\left|\phi_j^{(k+1)} - \phi_j^{(k)}\right|$ is chosen as the condition for termination, the accuracy of the solution depends on the targeted number of significant figures you wish to obtain for the temperatures. The smaller the acceptable error, the higher the number of iterations but this will achieve greater accuracy.

5.6 Fluid-Structure Interactions (FSI)

5.6.1 FSI in Computational Haemodynamics

The fluid structure interaction approach can provide advanced haemodynamic modelling of diseased vessel structure. A number of computational approaches to simulate the cardiovascular flow can be categorised into three types as follows (Cheng et al. 2005):

- *Geometry-prescribed CFD* constitutes prescribed moving meshes of a cardiovascular structure that is reconstructed from medical imaging modalities such as magnetic resonance imaging and computed tomography based on different time phases of a cardiac cycle. It does not consider the feedback from the structure to the fluid medium, and hence is considered as a one-way FSI approach.
- *Fictitious Fluid-Structure Interaction* method is based on the immersed boundary method that simplifies the structure to elastic fibers represented as chains of points immersed in a fluid. Then, a Dirac delta based interpolation function is used to compute the interaction of the fluid with the structure.
- *Realistic FSI* models the structural mesh using Finite Element Method and then simulates the fluid flow by CFD and the two models coupled via algorithms to achieve a fluid-structure interaction. By resolving the complex properties of the

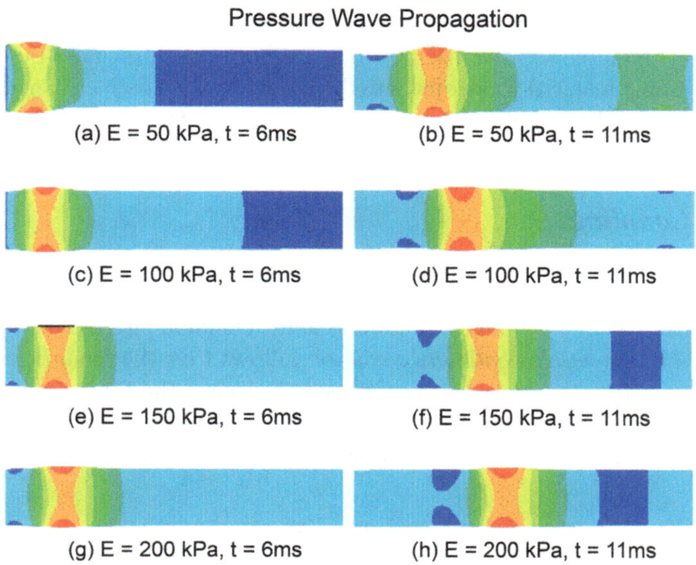

Fig. 5.39 Numerical study on the deformation of vessel wall and its associated blood flow structure using the FSI algorithm. In the default configuration, which is mainly used to as a reference to compare with other cases, the vessel radius $r=0.005$ cm, vessel length $L=0.06$ m, vessel thickness $h_s=0.001$ m, solid density $\rho_s=1000$ kg m^3, Young's modulus $E=750{,}000$ Pa, Poisson ratio $\upsilon=0.5$, fluid density $\rho_f=1000$ kg m^{3}and fluid dynamic viscosity $\mu_f=0.01$ Pa · s. At initial condition, fluid is set to be at rest and pressure pulse with peak of 2000 Pa is imposed at the inlet

cardiac structure and blood in two separate modeling platforms, a physiologically realistic simulation can be outputted.

Fluid structure interaction for large cardiac wall deformation presents a difficult problem. The fluid structure interaction must satisfy three important conditions: (1) The structural solver describes nonlinear, anistropic and inhomogeneous cardiovascular tissue characteristics; (2) The CFD code resolves large deformations on the flow domains and updates the mesh accordingly; and (3) coupling of the fluid and structural solver is needed to achieve convergence. For the fluid modeling, a finite volume method can be used to discretize the Navier–Stokes equations, and for the structural modeling, the nonlinear finite element method can be implemented.

As an example of vessel wall deformation such as the type is shown in Fig. 5.39. The combination of CFD with FEM is used to investigate the dynamic interaction of blood flow with the vessel wall solid structure due to the a pressure wave propagation. The FSI simulations pertain to either a stable or oscillatory structural interaction with its flow. One example would be the oscillatory deformation of heart valves whereby the induced structural strain causes the valve to assume a state of reduced strain, and before returning to its former state repetitively.

In today's commercial CFD software, FSI can be applied with multi-physics components required to solve haemodynamics problems. It includes the nonlinear and anisotropic material properties into the model by user-defined functions. This allows flow simulations of the distensible cardiovascular vessels.

5.6.2 Coupling

The interaction between the fluid and structural fields is achieved by enforcing kinematic and dynamic continuity conditions at the common interface between these domains. The *kinematic continuity condition* follows from the requirement that no mass flows across the interface. For a general viscous fluid, a no-slip boundary condition for both normal and tangential components of displacements and velocities are imposed at the interface (Donea et al. 2004; Gerstenberger and Wall 2008):

$$\begin{aligned} \mathbf{d}^f &= \mathbf{d}^s \\ \mathbf{u} &= \dot{\mathbf{d}}^s \end{aligned} \tag{5.84}$$

where $\mathbf{d}^f$ and $\mathbf{d}^s$ represent the fluid and structural displacement vectors at the interface, $\mathbf{u}$ and $\dot{\mathbf{d}}^s$ are the fluid and structural velocity vectors, and $\mathbf{n}$ represents the outward unit normal vector of the interface. The *dynamic continuity condition* is based on force equilibrium across the interface according to principle of action and reaction. Since the interface area is equal, this reduces to equality of traction between both fluid and structural fields as:

$$\sigma^f \cdot \mathrm{n} = \sigma \cdot \mathrm{n} \tag{5.85}$$

where σ represents the stress in three coordinates, the Cauchy stress tensor for structural field; and σ^f is the stress tensor for the fluid field, made up of the pressure and shear stress terms.

Solution strategies for FSI Coupling are mainly divided into "Monolithic" and "Partitioned" methods. The differences between them are the method for solving the governing equations and the interface conditions. The monolithic method completely couples the solid and fluid domains into one system of equations and solves them directly while the partitioned approach separates the fluid and solid domains and then solves their governing equations iteratively using two different solvers until the solutions converge (Fig. 5.40).

There are benefits and deficiencies to both the monolithic and the partitioned approaches. The obvious benefit to the monolithic approach is that, if solved, it always produces a fully-coupled solution in a single iteration per time step. On other hand, the biggest issue to this approach is the relatively higher computational cost. Michler (2004) showed that in a simple one dimensional FSI problem, the

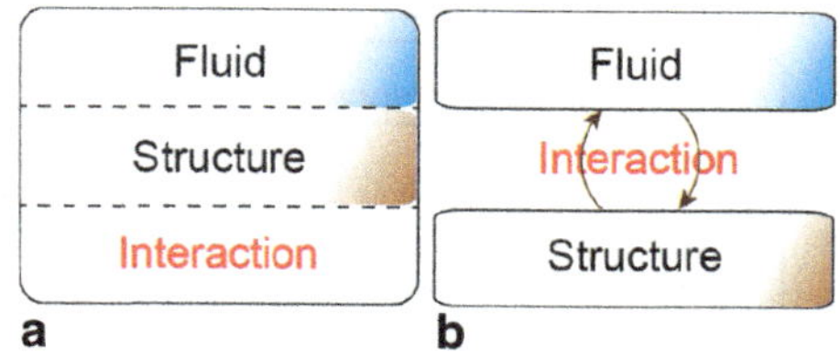

Fig. 5.40 Solution strategies for FSI simulations. **a** Monolithic methods **b** Partitioned methods

computational cost of the monolithic procedure is three to four times the partitioned procedure; however, the monolithic approach reached accuracies much greater than the partitioned approach when weakly coupled with respect to calculated structural displacements (Wong et al. 2013a). Michler (2004) hypothesized that for more complex problems the monolithic approach would be superior in a comparison of computational cost to accuracy. One of the benefits to the partitioned approach is its relative simplicity in implementation and solution. The interface conditions for the partitioned case can be applied directly, while in the monolithic case they must be added to the system of equations as extra Jacobian terms. Also, each set of governing equations for the partitioned case can be solved independently, requiring fewer unknowns per solve than for the monolithic case. Due to its relative ease of implementation and lower computational cost, partitioned schemes have dominated the majority of FSI research.

Therefore, this chapter will mainly focus on partitioned methods. To match both kinematic and dynamic conditions simultaneously in both fluid and solid equation solvers, a successive iteration method is applied. In this partitioned approach, segregated solvers for each fluid and structural field are employed and the interacting quantities at the interface from each field are exchanged sequentially. A flow chart demonstrating this process is given in Fig. 5.41.

Within each component of iterations, the solutions must reach converged results before moving to the next set of iterations. The steps are detailed as:

1. **Fluid Solver:** The fluid variables, (u, v, w, P) are solved based on initial or current geometrical configuration, i.e. based on the current displacement at the interface, $\mathbf{d}_{k-1}^{n}$ at time t^n and coupling iteration $k-1$. The fluid pressure and shear forces at the interface are resolved in all three components, x, y, and z.
2. **Structural Solver:** The forces and boundary constraints produced from the fluid equations are applied to the structure at the interface. The structural deformation is determined, giving the current interface displacements $\mathbf{d}_{k}^{n}$.
3. **Mesh Adapt:** The interface displacement is interpolated across the fluid interface mesh nodes which are used to alter the mesh deformation on the fluid domain.
4. The fluid equations are solved again for the unknown fluid variables based on the new geometrical configuration caused by $\mathbf{d}_{k}^{n}$.
5. **FSI Coupling**: The process is repeated until the difference in nodal deflection and forces exchanged in steps 3 and 1, from current and previous coupling iterations, are within a specified tolerance—suggesting that at a given timestep, the kinematic and dynamic continuity are satisfied at the interface.

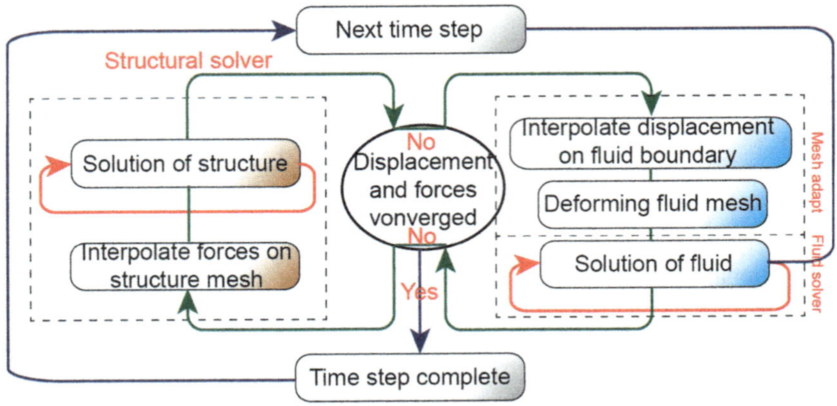

Fig. 5.41 Fluid-structure coupling flowchart

Transfer of the fluid flow variables (e.g., velocity, pressure, force, temperature) from the fluid solver to the structural solver, without the return influence from structure to fluid is referred to as a one-way FSI coupling. One-way transfer is appropriate when displacement and temperature differences calculated in the structural application are small enough to have insignificant impact on the fluid analysis. In two-way coupling, deformation of the fluid-structure interface on the structural side is transferred to the fluid solver to deform the fluid mesh. Therefore, the two-way FSI coupling solution is more needed for cases with larger deflections where the fluid field is strongly influenced by the structural deformation.

Using partitioned methods, FSI coupling solutions are separated into fluid dynamics, and structure dynamics. At interface, information for the solution is shared between the fluid and structural solver. The information exchanged is dependent on the coupling method. For one-way coupling, only the fluid pressure acting at the structure is transferred. For two-way-coupling calculations, the displacement of the structure is also transferred to the fluid solver.

A structural solver will most probably use finite elements on unstructured grids, while a fluid solver might use a finite volume approach on Cartesian grids. Therefore, interpolation methods plays a key role in mapping data from one mesh to the other. In addition, a fluid with low viscosity might be highly turbulent and require very small time steps to stay stable, while the structure could advance with much larger steps. Sub-cycling might improve the situation by allowing the fluid solver to accumulate several small time steps on its own before transferring the variables to the structural solver.

One argument for partitioned approaches is their flexibility. Flexibility in both the choice of coupling algorithms, and the choice of software for solving the equations. Since the solvers are separated, they use their own discretisation method, leading to nonmatching grids on both sides of the coupling interface between fluid and structure. In order to transfer loads across a dissimilar mesh interface, the nodes of one mesh must be mapped to the local coordinates of an element in the other mesh. The fluid nodes must be mapped to the solid elements to transfer displacements. Likewise, solid nodes must be mapped to the fluid elements to transfer stresses.

FSI coupling is characterised by the amount of repeated iterations between the fluid and solid domain. Computationally, this refers to how many iterations are used to reach convergence. A weak coupling is one that uses a minimum iteration while a strong coupling uses a maximum. Physically this means that a weakly coupled FSI is suitable for cases with minimal deformation between the fluid and structureal domain, while a stronger coupling is needed for cases with more deformation.

5.6.3 Stability and Convergence

Stability at the fluid–solid interface is a big concern with FSI simulations, particularly when the structure is very flexible and exhibits instabilities at the interface. In two-way coupling, the fluid mesh deformation of is calculated by the displacement diffusion algorithm. The connections between the nodes are modelled as springs with different stiffness. This stiffness can change from node to node, and it is set very high for nodes near boundaries. In the normal direction from the FSI interface, the first ten nodes have displacements in the range of the boundary displacement. This provides high quality meshes even in the boundary layer. Figure 5.42 shows a comparison of non-deformed and deformed mesh for an idealized arterial vessel, and in both cases, the mesh has a satisfactory quality.

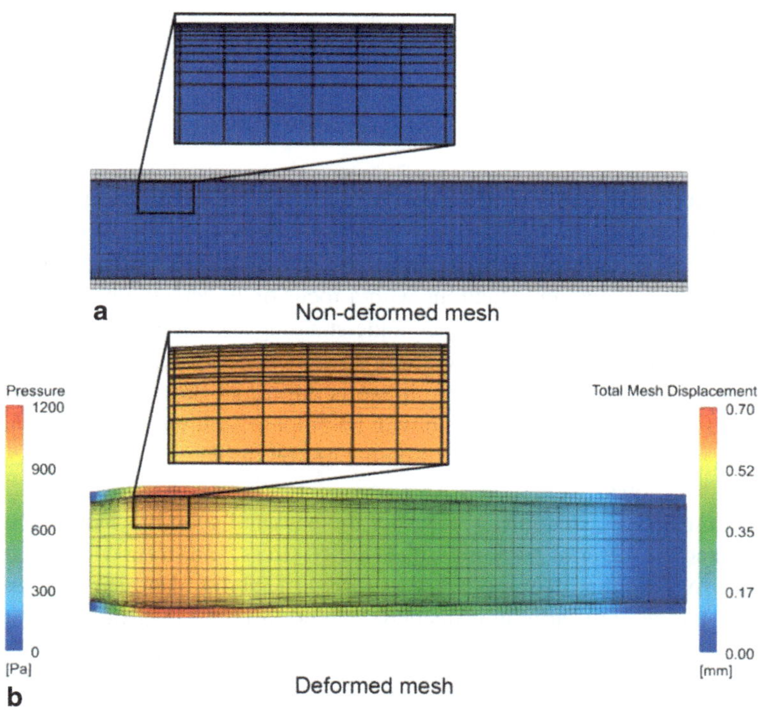

Fig. 5.42 Comparison between non-deformed mesh and deformed mesh for an idealized arterial vessel

In some cases, applying peak values on the target side of the interface will initiate oscillatory convergence or even divergence within and between the coupled solvers. Therefore, the target side data may be ramped from the final value observed in the previous coupling step to the peak value during the initial coupling interactions within the current step by using under-relaxation factors.

The order of which solver is initiated first also affects the FSI solution stability. In general, the fluid analysis should be processed first since the it causes the structure to deform To check the convergence the structural solver, the fluid solver and the data exchanged across the interface have to all converge to the a prescribed residual limit. For a full transient simulation, convergence should take place at every time step for the results to be accurate.

Sometimes, it is difficult to get a coupled simulation to initiate, or you want to check the setup before conducting a multi-field simulation. Then it is recommended to run the fluid solver and the structure solver separately for trial simulations. For example, if you want to get an impression or preliminary results of the structural deformation, a characteristic pressure load roughly reflecting the real pressure exerted by the blood flow can be loaded to the same location where you have defined the FSI interface. Then, run the structural solver to solve just the mechanical component with this dummy pressure load. To test the fluid component of the FSI coupling system, a fixed or time-dependent mesh displacement wall boundary can be used to replace the mesh displacement settings at the FSI interface. Then run the fluid solver separately to check the fluid solver's setup and perform debugging if necessary.

5.7 Summary

Fluid structure interaction modelling is the merging of fluid dynamics and solid mechanics. It is therefore important to be familiar with both sets of equations that govern fluid flow and structural deformation. This chapter presented mass conservation, and the momentum equations with a focus on understanding the individual terms in the equations. This includes understanding the fundamental principle of mass conservation, the local acceleration, advection, pressure, diffusion, and body forces that act on the blood and that it exerts during its flows. An introduction to turbulence modelling was also given with an emphasis on understanding the characteristics of turbulence and practical guidelines on turbulence model selection for internal blood flows.

For the structural equations a review of the basics of solid mechanics was given which linked the material properties of elasticity, stress, and strain with its structural dynamics to yield the three fundamental forces of inertia, damping, and stiffness. The composition of the arterial walls plays a major role in its elastic properties. The relative tissue composition of endothelium, elastic tissue, smooth muscle, and fibrous tissue was shown for different artery types and how this influences the ability to deform during pressure loadings from the blood flow.

Conversion of the fluid and structural set of equations to discrete nodal locations is referred to as discretisation. The fluid equations are typically discretised using finite difference and finite volume which are based on the Taylor series expansions of the differentials. Although the finite element method may also be used for the fluid equations, its use for structural equations is more prevalent. A characteristic stiffness matrix is used to connect each node for the finite elements. This was shown for a simple one-dimensional spring-mass system and the plane stress for a two-dimensional triangle was presented.

After discretisation, a set of algebraic equations is formed which are solved numerically by either an iterative or direct solution method. For complicated geometries the matrices of equations can be extremely large which take some time to solve. Simple examples were provided to highlight the nature solving the matrices.

If the pressure exerted by blood flow onto the vessel artery wall is accounted for, then a one-way FSI coupling has been achieved. Typically if the wall deformation is small and has little bearing on the blood flow, then the coupling is sufficient. However in cases where large deformations are found, then two-coupling is required. The interface boundary separating fluid from structure is used for matching mesh nodes to transfer the force loadings between the two domains. FSI is very prone to unstable simulations where large deformations cause problems in the mesh which is related back to the discretised domain onto nodal points. Strategies to help convergence, such as under-relxation, and time step selections were discussed.

This chapter provided an introduction into the numerical aspects of the computational modelling required to satisfactorily achieve FSI simulations. Following on from an understanding of the discretisation of the physical domain into discrete nodes, the next chapter continues this theme by presenting methods for producing a computational mesh that contain the discrete nodes.

5.8 Review Questions

1. Write down the relation for the total mass flow rate for the four artery outlets shown in the figure on the right. If the mass flow fraction for $\dot{m}_1$ and $\dot{m}_2$ is 0.4, what does this mean for the flow in $\dot{m}_3$ and $\dot{m}_4$?
2. What factors contribute to the mass flow?
3. In a converging nozzle, the flow accelerates due to the narrowing geometry. Discuss the changes in the velocity gradients $\frac{\partial u}{\partial x}$ and $\frac{\partial v}{\partial y}$ during the flow (assume a constant density). What is its corresponding pressure gradient?
4. Write a force balance equation for all the forces acting on a 2D differential control volume. What is its equivalent 3D form?
5. Discuss the difference between the local and advection acceleration in the fluid momentum equation

$$\underbrace{\frac{\partial v}{\partial t}}_{\text{local}} + \underbrace{u\frac{\partial v}{\partial x} + v\frac{\partial v}{\partial y}}_{\text{advection}}$$

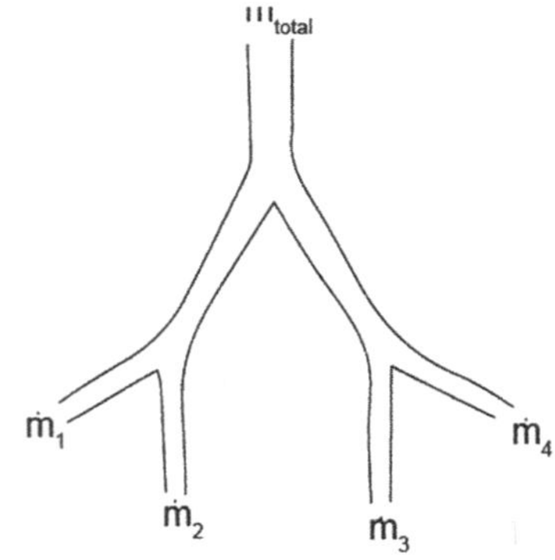

6. For flow in a stenosed artery what is the significance of the diffusion term $\left(\mu\left(\frac{\partial^2 v}{\partial x^2}+\frac{\partial^2 v}{\partial y^2}\right)\right)$ on the reattachment of the flow after the stenosis?
7. During deformation of an elastic material, describe the relationship for stress, strain, and Hooke's Law.
8. What are three opposing forces in response to an applied force on a vessel wall?
9. What are the four types of tissues that make up an arterial wall? Which tissue type is most abundant? Does it contribute towards the artery elasticity?
10. Which of the following, *forward difference*, *backward difference* and *central difference*, is most accurate and why?
11. How does the finite difference and the finite volume approach to discretisation differ?
12. Describe what false diffusion means.
13. What is the significance of the stiffness matrix in a finite element discretisation?
14. What is the difference between using a direct method and an iterative method to solve the discretized equations?
15. Is the direct method or iterative method more suitable in solving for a large system of non-linear equations?
16. Why does the Gauss-Siedel iterative method converge to a solution quicker than the Jacobi method?
17. What is the Gaussian elimination method based on? Can this method be used to solve a system of nonlinear algebraic equations?
18. Solve the following set of equations by Gaussian elimination:
19. What are the two approaches of FSI coupling?
20. Solve the following set of equations by Gaussian elimination

$$\begin{bmatrix} 100 & 100 & 0 \\ 200 & 100 & - \\ 300 & - & - \\ - & 200 & 300 \end{bmatrix} \begin{bmatrix} T_1 \\ T_2 \\ T_3 \\ T_4 \end{bmatrix} \begin{bmatrix} 400 \\ 100 \\ -300 \\ 400 \end{bmatrix}$$

Chapter 6
Generation of Computational Mesh for Haemodynamics Analysis

6.1 Introduction

A mesh can be viewed as a number of small elements or grid cells that overlays an entire domain geometry. In general the set of fluid and structural equations described in Chap. 5 are applied onto each cell, or finite element. These discrete equations, which calculate the changes or interpolations between cells are solved to yield the corresponding *discrete* values of the flow and field variables such as velocity, pressure, temperature, and deformation.

The type of mesh created is based on the level of fluid-structure-interaction involved in the haemodynamic flow. Coupling of the fluid dynamics and structural mechanics leads to hydrodynamic forces exerting onto vessel walls or vascular structures causing deformation and translation of the structure. A high quality mesh improves numerical stability, and increases the likelihood of attaining a reliable solution. It must allow for deformation while maintaining the integrity of the discrete mathematical equations.

For the fluid domain, the most common approach is the finite volume or finite difference methods. Although the finite element method may also be used for the fluid domain, it is most widely used for the structural domain.

In terms of the fluid-structure interaction, the two main approaches are: (i) the Arbitrary-Lagrangian–Eulerian (ALE) formulation, or (ii) immersed methods, where there is a fixed Eulerian formulation for fluid, and a moving Lagrangian formulation for the structure. In this chapter we present the main ALE method, which is more applicable to blood and its vessel interactions. In the immerse method, the Lagrangian mesh of a solid moves through the fluid mesh and is more suited for blood cell transport and its interactions with its environment.

Mesh generation is also referred to as grid generation, and is a separate discipline by itself, in a very active area of research and development. This chapter does not intend to elaborate on all the various methods of grid generation. There are many dedicated books that focus on this subject alone, such as Thompson et al. (1985), Arcilla et al. (1991), and Liseikin (1999). We also note that many existing commercial codes in the market have their own in-built powerful mesh generators and there

J. Tu et al., *Computational Hemodynamics—Theory, Modelling and Applications*,
Biological and Medical Physics, Biomedical Engineering,
DOI 10.1007/978-94-017-9594-4_6

is also a choice of a number of independent grid generation packages which is given in the Appendix. Although, a number of mesh-generation packages are designed with very user-friendly interfaces, these software packages still rely on the reader's aptitude to operate them.

Generating a mesh is as much an art form as it is science. One has to decide on the arrangement of discrete points (nodes) throughout the computational domain, and the type of connections of each point, which leads to either great success or utter failure of the numerical solution. In this chapter, guidelines and best practices are given for developing quality meshes. Furthermore the material presented will give the reader an introduction into mesh construction.

6.1.1 Meshing Topology

A computational mesh topology has a hierarchical system where higher topology assumes the existence of the topologies underneath it. For example, at the top of the hierarchy is a volume cell, which implies the definition of its lower constituent topologies—surfaces, lines, and vertices—that are used to define the boundaries of each higher topology. A vertex is defined by its spatial coordinates (x, y, z); two vertices define a line; three lines enclosing a surface defines a triangle, while four enclosing lines define a rectangle; and three or more surface faces define a 3D volume (Fig. 6.1). Three-dimensional volume elements can be hexahedra, tetrahedra, square pyramids pyramids, extruded triangles (wedges or triangular prisms) or polyhedra, while in 2D the elements may be quadrilaterals or triangles.

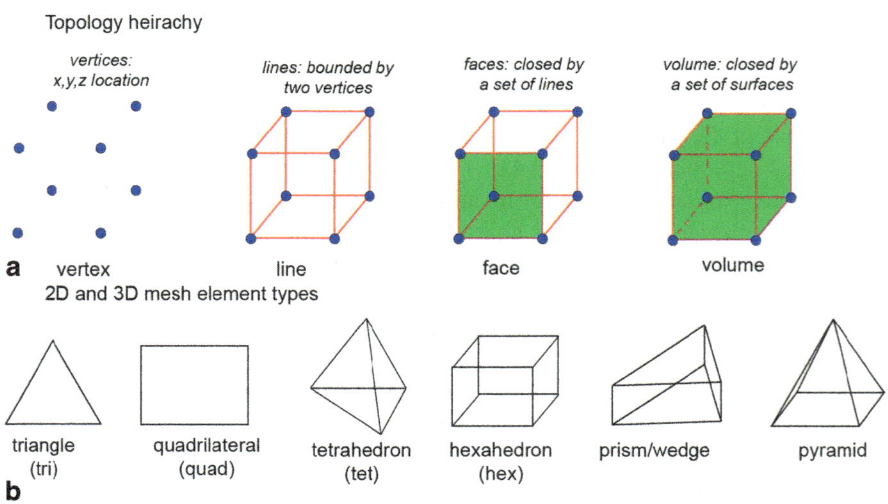

Fig. 6.1 Mesh topology hierarchy from lowest (*left*) to the highest (*right*)

Based on the topology, mesh generation is performed as a bottom-up or top-down approach. The bottom-up approach creates low dimensional topologies first and builds on top higher dimensional topology, e.g. create vertices, connect these to form edges, connect the edges to create faces, and combine the faces to create a volume. The top-down approach creates upper dimensional topology first and uses Boolean operations to decompose it down to lower topology.

The discretisation of the fluid equations in Chap. 5 was performed by Finite Volume and Finite Difference and Finite Element Method, which all produce a different mesh structure. In the next section we present an introduction to mesh terminology and different mesh designs for the fluid and structural set of equations; to assist the reader in understanding some of the issues related with creating a usable mesh for a haemodynamics analysis.

6.2 Mesh Configurations

6.2.1 Structured Mesh

By definition, a structured mesh is one that contains cells having either a regular-shape element with four-nodal corner points in 2D or a hexahedral-shape element with eight-nodal corner points in 3D. It is characterised by regular connectivity and is a straightforward prescription of an orthogonal (90°) mesh in a Cartesian system. This meshing should always be applied to geometries that exhibit orthogonal shapes. Figure 6.2 shows a uniform and non-uniform structured mesh on a simple rectangular domain in 2D and 3D. For a uniform mesh, the spacing of each cell is a single representative value in all directions (i.e. $\Delta x_i = \Delta y_j = \Delta z_k$). For non-uniformly distributed grid points, the spacing of either Δx, Δy or Δz can take any values, however a prescription of a four-sided surface face is maintained. The non-uniform rectangular mesh in Fig. 6.2b is regarded as a "stretched" mesh where the grid points are biased towards the wall boundaries.

6.2.2 Body-Fitted Mesh

Applying an orthogonal mesh to a non-orthogonal geometry, such as a 90° bend in the aortic arch, produces simplifications around the curved regions and staircase-like steps are found (Fig. 6.3). Developing an approximate boundary description around the curved edges is difficult as the steps at the boundary introduce errors in computing the wall stress, heat flux values, and resolving the boundary layer. Furthermore a fine mesh is required to match the curve more closely which resolves some of the flow approximations, however this comes at a cost of high computational demands.

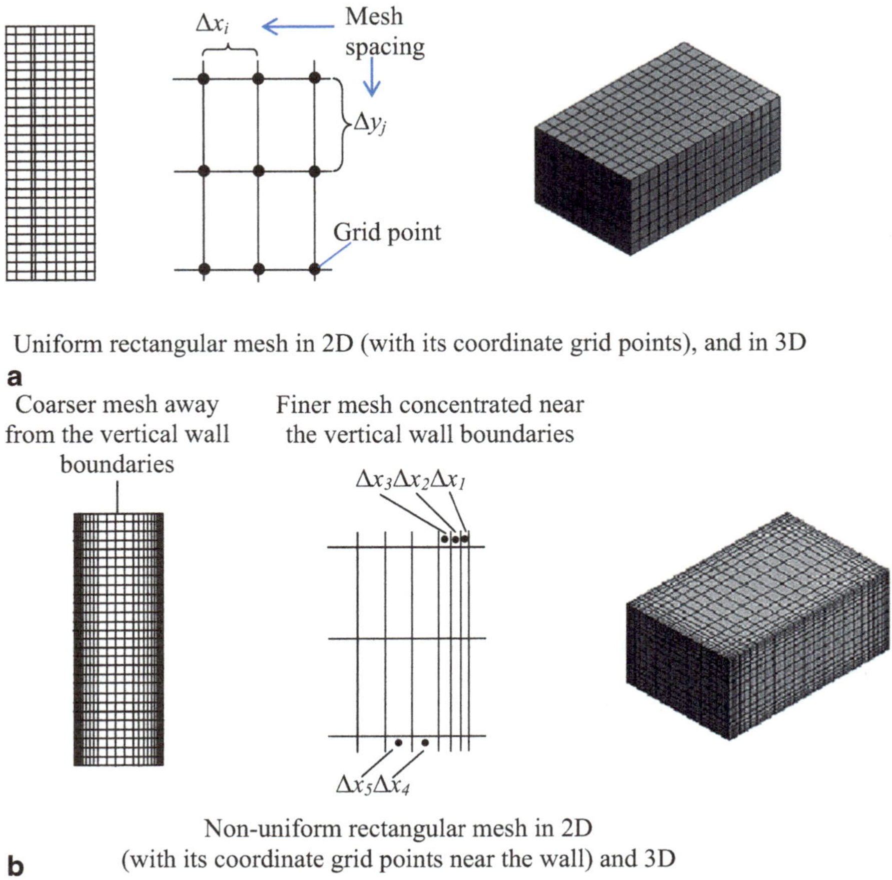

Fig. 6.2 **a** A uniform and **b** non-uniform structured mesh in 2D and 3D for a simple rectangular geometry. Note the evenly distributed grid points in the uniform mesh in contrast to the biased concentration of cells near the wall for a non-uniform structured mesh

As an alternative a similar type of structured mesh called a *body-fitted* mesh can be applied. This approach centres on mapping the distorted region in physical space onto a rectangular curvilinear coordinate space through a transform coordinate function. Applying a body-fitted mesh to the 90° bend geometry, the walls coincide with lines of constant η (see Fig. 6.4). The path length from the vertices A to B, and D to C, then correspond to specific values of ξ in the computational domain. In this example we see that η is constant but there is a stretching of ξ in the curved region.

A transformation must be defined such that there is a one-to-one correspondence between the rectangular mesh in the computational domain and the curvilinear mesh in the physical domain. The algebraic forms of the governing equations for fluid flow are carried out in the computational domain which has uniform spacing of $\Delta\xi$ and uniform spacing of $\Delta\eta$. Computed information is then directly fed back

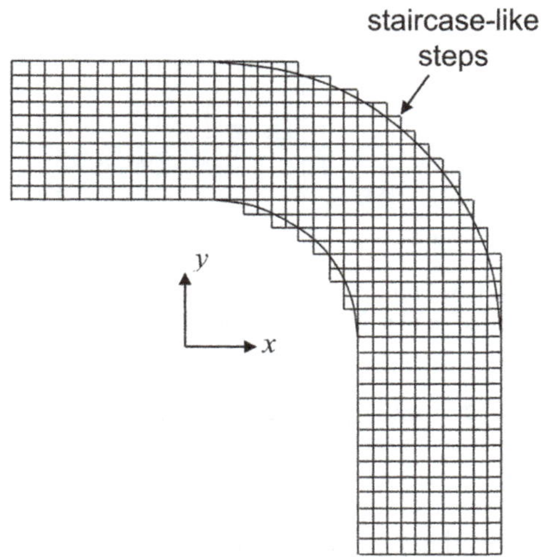

Fig. 6.3 An example of a mesh using staircase-like steps for a simplified 90° bend in the aortic arch

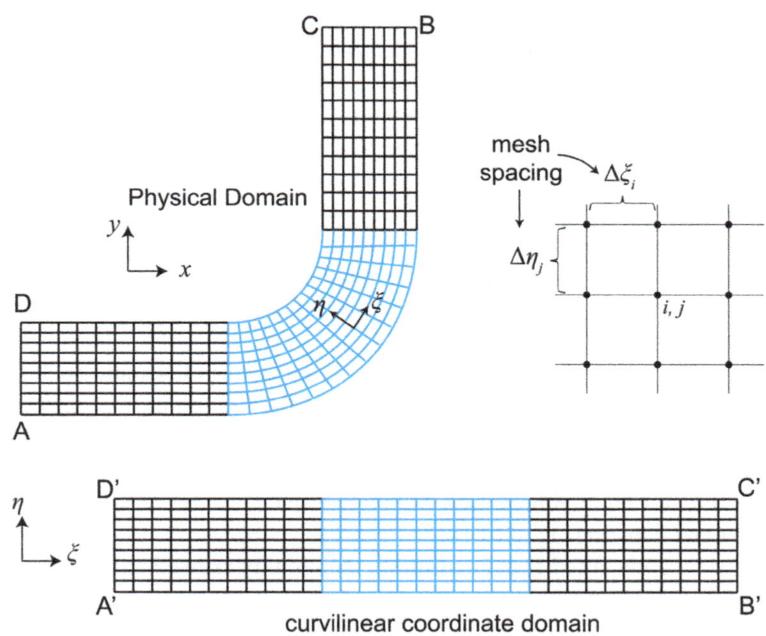

Fig. 6.4 An example of a body-fitted mesh for the 90° bend geometry and corresponding computational geometry

to the physical domain via the transformation of grid points. Because of the need to solve the equations in the computational domain, they have to be expressed in terms of the curvilinear coordinates rather than the Cartesian coordinates, which means that they must be transformed from (x, y) to (ξ, η) as the new independent variables.

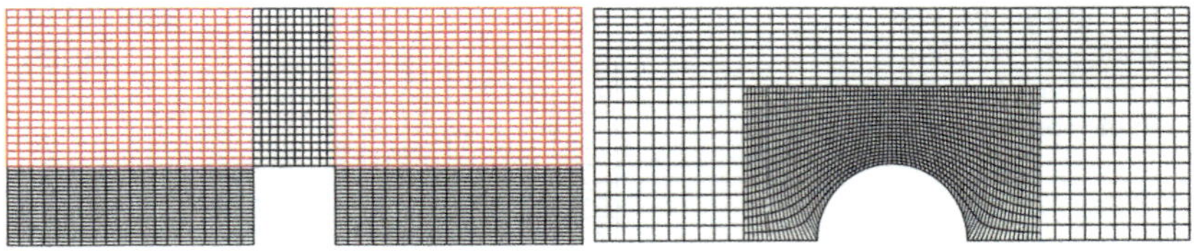

Fig. 6.5 Multiblock-structured mesh with matching and nonmatching cell faces at each block interface **a** matching cell faces (conformal) **b** non-matching cell faces (non-conformal)

6.2.3 Multi-Block Mesh

Block-structured or *multiblock mesh* is another case of a structured mesh. This is particularly effective for complicated shapes where it is difficult to apply a single block. The multiblock mesh allows the computational domain to be subdivided into topological blocks. Figure 6.5 shows mesh assembled from a number of structured blocks connected to each other. The attachments of each face of adjacent blocks may be regular (i.e. having matching cell faces, also referred to as *conformal mesh*) or arbitrary (i.e. having nonmatching cell interfaces, also referred to as *non-conformal mesh*). Generation of grids especially with nonmatching cell interfaces is much simpler than creating a single-block fitted to the whole domain. This approach provides flexibility in selecting the best grid topology for each of the subdivided block regions.

Special types of topologies for each block include structured *O*- and *C*-grids. These are in conjunction with the usual structured orthogonal mesh described earlier, as well as any unstructured mesh (discussed in Sect. 6.3) of triangle or tetrahedral elements.

An O-grid is a series of sub-blocks that arrange the grid lines into an "*O*" shape where the last point on a grid line joins up with the first point of that line, hence creating a circular '*O*' shape. Similarly a *C*-grid, which is half an *O*-grid, has lines that bend in a semicircle shaped like the letter '*C*'. Each series of sub-blocks can be found within the O- or C-block itself and represented in Cartesian coordinates in the computational domain (Fig. 6.6).

If we use a body-fitted structured mesh, then highly skewed and deformed cells are found at the perimeter of the geometry (Fig. 6.7), since the interior of the domain must satisfy the geometrical constraints imposed by the circular boundary. This mesh generally leads to numerical instabilities and deterioration of the computational results. Hence, it is better to mesh the geometry with an O-grid. In this operation, the O-grid fills the external circular cross-sectional conduit, while a square block fills the centre.

Finally, grids may be overlapped on top of one another to cover an irregular flow domain referred to as Chimera or overset grids. This differs from a single flat block-structured mesh which connects neighbouring blocks together. Chimera grids allow rectangular, cylindrical, spherical, or non-orthogonal grids to be combined with the

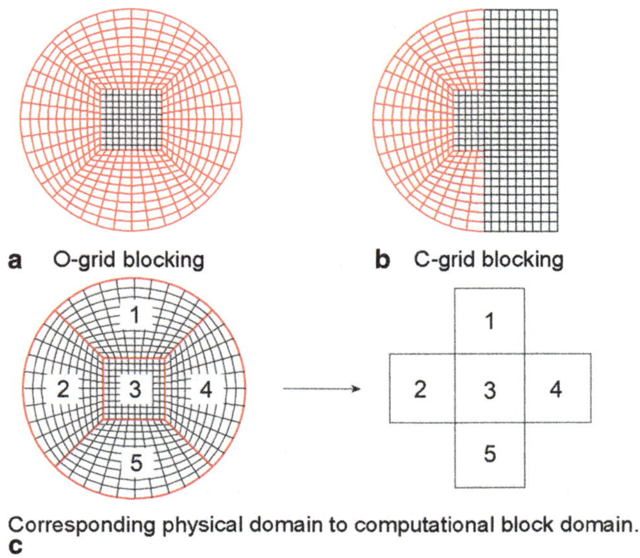

Fig. 6.6 Example of a structured **a** O-Grid and **b** C-Grid blocking mesh and **c** its corresponding computational block domain, for a circular shape

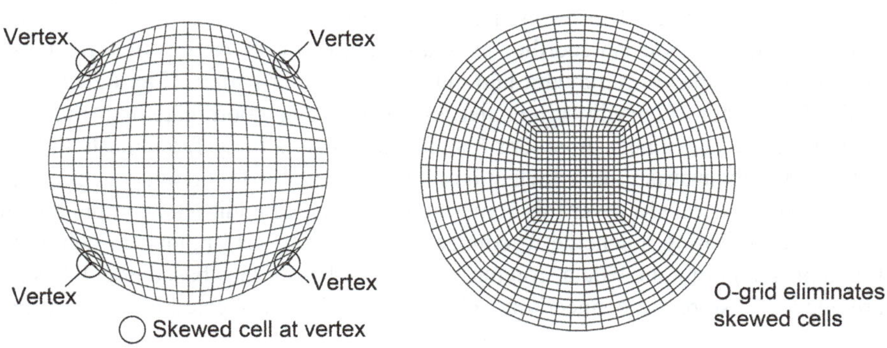

Fig. 6.7 Generation of body-fitted mesh, unstructured mesh and O-grid for the circular cross-sectional conduit

parent Cartesian grid in the solution domain. An example of an overlapping grid for a cylinder in a channel with inlet–outlet mappings is shown in Fig. 6.8.

Structured mesh blocks are placed freely in the domain to fit any geometrical boundary. The overlapping structured grids are periodically updated and exchange boundary information through interpolation. These grids can handle complex domains and large displacements found in dynamically moving geometries in stagnant surroundings. The Cartesian structured mesh provides computational efficiency, and also good resolution for boundary layers. Some examples can be found in Tu and Fuchs (1992) and Hubbard and Chen (1994; 1995). The disadvantages of these

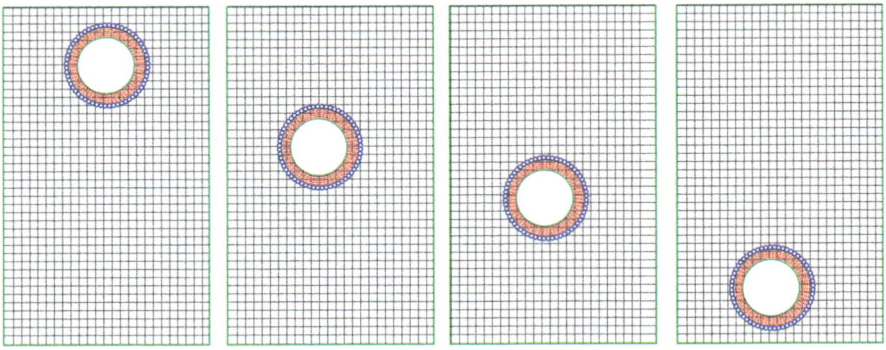

Fig. 6.8 A structured overlapping grid for a rigid blood cell in a vertical vessel

grids are that conservation is usually not maintained or enforced at block boundaries, and the interpolation process may introduce errors or convergence problems if the solution exhibits strong variation near the interface.

6.2.4 Unstructured Mesh

Geometries encountered in haemodynamics are typically non-orthogonal and do not fit exactly in Cartesian coordinates of a structured mesh. An unstructured mesh containing triangles or tetrahedra laid out in an irregular pattern (Fig. 6.9) is able to fill the interior of curvilinear geometries. There are no coordinate lines that correspond to the curvilinear directions ξ and η such as in a body-fitted mesh and the cells are totally unstructured which makes it highly suitable for complex geometries such as the arteries. However this comes at an expense of additional computational costs in establishing a scheme to connect neighbouring grid points. In computational programming, new data structures (e.g. edge based, face based or cell based) are needed to store the connectivity scheme which is complicated by the fact that

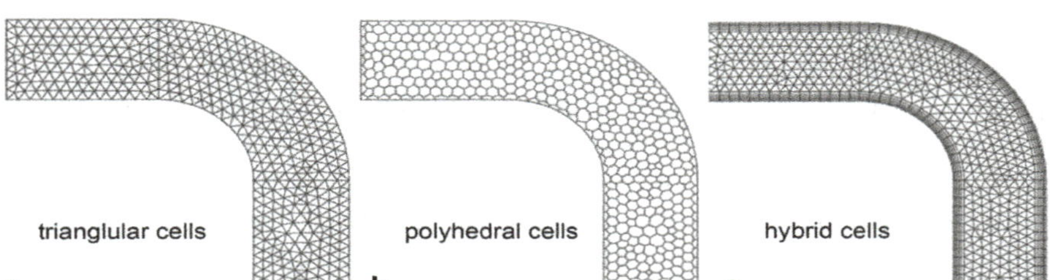

Fig. 6.9 Simplified 90° bend in the aortic arch filled with unstructured mesh comprising of **a** triangular cells, and **b** polyhedral cells. **c** A hybrid mesh combining quadrilateral cells at the wall boundaries and triangular cells in the centre

a node of one cell can be connected to any number of nodes from any number of other cells. The descriptions of some of these connectivity algorithms (Delauney, Quadtree/Octree, and Advancing front) are briefly described in this section.

In addition a large number of grid generation software and open source codes exist and a list is provided in the Appendix. Using a bottom-up approach a meshing process involves meshing the edges first, followed by filling the enclosed face (Advancing Front method), and for a 3D geometry filling the interior volume (Delauney method). Recently *polyhedral* cells have been used to fill the interior domain as an alternative to triangular cells. A benefit of applying a polyhedral mesh is that it allows the flexibility of an unstructured mesh to be applied to a complex geometry without the large computational demands associated with a large tetrahedral mesh. The application of such cells is relatively new. Nevertheless, polyhedral meshing has shown thus far to have tremendous advantages over tetrahedral meshing with regard to the attained accuracy and efficiency of the numerical computations (Kalvin and Taylor 1996; Spiegel et al. 2010).

Unstructured meshes can also involve the use of hexahedral, pyramid, and wedge cells in combination with tetrahedral cells, whereas a structured mesh is reliant on hexahedral cells or the use of block-structured mesh only. A mesh combining different cell types is a *hybrid mesh* and typically involves allocating cell elements that match the boundary surface. For example structured elements are used for wall boundaries instead of triangular unstructured cells as can resolve the boundary layer flow field gradients much better. The triangular mesh are difficult to cluster in the lateral direction due to the underlying triangular structure. In almost all cases, triangular elements are used to fill the surface face (in 2D) or tetrahedral elements are used to fill the remaining volume (in 3D). Many codes try to automate the generation of prismatic meshes by allowing the user to define the surface mesh and then marching off the surface to create the 3D elements. While very useful and effective for smooth shapes, the extrusion process can break down near regions of high curvature or sharp discontinuities. Hybrid grid methods are designed to take advantage of the positive aspects of both structured and unstructured grids with the structured grid in local regions while using unstructured grid in the bulk of the domain.

6.2.5 Delaunay Triangulation

Triangle and tetrahedral meshing are the most common forms of unstructured mesh generation as they allow maximum flexibility in matching cells with curved boundaries and increased resolution in flow regions where they matter most. A common method for triangle and tetrahedral meshing is the Delaunay triangulation criterion which states that no node must be located inside the circumcircle—a circle circumscribing the mesh element (Shewchuk 2002). In 2D, a circumcircle of a triangle is the unique circle that passes through all three vertices of its triangle whereby nodes from other triangles do not exist within the circle. Figure 6.10 shows that the circumcircle of every Delaunay triangular cell within the mesh generated is free from other nodes.

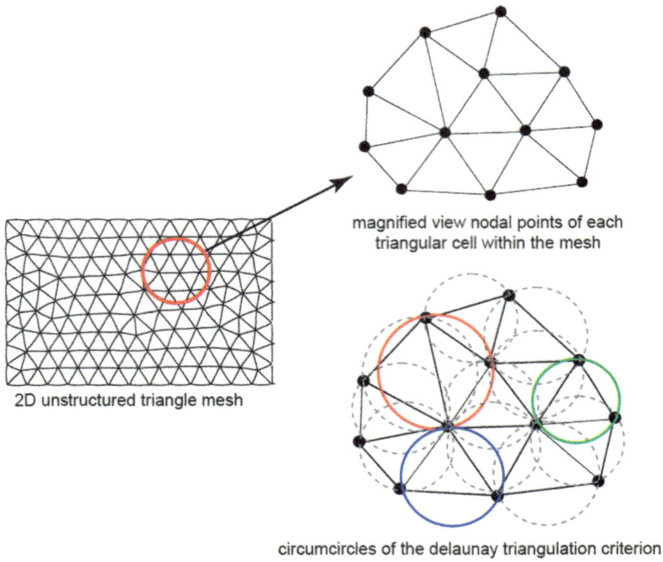

Fig. 6.10 A triangular mesh based on Delaunay triangulation. Coloured circumcirlces highlight the relationship between the triangle points and the absence of any external nodal points inside the circumcircle

To produce Delaunay triangulations an algorithm to detect when a grid point is within a triangle's circumcircle and an efficient data structure for storing triangles and edges is required. An approach is to first mesh the boundary edges, then repeatedly add one node at a time inside the surface face; each time re-triangulating (or redefining) the affected triangles locally to maintain the Delaunay criterion. In 3D the Delaunay triangulation for 2D is extended naturally to three dimensions by considering the *circumsphere* (*circumscribing sphere*) associated with a tetrahedron. For in-depth descriptions of this method, the reader is strongly encouraged to refer to the classical paper by Shewchuk (2002), a review paper by Mavriplis (1997) and a book by de Berg et al. (2008).

6.2.6 *Quadtree/Octree Subdivision*

Mesh generation by the quadtree (2D geometries) or octree (3D geometries) method is based on the divide-and-conquer principle. In quadtree, a 2D shape is recursively subdivided into 4 smaller regions or quadrants, and these regions may be rectangular or any arbitrary shape. Each smaller region is then subdivided until it meets some criterion, e.g. size of cell smaller or the number of cells greater than a certain value, or a homogeneous attribute. An example for a quadtree subdivision given for 90° bend is shown in Fig. 6.11. The vertices of the quadtree structure are used as nodal points, and the tree quadrants are divided into rectangular elements. The quadtree cells intersecting the boundary must be displaced or wrapped in order to coincide with the boundary.

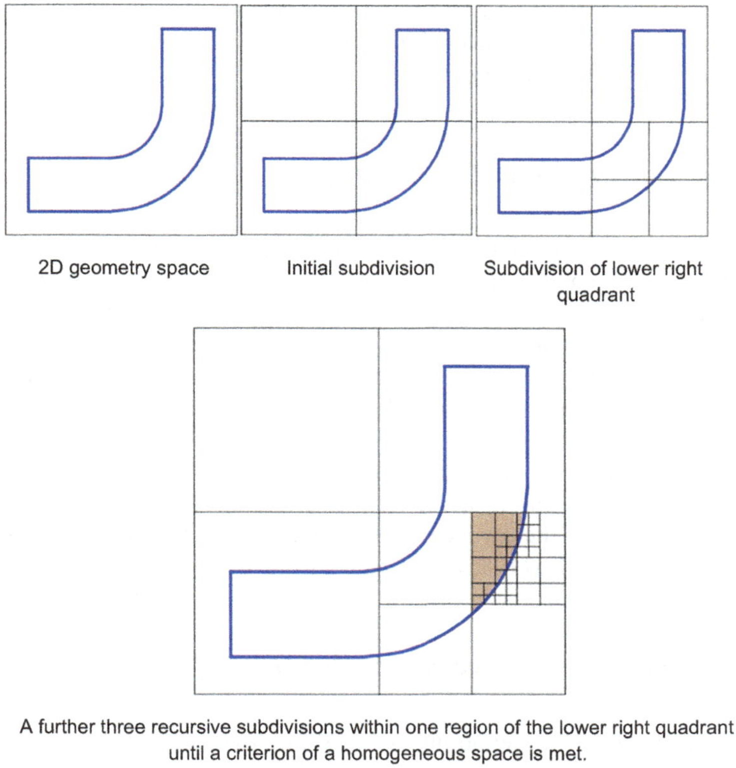

Fig. 6.11 A quadtree subdivision concentrating only on the bottom right corner region. A criterion of region filling by the interior of the 90° bend is used. Any region that does not contain any interior region is neglected. The interior region is shaded to highlight the selected region

Octrees are based on the same principle of divide-and-conquer as in quadtrees but instead applied to 3D geometries (Fig. 6.12). The geometry is subdivided in the three perpendicular planes. For each subdivision a criterion is used to determine whether further subdivisions are needed. Overall, the quadtree/octree method is relatively simple, inexpensive and produces good quality mesh in interior regions of the domain. One drawback of the method is that it has a tendency to generate

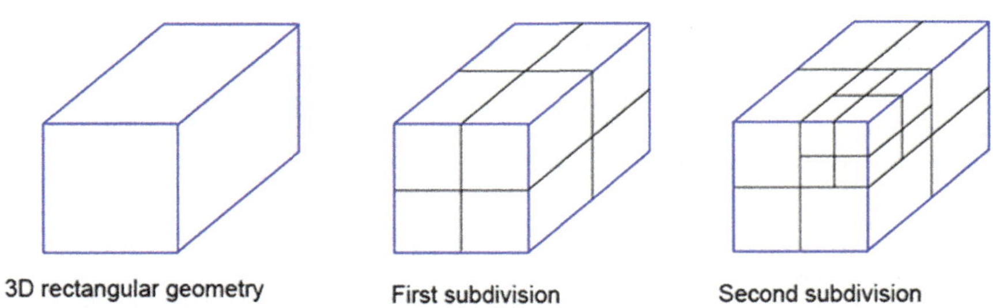

Fig. 6.12 Octree of a simple rectangular geometry

irregular cell distributions near boundaries. For detailed descriptions of quadtrees/octrees the reader is strongly encouraged to refer to the work by Yerry and Shephard (1984) and Shephard and Georges (1991).

6.2.7 Advancing Front Connectivity

The advancing front method for an unstructured mesh is created by adding individual elements inwards one at a time from an existing front of generated elements (typically a boundary or edge). New triangular cells are created by extending the front, with its growth constituting a stack, and the edges of the stack continuously added or removed. The process continues until all fronts have merged and the domain is entirely covered. For three-dimensional grids, a surface grid is first constructed by generating a two-dimensional triangular mesh on the surface boundaries of the domain. This mesh forms the initial front, and is then advanced into the physical space by placing new points ahead of the front and forming tetrahedral elements. For detailed descriptions of the advance front method the reader is strongly encouraged to refer to the work by Lo (1985), and Marcum and Weatherill (1995) (Fig. 6.13).

6.2.8 Comparisons Between Structured and Unstructured Mesh

The main advantage of using an unstructured mesh is its ability to conform onto complex geometries where it may be impossible to apply a structured mesh. With advances in mesh generation algorithms, combined with a user friendly graphical interface, an unstructured mesh can be created automatically or with minimal user interaction. In comparison, a structured mesh requires laying out a single mesh block that is then split, merged, or have O-, and C- grids applied. This process is much more difficult to perform and requires experience in the blocking strategy.

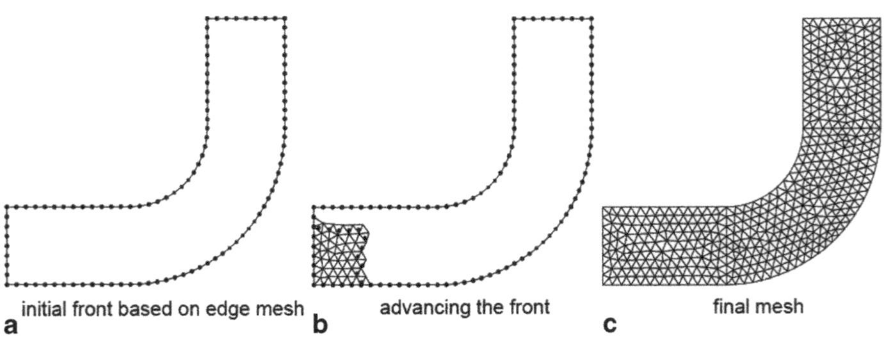

Fig. 6.13 Mesh construction based on advance front method on a 90° bend pipe. **a** Initial front is established on the edge mesh setting. **b** New triangular mesh element is added by inserting one new point to form a triangle and this continues to create a front. **c** Final completed mesh

Some drawbacks of unstructured mesh include the treatment of the nodal points which requires additional computational memory as they can't be treated simply by a double of indices (i, j) in two dimensions or a triple of indices (i, j, k) in three dimensions; additional computations are needed to connect up arbitrary neighbouring nodes. This typically results in increased computational times to obtain a solution and erode the gains in computational efficiency compared with a structured mesh.

An unstructured mesh is ineffective in resolving wall boundary layers since triangles and tetrahedrals do not deform (stretch or bend) during local refinement to make the cells very small. Tetrahedral cells are also prone to high aspect ratios which affect the skewness of the cell. Additionally these cells are difficult to align with the flow direction. These two problems can impede solution convergence and lead to artificial errors in the solution known as *numerical diffusion*. It is a common source of error and is also called *false diffusion* because it is a numerical error and does not represent a physical diffusion process. A few techniques may be applied to minimise the likelihood of false diffusion such as choosing higher order discretisation schemes (see Chap. 5) and increasing the resolution of the mesh.

6.2.9 Mesh Terminology

When developing a mesh, there are some terminologies that a new user will encounter. In this section we describe some of the terminology associated with meshing, and with examples shown in Fig. 6.14.

Structured Mesh

- any mesh containing an ordered set of locally orthogonal lines
- quadrilateral cells are used

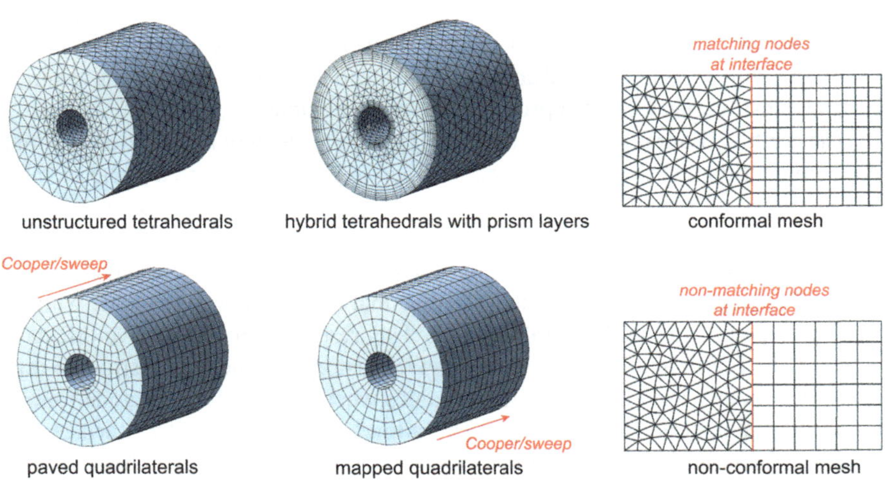

Fig. 6.14 Examples of mesh configurations using both structured, unstructured, and hybrid meshes

- a mapped mesh refers to a structured mesh where a Cartesian coordinate domain is mapped into a physical domain
- the grid lines are curved to fit the shape of the boundaries

Unstructured Grids

- any mesh containing unordered set of polygons, typically triangles or tetrahedrons
- main algorithms are: Delaunay triangulation, Advancing front, Octree
- quads used in unstructured meshing is called *paving*

Finally some additional examples include:

- a *sweeping mesh (*also called *Cooper mesh)* refers to using an initially meshed face (source face) and sweeping that source faced mesh through a path to a target face. A source face with the initial mesh is defined.
- a *hybrid mesh* is one that combines an unstructured triangle/tetrahedral mesh with horizontal prism elements at wall boundaries.
- when a domain is subdivided into multizones with each zone having its own mesh topology, then at the interface the meshing nodes (vertices) can match up (*conformal mesh*), or not match up (*non-conformal mesh*).

6.3 Dynamic Meshing

There are a number of methods to deal with a dynamic moving mesh. They range from simple zero coupling to full coupling. For example the simplest dynamic mesh is that of a moving object independent of the surrounding fluid forces. A step up from this is the immersed boundary method, where a solid object immersed in a fluid is subjected to the surrounding fluid forces, thereby influencing its geometry and net acceleration. Both methods are classes of fluid structure interactions, but for heamodynamics where the fluid and structures need to be fully coupled the arbitrary Lagrangian-Eulerian (ALE) approach is the most suitable.

The arbitrary Lagrangian-Eulerian (ALE) is a finite element formulation in which the computational system is not fixed in space (e.g. Eulerian-based finite element formulations) or attached to material (e.g. Lagrangian-based finite element formulations). ALE-based finite element simulations can alleviate many of the drawbacks that the traditional Lagrangian-based and Eulerian-based simulations have.

The Lagrangian frame is based on the idea that material points coincide with the same grid points during the whole motion. This results in no convective effects in calculation. The main advantage is it provides history-dependent constitutive relations. It is mainly used in structural mechanics where each individual node follows the associated material during motion. This allows easy tracking of free surfaces and interfaces between different materials, but it is not able to follow large distortions in the mesh without having remesh. Its drawback is its inability to follow large distortions of the computational domain without expensive remeshing operations.

The Eulerian frame, on the other hand, is based on the idea that material point moves with respect to fixed computational grid. This allows large distortions in the continuum motion to exist in computational domain. It is widely used in fluid dynamics where large movements in the continuum fluid is handled with relative ease, but generally at the expense of precise interface and free surface detection. It also leads to a decrease in the accuracy of solution.

The Arbitrary Lagrangian-Eulerian frame is developed to incorporate the desirable feature of Lagrangian and Eulerian frame. In this frame of reference, the computational grid can either be fixed or moved with continuum. This is the reason why ALE can prove to have large distortion and also the accuracy of solution is not degraded. When using the ALE technique in engineering simulations, the computational mesh inside the domains can move arbitrarily to optimize the shapes of elements, while the mesh on the boundaries and interfaces of the domains can move along with materials to precisely track the boundaries and interfaces of a multi-material system. ALE-based finite element formulations can reduce to either Lagrangian-based finite element formulations by equating mesh motion to material motion or Eulerian-based finite element formulations by fixing mesh in space. Therefore, one finite element code can be used to perform comprehensive engineering simulations, including heat transfer, fluid flow, fluid-structure interactions and metal-manufacturing.

6.4 Mesh Generation

6.4.1 Mesh Quality

Mesh quality depends on the cell shape based on its *aspect ratio*, *skewness*, *warp angle,* and *smoothness*. A quadrilateral cell having a mesh spacing of Δx and Δy and an angle of θ between the grid lines of a cell is shown in Fig. 6.15. The cell *aspect ratio* is defined as $AR = \Delta y / \Delta x$. Large aspect ratios should always be avoided in important flow regions (e.g. jets, flow separation, attachment and recirculation) as they degrade the solution accuracy and result in poor iterative convergence (or divergence) depending on the computational flow solver during the numerical computations.

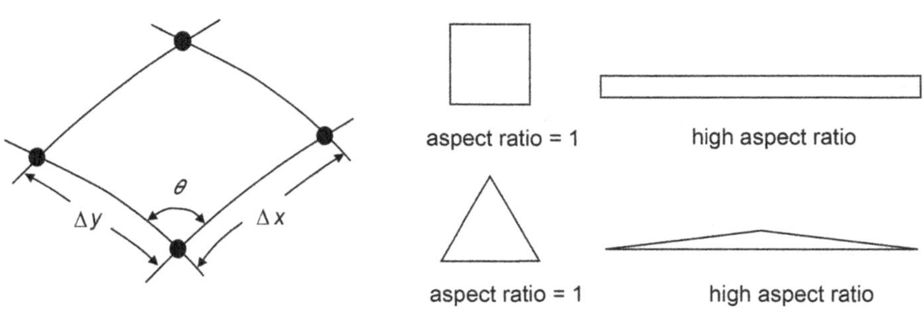

Fig. 6.15 Definition of mesh spacing, Δx and Δy and an angle of θ between the grid lines of a quadrilateral cell, and examples of high aspect ratio elements

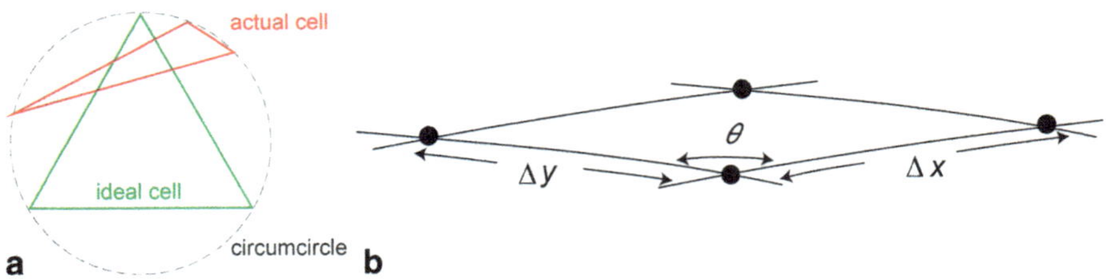

Fig. 6.16 An example of a highly skewed triangular cell, and a quadrilateral cell. **a** skewed triangular cell **b** skewed quadrilateral cell

The *AR* should be maintained within the range of $0.2 < AR < 5$ within the interior region, if possible. For near wall boundaries the condition for *AR* can, however, be relaxed. If the fluid flow is in the *x* direction, then the first mesh requirement is to resolve the velocity gradient in the *y*-direction because of the thin boundary layer. To avoid poor AR, the Δx mesh spacing should also be small enough to produce an AR within the suitable range. Such consideration can assist in possibly alleviating convergence difficulties and enhancing the solution accuracy especially where appropriately resolving the wall boundary layers is necessary.

Mesh *distortion* or *skewness* is measured by determining the angle θ between the mesh lines (Fig. 6.16). For triangles and tetrahedrals, the skewness can be quantified by using the ideal equilateral triangle as a reference by the equation

$$skewness = \frac{Ideal\ Cell - Actual\ Cell}{Ideal\ Cell}.$$

For quadrilateral cells, minimal distortion, mesh lines should be at an angle θ of approximately 90 degrees (orthogonal). If the angle is $\theta < 45°$ or $\theta > 135°$, the mesh becomes skewed and lead to deterioration in the computational results or numerical instabilities.

For an unstructured mesh *warp angles* measuring between the surfaces' normal to the triangular parts of the faces should be no greater than 75 degrees, indicated by the angle β in Fig. 6.17. Cells with large deviations from the co-planar faces can lead to serious convergence problems and deterioration in the computational results. This occurs when trying to cluster tetrahedrals in wall boundaries to resolve the thin fluid boundary layers. Whenever possible, tetrahedral elements should be avoided in wall boundary layers and instead prismatic or hexahedral cells are preferred. In any case, warp angle problems can be overcome by grid smoothing to improve the element warp angles.

Mesh elements should also exhibit gradual or smooth change in sizes from one to another. Adjacent cells should be less than 20% in size difference as any large sudden increase or decrease in adjacent cells causes difficult convergence of the solution.

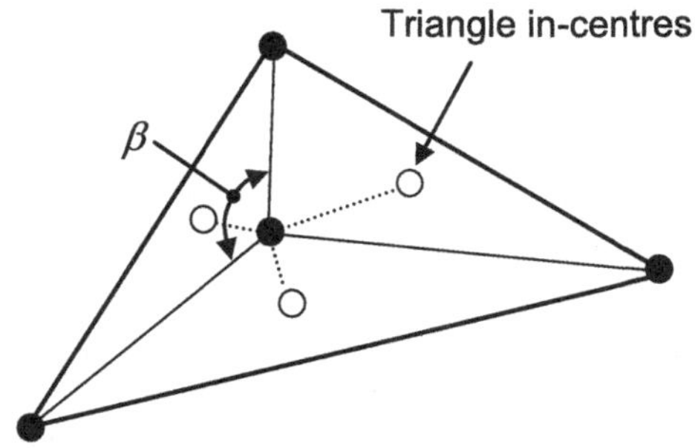

Fig. 6.17 A triangular cell having an angle of β between the surfaces normal to the triangular parts of the faces connected to two adjacent triangles

6.4.2 Mesh Design Strategy

A meshing strategy is to create an initial coarse mesh topology (i.e. use a small number of elements). This allows a quick solution which can be evaluated and the model then later refined. A suitable coarse mesh allows a number of "test-runs" to be carried out in quick turnaround time to assess the convergence or divergence behaviour of the numerical calculations, and physical models. When the numerical setup is correct and the solution is converging, mesh refinement is then undertaken to achieve a more accurate solution. If the solution is diverging, then the mesh or model setup needs to be debugged and investigated. Some possible sources of errors can be attributed to *physical modeling* and *human* errors, and not necessarily the numerical setup itself. During the testing phase, using a fine mesh could take hours or days to simulate only find the solution is diverging or that the physical model was applied incorrectly.

Since, there is no restriction on the use of particular cell type in an unstructured grid arrangement, a *hybrid* mesh that combines different element types has the ability to match appropriate cells with boundary surfaces and allocating cells of various element types in parts of the complex flow regions. For circular geometries such as arteries, grid quality is enhanced through the placement of quadrilateral or hexahedral elements in resolving the viscous boundary layers near the walls whilst triangular or tetrahedral elements are generated for the rest of the flow domain.

Finally, special grid design features such as O-grid or C-grid introduced earlier need careful consideration of block interfaces as this significantly improves the overall quality of a block-structured mesh. For example the presence of arbitrary mesh coupling, non-matching cell faces, or extended changes of element types at block interfaces should always be avoided in critical regions of high flow gradients or high shear. Wherever possible, finer and more regular mesh in these critical regions should be employed. In all cases, after preliminary testing the mesh should be re-checked in critical regions of high flow gradients and large changes to ensure the results are agreeing with validation data such as experiments or known flow behaviour.

6.4.3 Local Refinement and Solution Adaptation

To capture critical flow regions an adequate mesh resolution is needed. Such regions may include flows around obstacles that cause flow separation, attachment and recirculation, near wall boundaries, interface shear regions, and converging and diverging regions. These regions typically exhibit sharp flow property gradients (e.g. velocity, pressure, temperature gradients). This has enormous impact in the stability and convergence of the numerical procedure.

One local refinement technique we have seen is to refine the mesh close to obstacle boundaries and walls. For a viscous flow bounded with solid wall boundaries, clustering a large number of small cells within the physical boundary layer is important. As an illustration, the boundary at a surface wall is shown in Fig. 6.18 to highlight the need for near wall mesh refinements. In the real physical flow, a boundary layer develops at the wall growing in thickness as the fluid enters the left boundary, migrating downstream along the bottom wall of the domain. The local thickness of the boundary layer is given by δ, which increases with x and therefore $\delta = \delta(x)$. A uniform coarse mesh misses the physical boundary layer as it develops. In contrast, the stretched mesh with clustered nodes near the wall at the very least catches some of the boundary layer development. It is therefore not surprising that the accuracy of the computational solution is greatly influenced by the mesh distribution inside the boundary layer region.

When applying a stretched mesh care must be taken to avoid sudden changes in the mesh size to maintain a smooth mesh. The mesh spacing should be continuous and mesh size discontinuities should be removed as much as possible in regions of large flow changes. Discontinuity in the mesh size destabilizes the numerical procedure due to the accumulation of truncation errors in the critical flow regions. Making sure that the grid changes slowly and smoothly away from the domain boundary as well as within the domain interior will assist in overcoming divergence problems tendencies of the numerical calculations. It is also worthwhile noting that most mesh generators have the means to prescribe suitable mesh stretching or expansion ratios (rates of change of cell size for adjacent cells).

Local mesh refinement includes allocation of additional nodal points to resolve important fluid flow regions action or a removal of nodal points from other regions where there is little or no action. However, since mesh creation occurs *prior* to the solution of the flow field being calculated, where to perform local mesh require-

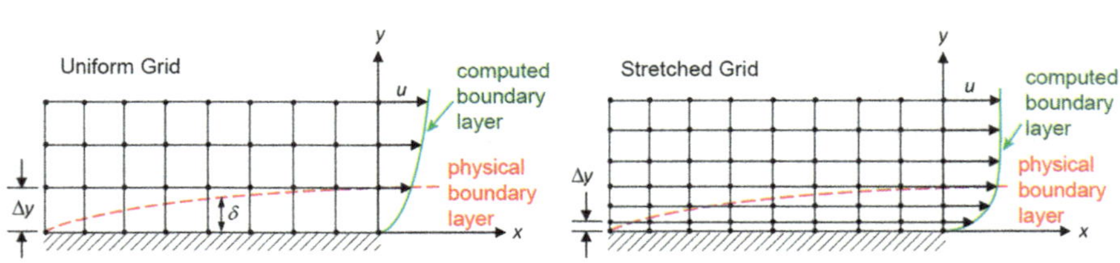

Fig. 6.18 Two illustrations demonstrating the influence of local refinement in the near vicinity of the bottom wall to resolve the physical boundary layer

ments is not always known. One method to overcome this uncertainty is to refine the mesh by *solution adaption*. This allows a mesh adaption based on both geometric and numerical solution data in regions where they are needed.

This employs the solution flow field to locate mesh nodes in the physical flow domain. During the solution process, mesh nodes in the physical flow domain *adapt* to the evolution of large flow gradients. Hence, the actual mesh nodes are constantly in motion during the solution and become stationary when the flow solution approaches a quasi-steady state condition. An adaptive mesh is therefore intimately linked to the solution and alters as the flow field develops unlike the stretched mesh described above where the mesh generation is completely separate from the flow field. For this purpose, unstructured meshes are well suited in automating the generation of elements such as triangular or tetrahedral meshes. Figure 6.19 shows an adaptive mesh for flow over two cylinders using a triangular mesh. For this flow problem, the wake region has been adapted to capture the formation and shedding of vortices behind the two cylinders.

6.4.4 Mesh Independence

A mesh independence study is performed to determine the suitability of the mesh in terms of its number of elements, and to yield an estimate of the numerical errors. In addition it is used to determine the minimum mesh resolution required to generate a solution that becomes independent of the mesh size used. Ideally, at least three significant different grid resolutions should be tested, where each subsequent mesh is approximately doubled in each direction. If this is not feasible, selective local refinement of the grid in critical flow regions of the domain can be applied. A flow

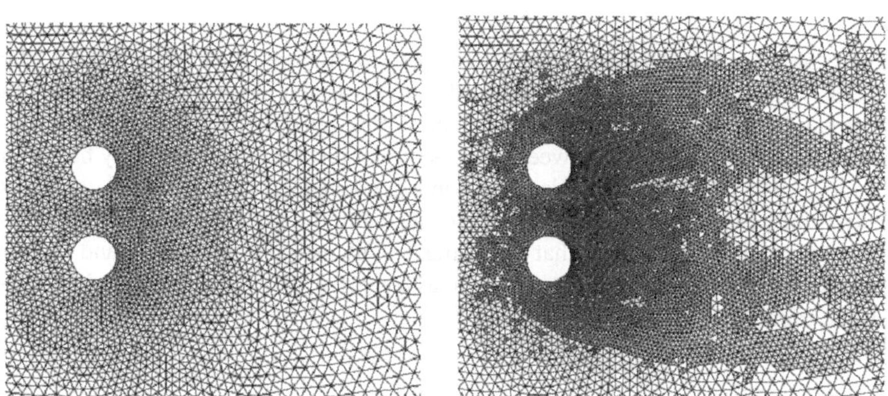

Fig. 6.19 A demonstration of solution adaptation through the use of triangular meshes for the fluid flowing over two cylinders

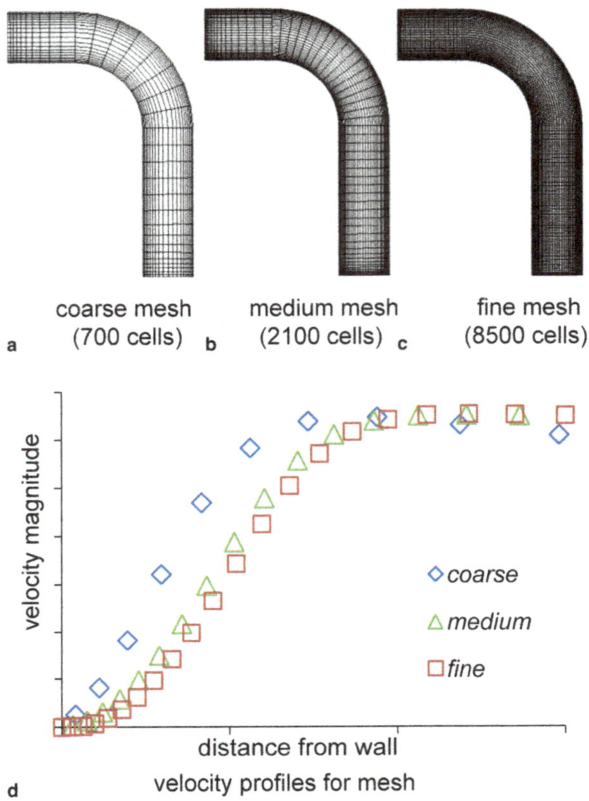

Fig. 6.20 Mesh independence study for three different mesh designs on a 90° bend. The velocity magnitudes are taken after the bend where the flow recovers. **a** coarse mesh (700 cells) **b** medium mesh (2100 cells) **c** fine mesh (8500 cells) **d** velocity profiles for mesh

field variable should be checked at each mesh to monitor its value or profile. The mesh is considered independent when the flow variable does not change with an increase in the mesh size. This leads to a mesh that has optimised and grid independent where a compromise between computational resources and accuracy has been made. Figure 6.20 shows a typical mesh independence test on a 90° bend geometry and its influence on a velocity profile taken just after the bend where the flow begins to recover. The results show that as the mesh is increased to a medium and further increased to a fine mesh, the results begin to converge, and hence reaching independence from the mesh density

6.5 Meshing Examples

In this section we first present practical examples, mesh designs, and guidelines that will accelerate a new student to mesh generation which is aimed at giving an insight into the practical strategies and techniques for mesh generation.

6.5.1 Flow in Blood Vessel Mesh

The nature of blood flow is inherently internal flow within circular pipes. Earlier it was shown that applying structured mesh elements onto non-orthogonal shapes required special treatment of the geometry. In particular the use of the O-block multi-block strategy provided excellent matching of mesh elements on the curved geometry. Figure 6.21 shows the different meshing patterns that are produced from different meshing strategies. The results show that the mapping scheme fails with highly skewed elements in local corners of the circle. The triangular and paved scheme can match elements to can fit the circular shape but refinement in the near wall region requires prism layers. The multiblock strategy can be used with a square internal block or with greater resolution an octagonal shape in the centre, where both allow prismatic layers in the near wall while maintaining near orthogonal elements throughout.

6.5.2 Blocking Strategies

Setting up the blocks to produce a structured mesh where possible can be difficult to see at first, however with experience then this visualization can be attained. The next section provides two examples in setting up the required blocks starting from a single block (Fig. 6.22). The first example is a bend in the blood vessel is a common feature found in the vasculature vessel network. Any curved bend and in particular

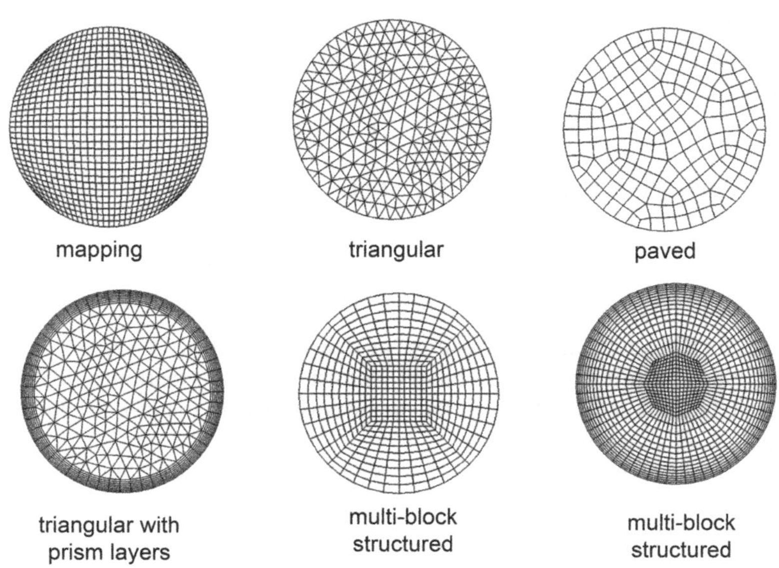

Fig. 6.21 Meshing outcomes using different mesh strategies on a cross-sectional slice of an artery vessel

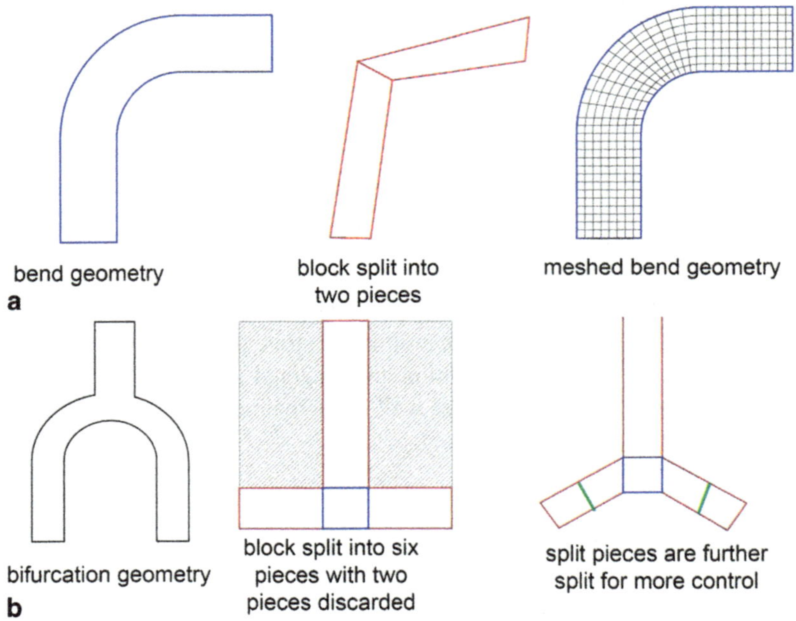

Fig. 6.22 Blocking strategies for bend geometry and bifurcating geometry

a 90°-bend found in the aortic arch can be treated as a single O-grid block topology, where the block is split into two sub-regions to allow for the bend. The second example is a bifurcating geometry where the initial block is split into six smaller blocks. The two outer blocks are then discarded to leave four smaller blocks that form a T-shape. This block is then re-shaped and split further for more control in the final mesh.

6.5.3 Stenosed Artery Step-By-Step

In this section we present a specific example of a representative straight artery vessel with a moderate stenosis and provide detailed step by step guide Fig. 6.23.

1. First, the fluid domain is enclosed by a single initial block.
2. This block is split into six sub-blocks (shown by the red markings in Fig. 6.23) that includes four sub-blocks at the stenosis site where the cross-sectional area experiences abrupt changes.
3. Each edge from the six sub-blocks is projected to the corresponding fluid domain surfaces and curves at the inlet and outlet. To optimize mesh quality, the block vertices are evenly spaced to minimize the average deviation of the edges from the surface curvature.

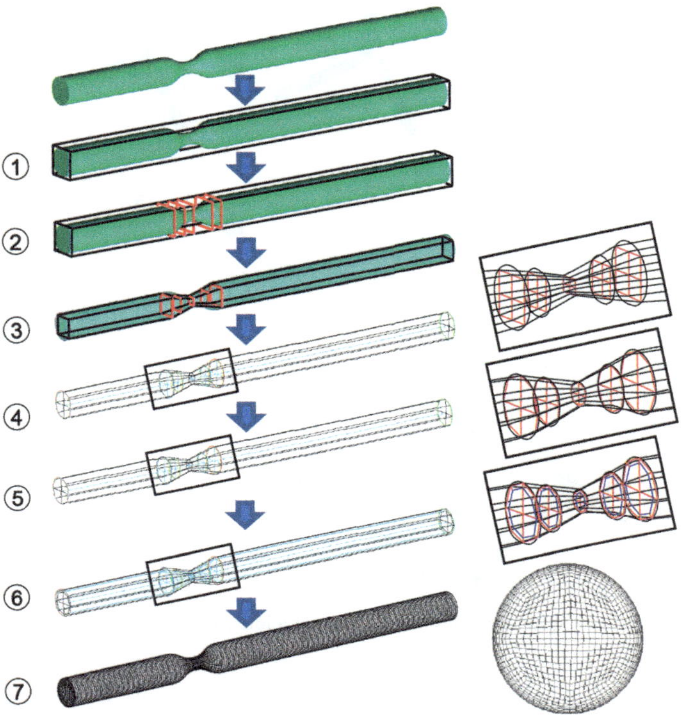

Fig. 6.23 Multi-Block structured mesh flow chart for a stenosed artery

4. To further fit the fluid domain, all blocks are split along the x-direction and y-direction, thus creating further blocking topology for the fluid domain.
5. Similar with step three, all new block edges are associated to their nearest geometry entities.
6. The existing hexa-grid blocking topology is converted to an O-Grid topology, where a prism layer is created at the near wall region.
7. Based on the dimension of the fluid domain, reasonable node distributions are defined at each block edge to produce the final mesh.

6.5.4 Left Coronary Artery Bifurcation Step-By-Step

Similar to the stenosed artery case study, a representative left coronary artery bifurcation is meshed, and detailed step instructions are provided in Fig. 6.24.

1. First, a single initial block is created to enclose the whole fluid domain.
2. The initial block is divided into six sub-blocks
3. Two blocks are deleted to form a T-shape as seen in Fig. 6.22.
4. The remaining blocks are associated to the nearest geometry entities, and a preliminary block topology is built to meet the bifurcating branches.

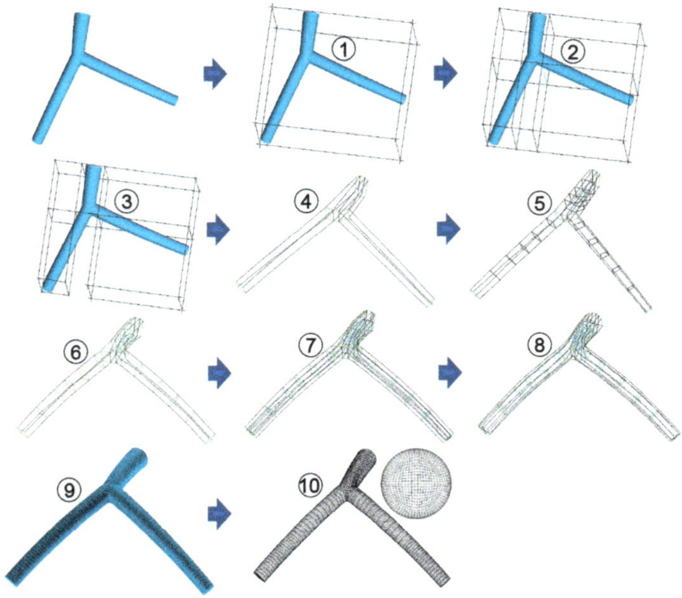

Fig. 6.24 Multi-Block structured mesh flow chart for a left coronary artery bifurcation

5. To better capture the branch curvature, all blocks are split into more sub-blocks along main branch and sub-branches.
6. The new blocks are associated to the nearest geometry entities, once again to attain the same curvature as the fluid domain.
7. All blocks are split along the vessel radius-direction to further fit the fluid domain, and all block edges are projected to the corresponding geometry entities again.
8. The existing hexa-grid blocking topology is converted to an O-Grid topology, where a prism layer is created at the near wall region.
9. Based on the dimension of the fluid domain, reasonable node distributions are defined at each block edge to produce the final mesh.
10. The final mesh results are produced and a close view of inlet mesh is shown.

Apart from idealized models, two selected locations of image-based aortic artery, aortic arch and Aortic Abdominal Aneurysm (AAA), were meshed using the same strategy. Figure 6.25 shows the outcome of applying the same steps.

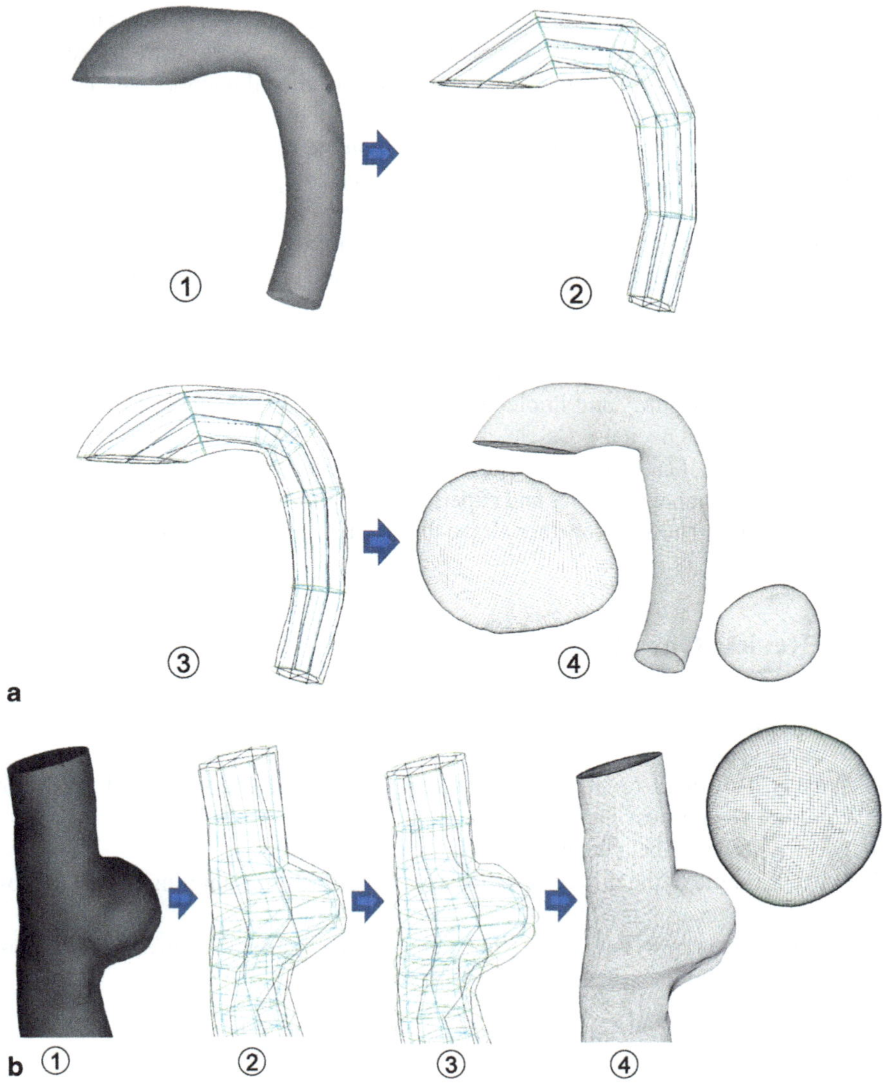

Fig. 6.25 Schematic mesh procedures for an image-based aortic artery: **a** aortic arch, **b** aortic abdominal aneurysm

6.6 Summary

The importance of generating a quality mesh in order to obtain reliable computational solutions was discussed. The hierarchy for a mesh element from the lowest to highest topology are points, lines, faces, and then volumes. Different mesh configurations including structured (body-fitted, and multi-block) and unstructured (Delaunay triangles), and meshing growth schemes (Quadtree, Octree, and Advancing Front) schemes were described. The different type of mesh element used (or its combination of) determines whether the mesh can be set up as a structured or unstructured mesh. Structured mesh has the advantage of smaller computational memory, and better solution convergence, however it can be difficult to apply to complex geometries such as the respiratory organs. An unstructured mesh on the other hand, has the ability to conform onto complex geometries but requires more computational memory because of the need to connect up an arbitrary number of neighbouring nodes.

Generating a quality mesh requires as much creativity as it does technical knowledge. Therefore different techniques and strategies are presented for both structured and unstructured mesh to provide the reader with some ideas that can be applied to generating a quality mesh for different flow problems. Practical guidelines are given which includes the importance of developing a coarse mesh to begin with for initial flow field testing, and implementing a mesh independence test to ensure an optimum mesh has been used.

Examples specifically related to haemodynamic flows were given in detail to accelerate the reader's understanding and experience into meshing. The next chapter presents case studies that wrap together all the theory presented in these first six chapters.

6.7 Review Questions

1. What are some of the benefits of a well designed grid?
2. What are some of the advantages of a structured mesh?
3. Why is it difficult to write CFD programs that involve a structured mesh for complex geometries?
4. Discuss some of the advantages of an unstructured grid.
5. What are some of the difficulties that arise regarding programming of CFD problems for an unstructured mesh?
6. For the geometry below discuss how using a block-structured mesh has advantages over a single structured or unstructured mesh.

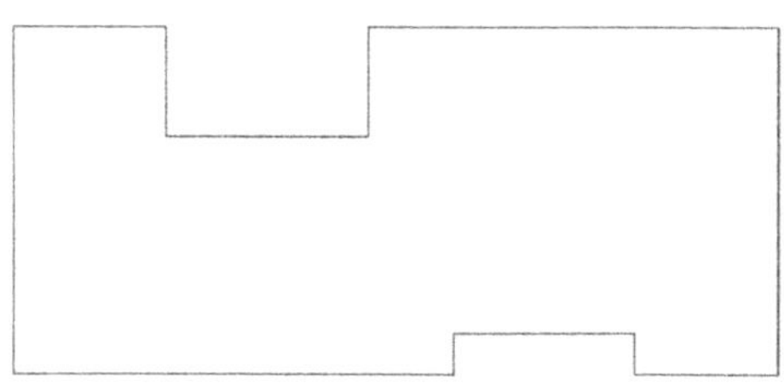

7. Why is it better to start off with a coarse mesh when you first begin to solve a CFD problem?
8. The stretched grid technique is an example of a local refinement technique that is applied before a solution is obtained. What is one problem associated with this?

9. Discuss why tetrahedral elements are a poor choice for meshing near walls and boundaries.
10. What is a mesh independence test?
11. For a circular cross-section representing a blood vessel, what kind of mesh would you apply? Sketch the approximate mesh onto the figure provided.

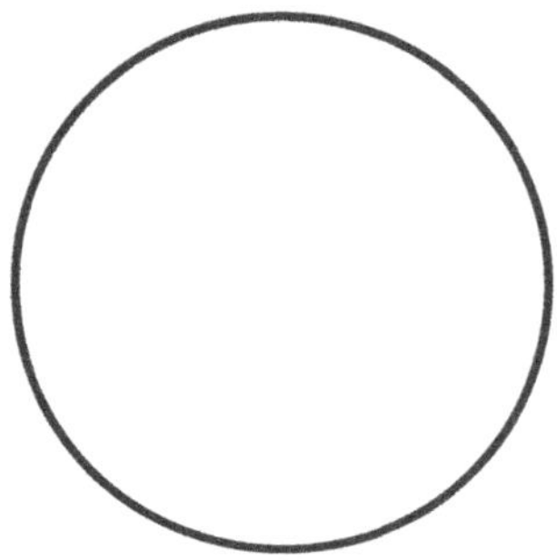

Chapter 7
Case Studies of the Human Cardiovascular System

7.1 Introduction

Having laid the groundwork to establish a strong theoretical base, this chapter culminates the foundational knowledge attained by applying the theory and putting them into practice through selected demonstrative applications. From a practical viewpoint, the selected detailed case studies in this chapter will provide the reader with insight and confidence in applying the techniques to a wide range of biomedical engineering applications. The computational model of a human atherosclerotic artery and the heart structures developed and described in Chap. 3 and 4 are used for the case studies herein. Analyses of the results are shown to inform the reader how to translate the contours, vectors, and line plots and what to look out for in case of any spurious results. Just like a medical practitioner reading CT or MRI scans needs the ability to look for abnormalities, the colourful CFD results needs a person who can interpret and distinguish it as a product of the numerical setup or as a new physical finding. Therefore, the important aims of this chapter are to:

- apply the theoretical knowledge attained from previous chapters through worked case studies,
- demonstrate different setup procedures that reflect the needs of individual cases,
- demonstrate how to convert raw data into presentable contour and vector maps, and line plots, and
- demonstrate how to analyse the results and what type of results should be expected under different cases.

7.2 Haemodynamics of a Stenosed Carotid Bifurcation

The human carotid bifurcation is an important component of the cerebral vascular system. Individual aspects of the carotid artery such as the geometry and flow rate collectively affect cerebral function by bringing oxygenated blood into

J. Tu et al., *Computational Hemodynamics—Theory, Modelling and Applications*,
Biological and Medical Physics, Biomedical Engineering,
DOI 10.1007/978-94-017-9594-4_7

the brain, thereby ensuring its survival and functioning. To better understand the physiological aspects of the carotid artery, detailed blood flow patterns from the entrance of the common carotid artery (CCA) through to the internal carotid artery (ICA) and external carotid artery (ECA) can provide data that is pertinent to the prediction of atherosclerosis. Because of the inter-subject variations inherent between carotid arterial geometries, the numerical results herein are compared with experimental data available in literature to complement and reinforce the current knowledge base.

The physics, numerical and CFD setup details that are needed in order to study the effects of morphological differences in the patient-specific carotid arteries on the blood flow and their fluid mechanical properties are discussed. Geometrical differences are also compared with available data in the literature. Additionally, variations in the flow patterns and flow features such as pressure drop, wall shear stress, velocity, and flow distribution between the internal and external carotid arteries, as well as the different geometries are also presented. The flow in the arterial region is studied in particular detail to provide better insight into its biofluid mechanical properties.

This case study is composed of three stages: three-dimensional artery reconstruction, downstream vascular flow resistance modelling, and haemodynamic analysis of the carotid artery. We shall examine each of these stages sequentially.

7.2.1 Physiologically Realistic Geometrical Reconstruction from MRI

The medical image reconstruction for modelling a blood vessel to use in CFD has rapidly developed in recent decades. With the development of modern imaging technology, especially magnetic resonance imaging (MRI) and computed tomography (CT), it is now possible to quantify arterial blood flow in subject-specific physiologic models (Philips Medical Systems Clinical Education 1984; Powell et al. 2000). For three-dimensional numerical studies, CFD models of the carotid artery can be constructed from MRI or CT images (Merrifield et al. 2004b). Imaging data can be stored in the Digital Imaging and Communication in Medicine (DICOM) format. The DICOM format file contains two parts: the header which stores detailed information about the patient such as name, type of scan, dimension of the image, image position, and so forth. The second data set contains information of each scanned image.

Segmentation of the MRI was required to extract the geometry of the carotid vessels. The segmentation process includes thresholding and region growing (Gonzalez and Woods 2002), followed by 3D anatomical reconstruction (Chandran et al. 2006) to obtain a very coarse solid model. During thresholding, a range of gray scale values are selected such that the region to be selected is of best contrast. A schematic diagram depicting the segmentation of magnetic resonance images at various locations of the carotid bifurcation is shown in Fig. 7.1. After the regions of

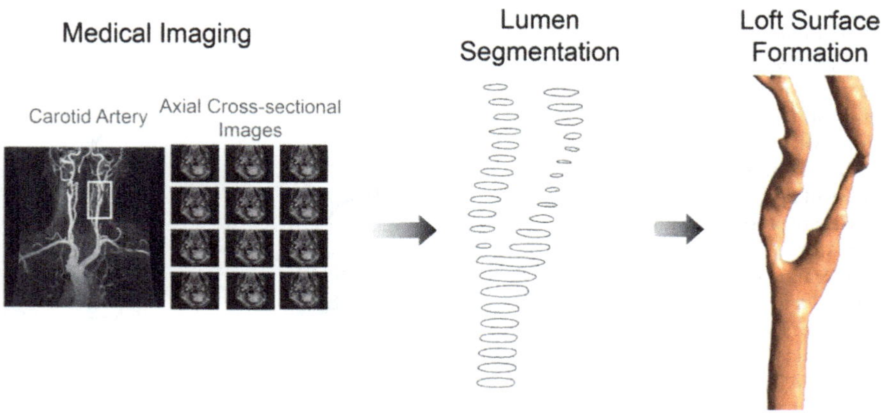

Fig. 7.1 Three-dimensional reconstruction of carotid bifurcation. Magnetic resonance imaging was carried out on the neck region of the patient whose carotid artery can be imaged at axial orientation for multiple slices. Segmentation based on the threshold of the blood vessel at various slices is performed in the initial stage. The segmented voxels can be grouped to form a three-dimensional anatomy and a mesh reconstruction based on the contours of these segmented regions is carried out

interest are extracted, the voxels are grouped together to form a 3D geometry. Then the reconstruction of carotid bifurcation anatomy into a Computer Aided Design (CAD) model is performed based on the segmentation information. The reader may refer to the detailed procedure of medical image reconstruction from Chap. 3.

High resolution magnetic resonance imaging of a stenosed carotid bifurcation was acquired through scanned images performed on a 42 year-old male using a 1.5-T General Electric scanner. A total of 112 contiguous slices were generated from the high-resolution T-1 weighted spoiled gradient echo with parameters as follows: TR, 35 ms; TE, 7 ms; flip angle, 35°; field of view, 24 cm; voxel size 0:63 mm × 0:73 mm × 0:63 mm. An automated detection algorithm was then applied to the images to generate a conservative segmentation of stenosed carotid bifurcation using a two-dimensional watershed transform form markers applied to each slice (Meyer and Beucher 1990a). Based on segmentation of different adjacent planes, a high resolution three-dimensional computer artery model was created and the data was stored in a stereolithography (STL) file format. To enhance the resolution of PIV measurement, the flow phantom was enlarged to 10 times of its human counterpart. Figure 7.2 shows the reconstructed carotid geometry employed in this case study. Based on the geometrical data from the STL file, a negative model was created by a Rapid Prototyping (RP) printer (model ZCorp 3D) using a water-soluble plaster. Following the method adopted by Hopkins et al. (2000) five layers of water-soluble glue were painted on the plaster model to smooth and seal the pores on its surface.

The painted prototype was then encased in a Plexiglas box where clear silicone (using Dow Sylgard 184) was casted around the plaster model to create an optically clear positive model. After the silicone has been cured, the plaster model was then dissolved out with cold water, leaving a patient-specific replica of the stenosed carotid

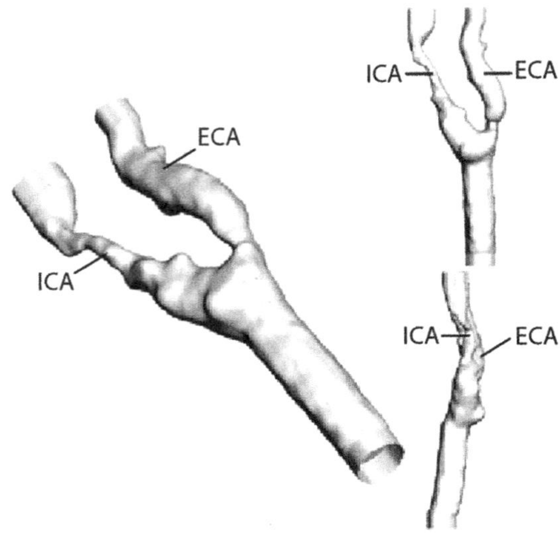

Fig. 7.2 Three dimensional views of the stenosed carotid bifurcation. Anatomical geometry of the carotid bifurcation is reconstructed using MRI data, and output in the STL format. Three views of the patient specific vessel are presented with the labels ECA, ICA, and CCA representing the external, internal and common carotid arteries respectively. We note the location of the stenosis at the ICA has a relatively smaller cross-sectional area compared to the ECA

bifurcation. Finally, three pieces of reinforced flexible plastic tubes were attached to the inlet and outlet of the phantom as connectors to the flow loop in the PIV experiment.

7.2.2 Computational Mesh Generation

After extracting the cardiovascular structure, the computational file can be saved as an IGES, STL, or STEP to be cross-compatible with 3D modelling and meshing programs. The meshing procedure can begin by applying a simple unstructured tetrahedral mesh all over which produces a single contiguous mesh. However for easier post processing of local flow variables, the computational model may be split into smaller sub regions during the CAD surface and volume generation stage, prior to meshing. While the process of sub dividing the computational model into smaller regions can be performed within some CFD packages, it is not always an easy task, and therefore it is recommended to be performed in CAD packages that have NURBS functionality.

For meshing of an atherosclerotic artery, surfaces were created and stitched to create a computational mesh. An initial model with 129,182 cells was created and refined by cell adaptation techniques that included refining large volume cells, cells that displayed high velocity/pressure gradients and near wall refinements. This process was repeated twice, with each repeat producing a model with a higher cell count than the previous model. A model containing 786,712 mesh cells is shown in Fig. 7.3.

Prism meshes were used to resolve the thin boundary layers present. The first mesh element, adjacent to the wall is a very thin layer and subsequent mesh elements

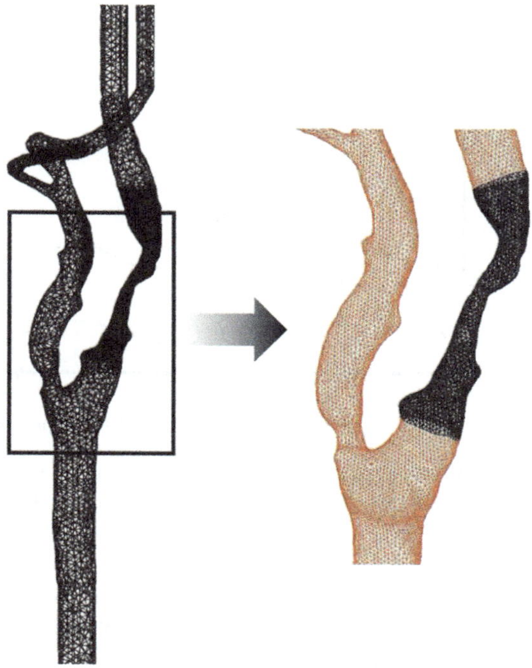

Fig. 7.3 Mesh configuration of the carotid artery model. The mesh construction for the stenosed artery is based on the medium scale selection. The element numbers of the stenosed artery is 786,712. Increased mesh density is performed at the stenosis region. Note that the density of grid is set to be higher at the region of interest, i.e. the stenotic region (coloured as *black*)

above the first mesh element become progressively thicker until the layers cover the distance of the boundary layer.

7.2.3 Computational Fluid Modelling

The blood was assumed as incompressible and isothermal fluid in this work and the blood flow was governed by the following Navier-Stokes equation and continuity equation:

$$\rho\frac{\partial u}{\partial t}+\rho(u.\nabla)u=-\nabla P+\rho g+\mu\nabla^2 u, \tag{7.1}$$

$$\nabla . u=0, \tag{7.2}$$

where ρ represents blood density, μ represents velocity, P stands for pressure, g stands for gravity constant and μ is the dynamic viscosity of the blood.

A group mean blood flow waveform as shown by Fig. 7.4, which is based on flow rate, Q (ml/s), for the ICA, ECA and CCA (Marshall et al. 2004a) was introduced as the flow boundary condition for the carotid artery object of interest.

Experimentally derived volumetric flow rates were set at one inlet and two outlets. To minimize artificial effects from all openings, extended regions were

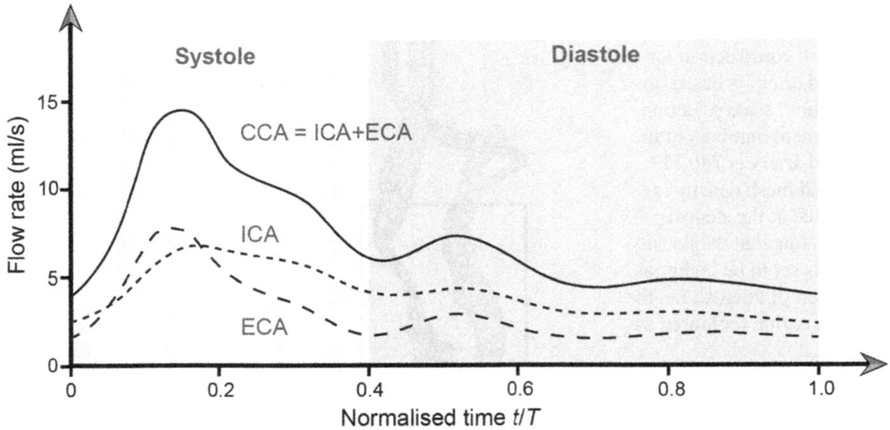

Fig. 7.4 Physiological waveforms used in simulation. This fluid mechanical property, which is based on flow rate, Q (ml/s), serves as the inlet boundary condition for the carotid artery used in CFD simulation

incorporated at both inlet and outlets of the computational model. Vessel walls were treated as rigid, and this will not affect physiological accuracy significantly because previous research has shown that the effect of blood wave propagation on vessel compliance can be generally regarded as secondary (Yiemeng Hoi et al. 2010). The outlet pressure was set at 20 mmHg, which represents the capillary pressure, and the SST model was applied in the simulation.

The generated mesh files were imported into a CFD package—ANSYS-CFX. In this case study, 3D unsteady incompressible Navier-Stokes equations were solved by a Second Order Backward Euler method. The convergence criterion for the relative residual of all dependent variables was set to 1×10^{-4} to ensure the convergence of each time step, which is fixed at 0.004 s. To minimize the influence of initial flow conditions due to the zero-velocity initial settings, all simulations were carried out for four cardiac cycles based on a time period T of 4.0 s, and the results presented were extracted from the third cycle onwards. Having a parabolic inlet flow setting according to the Hagen-Poiseuille's law eliminates the need to add an extra fluid domain to generate a fully developed flow at the inlet of the carotid artery, which can reduce computational mesh element and expenses.

7.2.4 Experimental Validation

The overall methodology of an experimental validation study is depicted by Fig. 7.5. Validation is performed by comparing the haemodynamic parameters between the numerical predictions and experimental measurements. The experimental measurements provides a means for validation of the numerical simulated results and as a form of support for fluid mechanists to review the accuracy of their simulation platform. In addition, experimental data can be used to determine the boundary

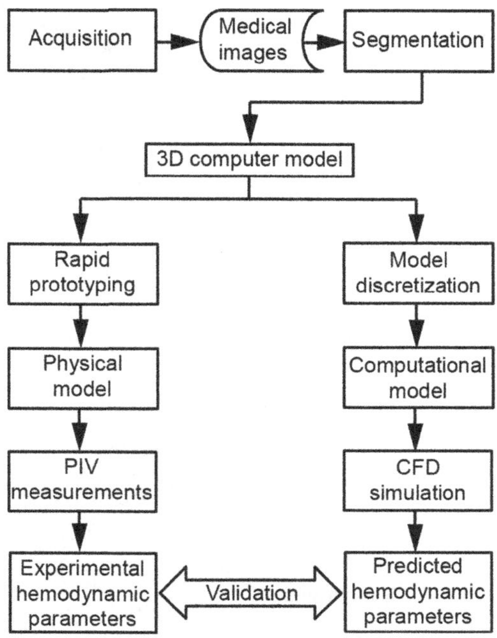

Fig. 7.5 Procedural flow chart for experimental and numerical studies. The presentation of a systematic approach to perform experimental and numerical data acquisition and comparison can allow us to compare the predicted and experimentally derived flow. Data retrieval and anatomical reconstruction based on MRI generates a geometrical model. Then, the rapid prototyping of the model that is followed by PIV measurement of flow within it, and numerical simulation using the same anatomical model are performed. The predicted data from simulation and the experimentally measured flows are examined in the final stage

conditions when setting the numerical framework. In the case of cardiovascular anatomies, velocity-encoded phase contrast MRI or ultrasound, are typically used to extract the flow values at the inlets and outlets of the anatomical structures to be used as boundary conditions for the computer models.

Figure 7.6 shows a schematic diagram of an experiment setup for PIV measurement. The experimental flow loop comprises of a silicone phantom, an elevated fluid tank, a flow meter and a suction pump. To eliminate refraction of the laser sheet, the index of refraction of working fluid was specifically chosen to match the refraction index of the phantom wall. The working fluid was a mixture of glycerol (55 % by mass) and distilled water (45 % by mass), which has a refraction index of 1.42 and kinematic viscosity of $\nu = 6.2 \times 10^{-6}$ m^2/s at constant temperature of 25 °C. In this case study the kinematic viscosity of human blood is taken as 3.4×10^{-6} m^2/s. Considering that the kinematic viscosity of the mixture and the phantom are scaled-up by 10 times, the flow rate of the working fluid can be established based on dynamic similarity. A flow Reynolds number (Re) of 485 was determined at the flow phantom inlet boundary, corresponding to a typical flow rate at the common carotid artery (CCA) (i.e. 12.17 ml/s) of a healthy adult at the peak of cardiac cycle (Tada and Tarbell 2005). The flow rate of the working fluid can be determined by

$$Q_{\mathrm{m}} = 10 \frac{\nu_{\mathrm{m}}}{\nu_{\mathrm{b}}} Q_{\mathrm{b}}, \tag{7.3}$$

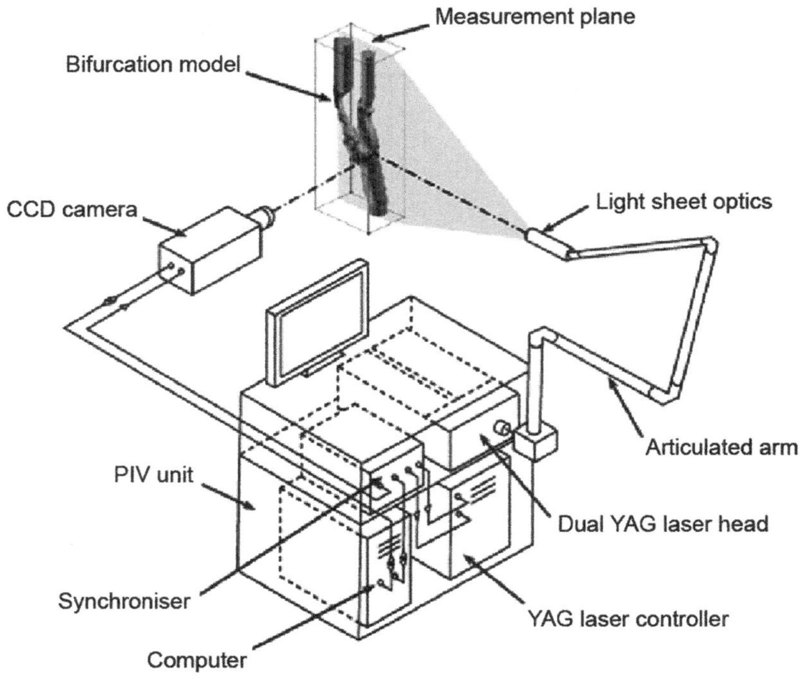

Fig. 7.6 The PIV apparatus adopted in the experiment. The experimental setup involves the use of PIV to capture flow within the carotid bifurcation silicon phantom. That comprises an elevated fluid tank, a flow meter and a suction pump to support the flow condition. A laser system and CCD camera allows the flow measurements optically. The PIV data is then post-processed and can be used as experimental information for validation of CFD simulation results

where Q_b is the volumetric flow rate and ν_b is the kinematic viscosity. A constant volumetric flow rate of mixture (i.e. $Q_m = 21.93$ ml/s) with kinematic viscosity ν_m is adopted throughout the experiment. Rhodamine B fluorescent particles with a mean diameter of 12.3 μm and a relative density of 1.1 kg/m^3 were adopted as the seeding particles. The mixture was circulated by a suction pump and entered the flow phantom through a reinforced flexible tube with a diameter about 25 mm. The working fluid then exited the flow phantom from the two outlets (internal and external carotid arties) and entered the elevated flow tank to eliminate cavitations that could occur within the flow system. The mock circulatory setup is displayed in Fig. 7.6.

The *in-vitro* flow validation experiments can be performed by implementing Particle Image Velocimentry (PIV) measurment of the carotid arteries and extract the velocity along a plane through the cross-section of the stenosed arteries longitudinally and for multiple planes axially (Cheung et al. 2010b).

An ILA (Intelligent Laser Applications GmbH, Germany) PIV system which consisted of a 1.3 Megapixel (1280 × 1024 pixels) 12-bit digital CCD camera was employed for measurements. A New Wave 120 mJ double-cavity Nd:YAG laser head that is synchronized with the CCD camera was installed on a horizontal traverser, which allowed the measurement plane to be altered by a repeatable and

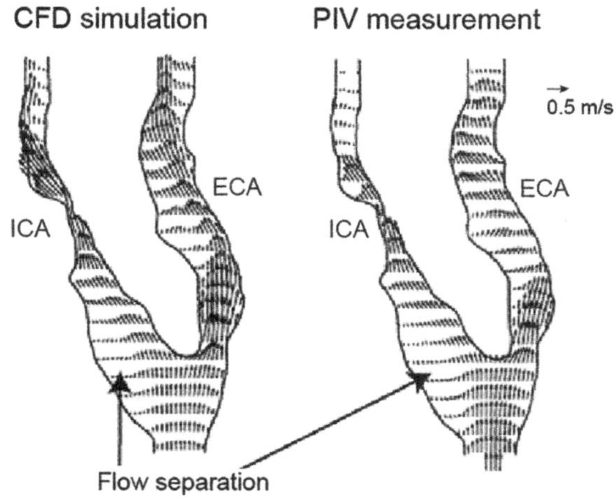

Fig. 7.7 Measured and predicted velocity distribution at the centre-plane of a stenosed carotid bifurcation. The longitudinal sectioning of the carotid bifurcation is performed to compare the measured and predicted flow fields. The PIV and CFD results show good agreement, which highlights the credibility of the numerical simulation in flow modelling of the patient specific anatomy

quantifiable amount. The laser beam was expanded to a 2 mm thick plane vertical light sheet that was directed through the replicated model. Measurements were taken in 3 mm slices in sagittal planes from left to right. The field of view of the CCD camera was 165 mm × 132 mm using 1280 × 1024 pixels of a CCD array.

Figure 7.7 presents the comparison between predicted and measured stream-wise flow patterns at the center-plane of the stenosed carotid artery. In general, both measured and predicted stream-wise flow patterns are in satisfactory agreement. Flow separation and re-circulation regions due to the abrupt cross-sectional area expansion at the sinus region[1] are successfully captured by the CFD simulation, including the highly skewed velocities caused by curvature and irregularity of both ICA and ECA. Nevertheless, stream-wise velocities in both ECA and ICA are found to be over-predicted by the CFD model.

Figure 7.8 shows the quantitative comparison of the measured axial velocity profiles at the centre-plane with the numerical simulation results at four different axial locations. Axial velocities are plotted against the dimensionless radial locations and are normalized by the total length of each cross-section lines. Overall, the predicted velocity profiles at all cross-section lines are in satisfactory agreement with measurements. In particular, the velocity profiles along the ICA that are depicted in Fig. 7.8b, c compared reasonably well with the experimental data. Axial velocities are over-predicted around 30–55 % in comparison to the measurements though the main trends are captured rather well for the ECA. Such error can be attributed to the probable onset of transitional turbulent flow in which laminar flow calculations that have been assumed are invalid. Without an appropriate transitional model or resolution of all turbulent scales (DNS), the CFD simulation underestimates the strength of turbulence being induced through the secondary flows which consequentially over-predicts the axial velocity magnitude. Nonetheless, it can be

[1] The *carotid sinus region* is at the bifurcation point of the carotid artery at the origin of the internal carotid artery (ICA). It exhibits a region of dilation in its cross-sectional area.

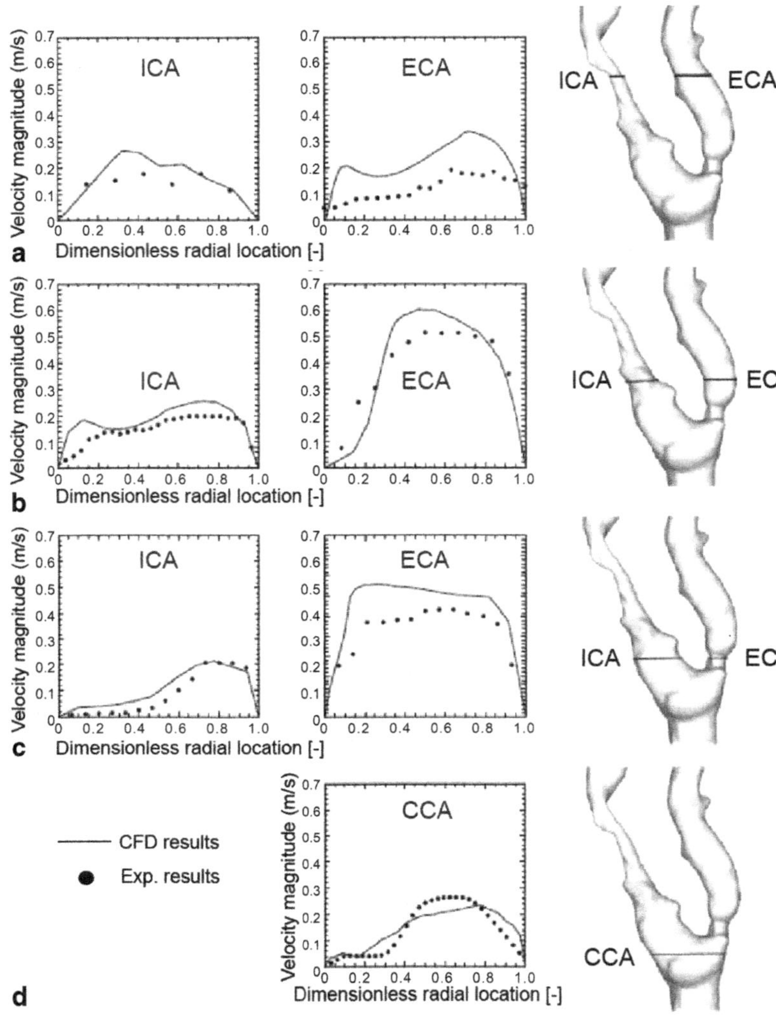

Fig. 7.8 Comparison of the measured and predicted local axial velocity profile. Velocity profiles at selected lines of the stenosed carotid bifurcation from **a** to **d** are examined using the predicted and experimental data. Axial velocities are over-predicted around 30–55 % in comparison to the measurements. It may be noted that the results are relatively accurate for flow in the ECA

noted that the prediction error for the present investigation is comparable to those that have been obtained through other researchers (Bertolotti et al. 2001; Lei et al. 2001; Zhang et al. 2008), who reported an error ranging from 1 to 47 % when compared to PIV or LDA measurements.

Figure 7.9 illustrates the predicted and measured vector plots of secondary flows at three selected cross-sectional planes of the stenosed carotid artery. The recirculation regions of secondary flows are successfully captured by CFD and the predicted vortex core locations compared favourably with measurements. The predicted

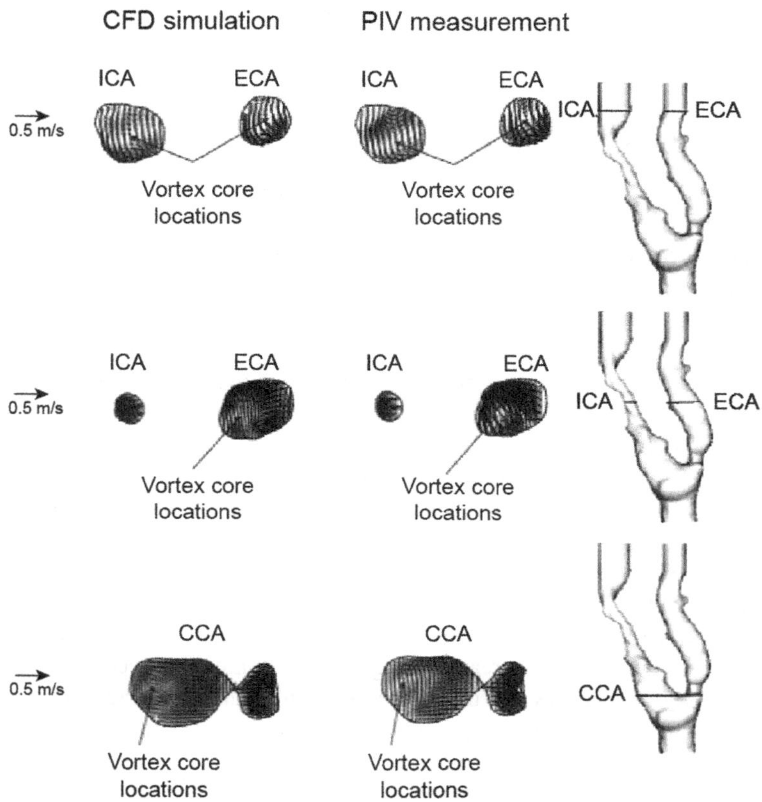

Fig. 7.9 Measured and predicted secondary flow pattern at the selected planes of a stenosed carotid bifurcation. Various sections of the carotid bifurcation from **a** to **c** are visualised based on the predicted and measured flow (in the *top-down* direction of the anatomy). The recirculation regions of secondary flows and the predicted vortex core locations simulated by CFD agree well with experimental measurements. The predicted flow differs slightly from the measured ones at the common carotid artery

secondary flows appear to be relatively weaker when compared to measurements. This further confirms the breakdown of the laminar flow calculations in attempting to capture the onset of turbulence, which subsequently underestimated the strength of the resultant secondary flows.

In general, the main flow characteristic can be successfully captured using CFD. The predicted vortex cores of the secondary flow compare favourably with measurement data. There is a limitation in the capability of the CFD model to accurately predict the transition of laminar flow upstream to weak turbulence downstream. This highlights the deficiency of the current CFD technique to properly capture transition turbulent flow, and therefore requires further studies to better understand these embedded chaotic flow structures. Experimental work may be carried out to measure blood flow structure under transient pulsatile flow situation, which allows additional validation to be performed on the CFD model in predicting actual blood flow behaviour. These studies can help to better understand the chaotic flow structure of transition turbulence, which is significant in the plaque weakening mechanism.

7.2.5 Flow Visualisation

Path streamlines that act as massless particle tracers to track the flow path of the blood, are released from the CCA inlet with a flow rate of 12.17 mL/s to provide qualitative visualisation of the flow field. The flow characteristic inside the stenosed carotid bifurcation is illustrated in Fig. 7.10. The streamlines in the carotid bifurcation at the stenosis, has a maximum velocity of approximately 0.28 m/s. In this region there is flow separation and reversed flow just posterior to the narrowest stenosis.

Contours of axial velocities are plotted on seven equally spaced slices normal to the flow direction (numbered 1–7). As depicted in slice 1, a Womersley flow profile with uniform streamlines is established at the upstream of the sinus region. After passing through this region, the flow enters both the internal and external carotid arteries (i.e. ICA and ECA). Velocity distributions are highly skewed toward the outer vessel walls. Similar findings have also been reported, where skewed flow structures are primarily caused by the misalignment of the mean axis of arteries (Vetel et al. 2009). Further downstream, strong flow separation and secondary flows are exemplified by streamlines. Owing to the rapid enlargement of the cross-sectional area, flow separation firstly occurs at the sinus region (see streamlines between slices 1–3). Similar flow separation is also found downstream of ICA as the flow continues through the stenosis (see streamlines after slice 7a). Streamlines between slices 4b–7b show a strong swirling secondary flow in the ECA. The inception of such flow is due to the action of centrifugal forces caused by the curvature of lumen,

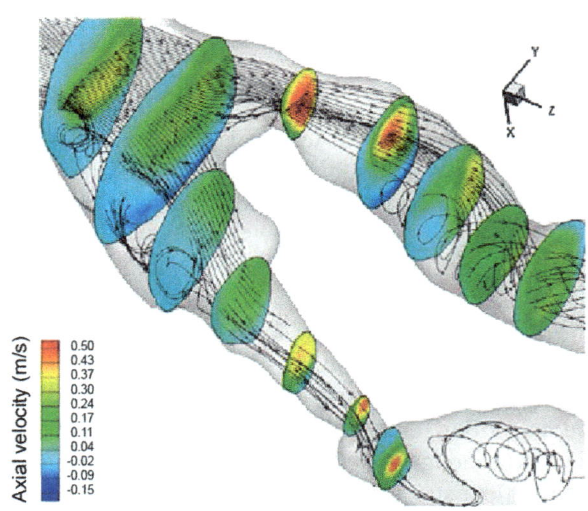

Fig. 7.10 Flow visualisation in a stenosed carotid bifurcation. The contour plots of axial velocities and streamlines traces in a stenosed carotid bifurcation are presented after extraction of the flow information within the anatomical geometry. This flow visualisation allows us to understand the flow condition within a diseased artery effectively

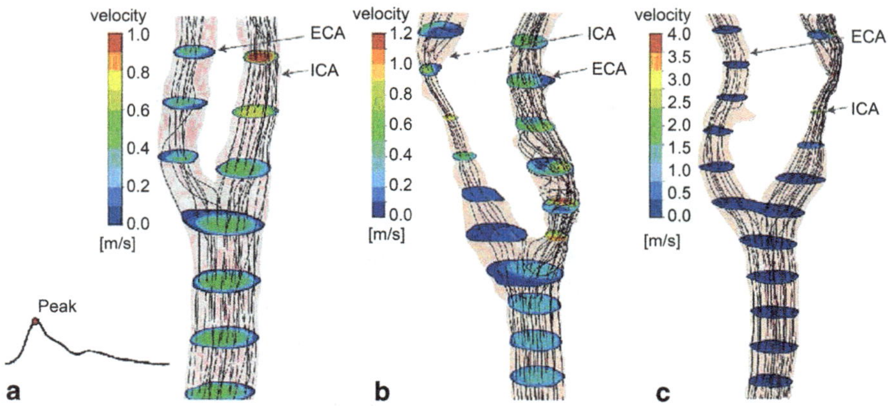

Fig. 7.11 Streamline plots pertaining to patient-specific carotid bifurcations. Streamline plots based on anatomical realistic geometries can demonstrate the regions of high velocity and also indicate the onset of turbulence from its laminar regrime. **a** shows a healthy normal carotid bifurcation, while **b** and **c** show stenosed carotid bifurcations at the ECA and ICA respectively

which redistributes the flow velocity and further intensifies the skewness of the flow structure (Vetel et al. 2009).

The velocity contours with streamlines passing through show that at the neck of the carotid stenosis, there is increased velocity and many streamlines converge through (Fig. 7.11). A higher flow velocity also corresponds to a higher Reynolds number, which initiates turbulence. Streamlines are useful for identifying the transition from laminar to turbulent flow. For example, laminar flow usually produces streamlines with tangential velocities aligned in their directions as shown by Fig. 7.11a, whereas turbulent flows are characterized by more random motion and indistinguishable paths. The latter usually occurs when there is a sufficiently high Reynolds number flow. For example, flow pertaining to the peak volume flow rate of the cardiac waveform through the stenosis of the carotid artery can result in turbulence downstream as shown in Fig. 7.11b and c.

In summary, path streamlines can be used to show that recirculation was prevalent downstream of the stenotic region where the flow experiences sharp changes in the flow conditions.

The complex asymmetric flow pattern within the stenosed carotid artery clearly depict the fluid flow transiting to a state of weak turbulence downstream while remaining laminar upstream (Lee et al. 2008). A closer examination of the iso-surface plot of the vorticity field by computational and experimental results is shown in Fig. 7.12. Upstream, the flow remained laminar and uniform in the CCA. Here, the Reynolds number ranges between 485 and 768. Through the sinus region, the flow in both ICA and ECA changed significantly and becomes more chaotic with pronounced vortical coherent structures and strong central vortex threading being formed through the stenosis. At the stenosed area of the ICA, the Reynolds number can reach as high as 2100, which indicates the probable transition to weak turbulent

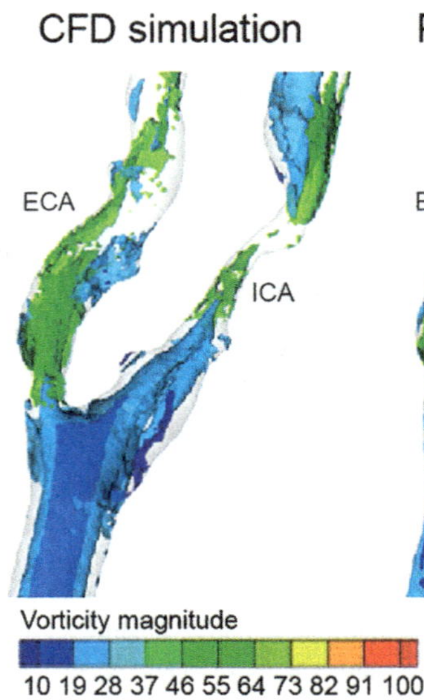

Fig. 7.12 Iso-surface plots of vorticity magnitude in a stenosed carotid bifurcation. The vorticity fields, which are derived from computational fluid dynamics and particle image velocimetry, are presented in three dimensional plots and enable the visualisation of rotational blood flow in the carotid bifurcation. The computationally predicted results and the experimentally derived flow measurements are observed to be relatively similar

flow. Break-down of vortex structure in the post-stenotic region can also be observed, which further ascertains the onset of transitional flow.

Velocity vector plot and axial profiles arising from computational haemodynamics was used to assess influence of the stenosis on the flow through a carotid bifurcation model (Zhao et al. 1999). The flow patterns within the carotid due to the geometry of the bifurcation are assessed by vector plot as shown in Fig. 7.13.

The axial velocity profiles in the bifurcation plane have a high degree of skewness near the bifurcation region. This is consistent with the results displayed in Fig. 7.8. Downstream of the bifurcation and along the ECA, axial flow accelerates due to the reduction in cross-sectional flow area as a result of the stenosis. However at the sinus bulb of the carotid artery the axial velocity is lower. It is worth mentioning that such consistently slow moving flow in this region may give rise to a higher susceptibility of atherosclerotic plaque deposits (Zhao et al. 1999).

The stenosis in the artery branch affects the flow field more in the ICA than in the ECA. At the ECA, the flow velocity is more consistent in terms of velocity profile upstream of the artery branch. The presence of the stenosis in the ICA does not affect the flow field in the ICA significantly due to its different branching.

Wall shear stress patterns in a patient-specific carotid bifurcation with different degrees of stenoses can be shown. Plots of wall shear stress can reveal the locations

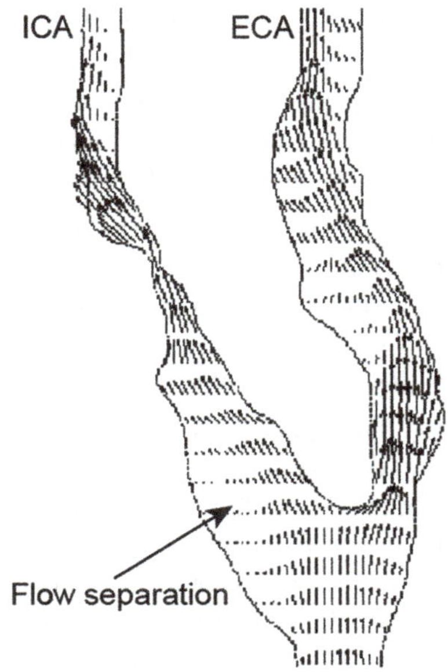

Fig. 7.13 Velocity vector plots of flow in atherosclerotic artery. Velocity vector plots of carotid bifurcation in the branching plane at t/T = 0.6 (systolic acceleration phase)

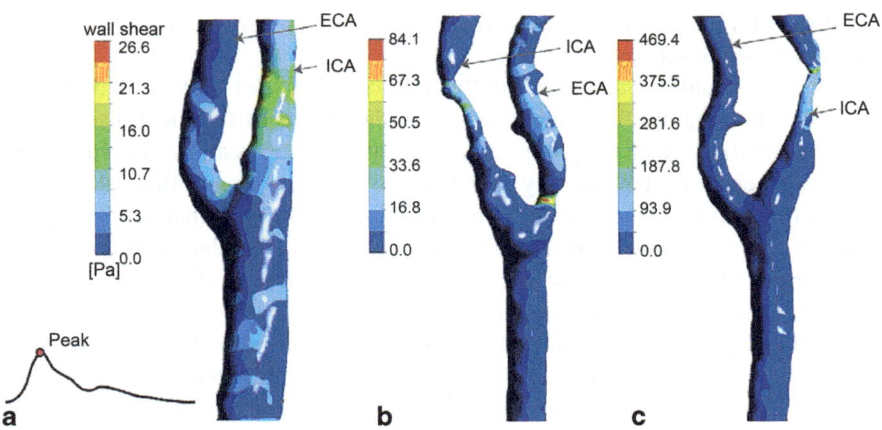

Fig. 7.14 Pattern of wall shear stress in patient-specific carotid bifurcations. Plots of wall shear stress based on anatomical realistic geometries can present the locations where the parameter is abnormally high. **a** shows a healthy normal carotid bifurcation, while **b** and **c** show stenosed carotid bifurcations at the ECA and ICA respectively

where the parameter is abnormally high. For instance, during the peak of the cardiac cycle, a higher level of wall shear stress at the stenotic regions with respect to other regions is found. This haemodynamic indicator can be used to detect the abnormal geometrical regions (Fig. 7.14).

7.2.6 Comments on Modelling Issues

The reconstructed human cardiovascular system model from Chap. 3 and 4 was used in this case study. Modelling the blood flow process raises some important conditions that need to be determined, such as:

- is the flow *laminar or turbulent*,
- *steady or unsteady*,
- additional equations that are needed to account for blood flow behaviour being *Newtonian or non-Newtonian*,
- the type of *inlet/outlet* conditions for the arterial branches.

Laminar flows are normally characterised by a smooth motion where the fluid's viscosity dominates allowing high molecular diffusion and dampening out any fluctuations in the flow. This leads to adjacent layers of fluid sliding past each other in an orderly fashion (like layers of lamina). However for an atherosclerotic artery, which has a complex geometry that is highly convoluted, flow separation and recirculation can exist especially in the stenosed region where a rapid decrease in the cross-sectional area of the artery is observed, and thereby enhancing flow instabilities. Although flow separation and recirculation are typical characteristics of turbulent flow, the presence of these characteristics does not necessarily assume the flow is turbulent, since its existence can be found in laminar flows with geometries that exhibit separation.

The dimensionless parameter, the Reynolds number (*Re*) is often used to determine the flow regime. For higher velocities the effects of turbulent disturbances become significant with the presence of velocity fluctuations in the flow field while for lower velocities the viscosity dominates to dampen out any fluctuations. The critical flow rate and therefore the *Re* number at which the flow changes from a laminar to a turbulent flow regime is difficult to succinctly define due to the complexity of the artery which has led to some debate concerning the type of blood flow regime to implement for numerical simulations. Table 7.1 summarises the model settings for recent atherosclerotic carotid bifurcation simulation studies. Experimental studies by Chen et al. (2006), Fan et al. (2009) and Long et al. (2000) have suggested that a laminar flow regime dominates. Simulation by Cheung et al. (2010), Botnar et al. (2000), Lee et al. (2008), Birchall et al. (2006) configured a turbulent flow model whereby at the stenosis, a disturbed laminar regime results.

In the context of CFD applications, existing uncertainties and difficulties remain in the modelling of embedded transition flow. The laminar flow assumption taken in the current CFD simulation is not strictly applicable in attempting to capture the onset of transitional flow. On the other hand, most existing turbulence models that have been developed primarily for fully turbulent flows cannot be directly applied to capture transitional flow.

Flow in a healthy and non-stenosed artery is dominantly laminar while turbulence is usually induced through the curvatures within the stenosed arterial geometry (Ku et al. 1985b). A number of numerical studies have also been performed to

Table 7.1 Summary of viscous model and flow setup found in selected atherosclerotic carotid bifurcation simulation studies

Researcher	Blood model	Viscous model	Steady/unsteady	Maximum Reynolds number	Maximum Mesh size
Cheung et al. 2010a	Newtonian	Turbulent	Steady	2100	1,517,434
Fan et al. 2009	non-Newtonian	Laminar	Unsteady	Not stated	71,312
Lee et al. 2008	Newtonian	Turbulent	Unsteady	1217	1,854,000
Chen and Lu 2006	non-Newtonian	Laminar	Steady	270	360,000
Birchall et al. 2006	Newtonian	Turbulent	Unsteady	Not stated	500,000–1,000,000
Tada and Tarbell 2005	Newtonian	Not stated	Unsteady	Not stated	60,000 (fluid), 36,000 (elastic wall)
Younis et al. 2003	Newtonian	Laminar	Unsteady	Not stated	56,209 (fluid), 43,334 (elastic wall)
Steinman et al. 2002b	Newtonian	Not stated	Unsteady	Not stated	100000
Long et al. 2002	Not stated	Laminar	Steady	330	41,600
Long et al. 2000	Newtonian	Laminar	Unsteady	Not stated	24,320
Botnar et al. 2000	Newtonian	Turbulent	Unsteady	Not stated	1,000,000

investigate the turbulent characteristic of stenotic flows using different turbulence models (Birchall et al. 2006; Lee et al. 2008; Younis and Berger 2004).

The pulsatile blood flow is clearly unsteady with an oscillatory motion. In order to assess the importance of the unsteadiness on the mean or average flow characteristics obtained through a steady solution, we make use of the Womersley number, α and the Strouhal number, S. The Womersley number is a ratio of unsteady forces to viscous forces named after John R. Womersley (1907–1958) and is defined as

$$\alpha = \frac{D}{2}\left(\frac{\omega}{T_g}\right)^{0.5}, \tag{7.4}$$

where D is the characteristic length which was taken as the inlet hydraulic diameter of the carotid artery equal to approximately 5 mm. Then, ν_g is the kinematic viscosity of blood (3.50×10^{-3} Pa·s), ω is the blood pumping frequency equal to

$\omega = 2\pi f = 6.28 s^{-1}$ and u_{ave} is the average velocity through the artery passage under the flow rate of approximately 10 ml/s or 10^4 mm^3/s, which is equal to 510 mm/s. When α is small (1 or less), the oscillatory effects are sufficiently low that the inlet conditions such as a parabolic velocity profile has time to develop during each cycle, and the flow will be very nearly in phase with the pressure gradient. When α is large (10 or more), the oscillation effects are sufficiently large that the velocity profile does not develop in time and the mean flow characteristics lags the pressure gradient by about 90 degrees, (Womersley 1955).

The Strouhal number is a ratio of the unsteady forces to the inertial forces named after Vincenc Strouhal (1850–1922) and is defined as

$$S = \frac{\omega D}{u_{ave}}, \tag{7.5}$$

where u_{ave} is the mean airflow velocity. For large S (> 1), the oscillations become important. For low S (<< 1), the contribution of the velocity dominates the oscillations. The calculated α and S numbers for the artery in this case study are 0.1059 and 0.0616 respectively. When α is smaller than 1, then the frequency is sufficiently low that a parabolic velocity profile has time to develop during each cycle. On the other hand, the low value for S suggests that the flow may be assumed to be quasi-steady.

Although the assumption of Newtonian behavior of blood is acceptable for high shear flows and also has been validated by several numerical studies, which indicated that the influence of shear thinning properties of blood are not significant (Cho and Kensey 1991; Perktold et al. 1991), it is not valid in low shear rate regions where the value is less than 100 s^{-1} (Chien 1982; Chua et al. 2005). The shear thinning behavior of the blood flow was incorporated by Carreau-Yasuda model (Bird et al. 1987) to predicate the flow pattern in the low shear rate region more accurately, and the correlation between blood viscosity and the shear rate variation is governed by

$$\mu = \mu_\infty + (\mu_0 - \mu_\infty)[1 + (\lambda\gamma)^a]^{(n-1)/a}, \tag{7.6}$$

where η_∞ is the viscosity for an infinite shear rate, and η_0 is the plasma viscosity at zero shear rate. λ, n, and a are fitting parameters, which are borrowed from the experimental data based on a well-tested blood-mimicking fluid (Gijsen et al. 1999a, b). As an example, these parameters can have the following values: $\eta_\infty = 2.2 \times 10^{-3}$ Pa s, $\eta_0 = 22 \times 10^{-3}$ Pa s, $\lambda = 0.11$ s, n = 0.392, $a = 0.644$, and $\rho = 1410$ kg m^{-3}.

Estimates of patient-specific haemodynamics parameters have been attained (Long et al. 2000; Steinman et al. 2002a). Some researchers employed non-Newtonian fluid models to aptly characterize the rheological effect in blood flows (Chen and Lu 2006; Gonzalez and Moraga 2005; Lee and Steinman 2007). However, many numerical investigations considered Newtonian fluid flows in rigid stenosed arteries (Farmakis et al. 2004; Giannoglou et al. 2005; Siauw et al. 2000; Zhao et al. 2002). In these studies, the fluid flow weres considered steady, isothermal, incompressible and Newtonian, consistent with experimental studies in the literature. Wall boundaries of the computational model were assumed rigid and impermeable. The blood was Newtonian with a density ρ of 1050 kg m^{-3} and a dynamic viscosity μ of 3.5×10^{-3} Pa•s (Ufuk Olgac et al. 2008).

Boundary conditions for the computational surfaces need to be defined. While the surface walls are easily understood in terms of its definition as a rigid or elastic wall boundary, the common carotid artery (CCA) inlet, as well as the internal carotid artery (ICA) and external carotid artery (ECA) outlets provide the user with more options for definition. In this case study, a uniform flow perpendicular to the CCA inlet was specified.

This is enforced as we are making direct comparisons with experimental results that used a fixed flow rate. It is important to maintain similar settings and to keep the flow rates the same between the computer and experimental model. A pressure boundary condition could have also been used for modelling the blood pumping cycle. Other boundary conditions commonly used are a velocity or mass flow rate boundary, if they are known. The cardiac cycle is caused by the blood pressure difference induced by the heart chamber dilating or contracting to increase/decrease the volume of the chamber. Therefore, it is natural to assign the CCA inlet, and the ICA and ECA outlets as pressure conditions. The inlet pressure condition is used to define the total pressure at the flow inlet. It is an ideal condition to use when the pressure at the inlet is known but the flow rate and/or velocity is unknown. The pressure outlets are used to define the static pressures at the flow outlets (and also other scalar variables, in case of backflow). However, the problem with the pressure outlets is that it is unknown in relation to the CCA inlet to induce the required flow rate. A modelling strategy to overcome this is to first simulate the flow field with the known mass flow rate in mind using a mass flow rate condition (Figure 7.15). Upon completion of the simulation, the pressure difference between the CCA inlet and ICA outlet as well as between the CCA inlet and ECA outlet will be known, and thus the pressure at the inlets and outlet can be defined.

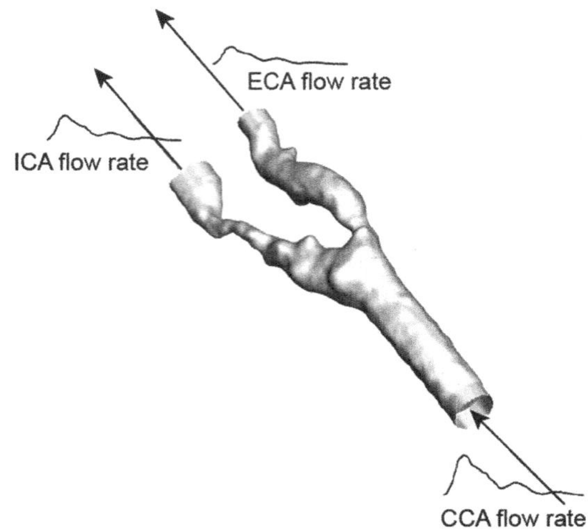

Fig. 7.15 Boundary conditions determination based on flow rates when the pressures at inlets and outlet are unknown

7.2.7 Downstream peripheral vascular impedance modelling

During past decades, various haemodynamic studies of carotid bifurcations have been carried out from the aspects of experimental and numerical modelling (Kleinstreuer et al. 1991; Ku et al. 1985b; Ojha 1993; Steinman and Rutt 1998; Yuan C et al. 1995). For most of these models, the inlet and outlet flow velocity waveforms from experimental measurements are able to establish accurate boundary conditions for a physiologically realistic model. For example, based on configuration of flow rates as boundary conditions, waveforms were imposed at the inlet artery of the stenosed model, and also with the fully developed velocity profiles applied at the CCA inlet and opening boundary conditions at the ICA and ECA outlets. However, such flow profiles are not readily available. The artery outlet branches are often set at specific pressures as outflow boundary conditions whenever the outlet volume flow profiles are not available (Steinman et al. 2000; Zhao et al. 1999). As such, velocity profiles may be poorly defined, not patient-specific, or lack clinical validation. The results generated by these configurations may not be accurate enough to reflect physiological realism, and further assumptions are required to approximate the modelling as close as possible.

The lack of measured inlet and outlet boundary conditions will jeopardise the accuracy of the modelling significantly, and it is important that sufficient assumptions such as accurate flow division are established to maintain the integrity of the modelling. Configurations of existing studies that utilized specific or measured flow division ratios are examined. The carotid bifurcation study by Steinman et al. (2000) utilised assumptions that involved a flow division of 56:44 between the ICA and ECA branches. It may be worthwhile noting that according to the statistics provided by such clinical studies, this ratio is not a good representative of the physiological condition, and does not come with experimental support all the time. For example, in a study by Zhao et al. (1999), fully developed flow profiles were set at both the

CCA inlet as well as the ICA and ECA outlets. In their settings, the flow division ratio of the ICA to ICA was assumed to be 7:3, which they referred to as an average physiological value under normal conditions (Ku and Giddens 1987). However, similar to Steinman et al. (2000) they model the resistance effect downstream of the arterial network by assuming specific flow division ratios. A more accurate setting by Long et al. (2002) implemented the 55:45 ratio based on magnetic resonance flow measurements, However, although this configuration is patient-specific, it is fixed for a steady flow while the real flow division ratio is transient and dependent on the flow rate profiles through the ICA and ECA. The study by Marshall et al. (2004a) took into account this transient property and reported the ECA/ICA flow division ratio of a healthy carotid artery with a time-averaged value of 0.39:1.

It is worth noting that the flow division ratio of ICA to ECA is not affected by the stenosis alone, but also by the downstream resistance of the arterial bed. It is important that models consider the downstream peripheral vascular impedance when simulating the carotid arterial intravascular flow. This is achieved by developing a downstream peripheral vascular model to incorporate its flow resistance. The downstream resistance model can be used for computational studies of idealistic geometries that relate to the carotid bifurcation, an artery with varying stenosis, or an artery that has lumen reconstruction based on virtual stenting (Cebral et al. 2001). Experimental measurements of the outlet boundary conditions for specific stenosed artery cases may be unfeasible, which necessitates the use of a downstream peripheral vascular model.

The development of a CHD model that incorporates the downstream peripheral impedance effect is used when the outlet flow conditions are not available. The proposed modelling accuracy can be verified by numerical simulations based on a healthy carotid bifurcation case study (Dong et al. 2013a). The haemodynamic differences between the proposed and fixed flow division ratio can be addressed based on an atherosclerotic carotid bifurcation. Finally, the influence of atherosclerosis on the intravascular blood flow can be accessed and evaluated based on the extracted local risk indicators.

In this example, two porous beds with different permeability configurations were connected with the ICA and ECA branches each. Darcy's Law (Batchelor 1967) was used to establish the correlation between the pressure difference across the porous bed ΔP, the volume flow rate Q, and the permeability k by the following equation:

$$Q = -\frac{kA}{\eta}\frac{\Delta P}{L_P} \tag{7.7}$$

where L_P is the length of the porous medium, A is the cross-sectional area to the blood flow, and η is the blood flow viscosity. To maintain numerical stability, L_p is chosen as 10 times the diameter of the extension (Figure 7. 16).

The pressure difference ΔP between large arteries and capillaries reported from Pries et al. (1995) is around 60 mmHg. The geometry details of the proposed porous medium are shown in Table 7.2.

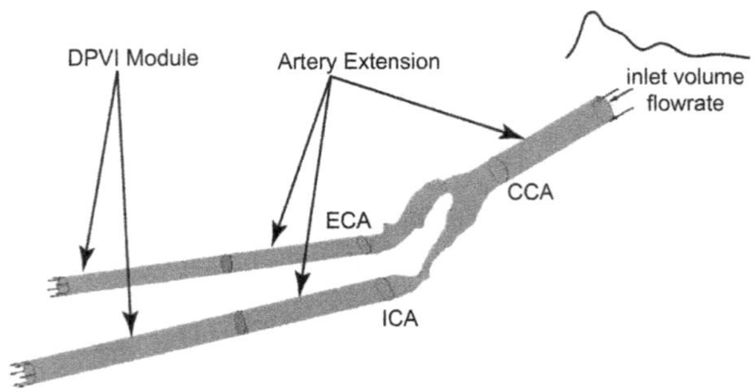

Fig. 7.16 Domains configuration for the computational haemodynamics of carotid bifurcation model. In order to eliminate the local fluid dynamic effects on the reconstructed fluid domain, flow extensions were attached to the carotid bifurcation model; two DPVI modules with difference transient porosities were connected downstream of the artery model to represent downstream flow resistance

Table 7.2 Geometrical parameters of the introduced porous mediums

Patients		Diameter (mm)	Cross-sectional area (mm2)	Length (mm)
Healthy Carotid Bifurcation	ECA Porous Medium	4	12.6	40
	ICA porous Medium	4	12.6	40
Atherosclerotic Carotid Bifurcation	ECA Porous Medium	4.6	16.6	46
	ICA porous Medium	5.6	24.6	56

Based on Equation (7.7) the permeability variation of the porous medium (kA/ηLp) reflects the downstream vascular impedance $R=-\Delta P/Q$. Therefore, the downstream ICA and ECA porous domain permeability can be used for other carotid bifurcation models to approximate the vascular bed impedance, and a standard peripheral flow environment can be established for the carotid bifurcation haemodynamics.

The porous domain is then connected to the carotid bifurcation. The 3D unsteady incompressible Navier-Stokes equations were solved by a second order implicit backward Euler method, and the SIMPLE pressure-velocity coupling was used. The Wilcox low-*Re* k–ω turbulence model (Wilcox 1998) was chosen for all the simulations to predict low-*Re* effects on the turbulence field at the near-wall regions.

Vessel walls were assumed rigid for all models (Hoi et al. 2010), and the convergence criterion for the relative residual of all dependent variables was set to 1×10^{-4}. Two full cardiac cycles were required to damp the initial transient errors. The third cycle was used for data analysis (Ghalichi and Deng 2003). Studies of grid independence based on scales of 0.8, 0.6, 0.4, and 0.2 mm and time-step size dependence based on time-steps of 0.01, 0.008, 0.005, and 0.002 s were performed based on the healthy carotid bifurcation model using the prescribed inlet and outlets boundary conditions. The investigation results are shown in Table 7.3. Accounting for the computational expenses, the most efficient grid scale L_G and time-step size t settings are $L_G=0.4$ mm, $t=0.005$ s.

Table 7.3 Comparison of angles based on ten carotid bifurcation case studies. We note that relatively larger angles between ICA and CCA exist for patients (4/L) and (4/R). The angles between ECA and CCA for patients (1/R) and (6) are also observed to be greater than the rest of the other patients

Comparison of carotid bifurcation angles										
Patients	(1/L)	(1/R)	(2/L)	(2/R)	(3)	(4/L)	(4/R)	(5/L)	(5/R)	(6)
Angle between ICA and CCA	24°	31°	30°	40°	10°	57°	65°	23°	30°	37°
Angle between ECA and CCA	13°	33°	25°	19°	25°	19°	0°	14°	10°	29°

Three commonly adopted flow indicators to evaluate the total shear stress exerted on the wall throughout a cardiac cycle are the time-averaged wall shear stress (TAWSS), and the oscillatory stress index (OSI), and the relative residence time (RRT), which can be obtained from the following equations:

$$\mathrm{TAWSS} = \frac{1}{T}\int_0^T |\tau_w| dt \tag{7.8}$$

$$\mathrm{OSI} = \frac{1}{2}\left[1 - \frac{\left|\frac{1}{T}\int_0^T \tau_w dt\right|}{\frac{1}{T}\int_0^T |\tau_w| dt}\right] \tag{7.9}$$

$$\mathrm{RRT} = \frac{1}{(1 - 2\times\mathrm{OSI})\times\mathrm{TAWSS}}, \tag{7.10}$$

where T is a cardiac cycle period and τ_w is instantaneous wall shear stress (WSS). As defined in Eq. 7.9, the OSI can be regarded as the fraction of angle and magnitude change between the instantaneous WSS and the time-averaged WSS. It ranges from 0 to 0.5, where 0 represents a total unidirectional WSS and the theoretical maximum value 0.5 describes a purely unsteady, oscillatory flow with zero WSS. Areas of high OSI are predisposed to endothelial dysfunction and atherogenesis (Taylor et al. 1998, 1999; X.J. He and Ku 1996). TAWSS values (lower than 0.4 Pa) (Malek et al. 1999), and high RRT (higher than 10 m^2/N) (Lee et al. 2009; Morbiducci et al. 2010) are also known to promote an atherogenic endothelial phenotype, while abnormally high TAWSS (higher than 40 Pa) values can cause direct endothelial injury and increase the risk of thrombosis (Fry 1969; Malek et al. 1999).

The results from the proposed downstream peripheral impedance model are shown in Fig. 7.17, and a TAWSS comparison was also carried out on the same carotid model using prescribed boundary conditions. According to the local flow indicator thresholds, the outer-ICA wall of the healthy carotid bifurcation shows low TAWSS, high oscillatory shear and long relative resident time (indicated by red boxes). Therefore this region is vulnerable for atherosclerosis in the long term.

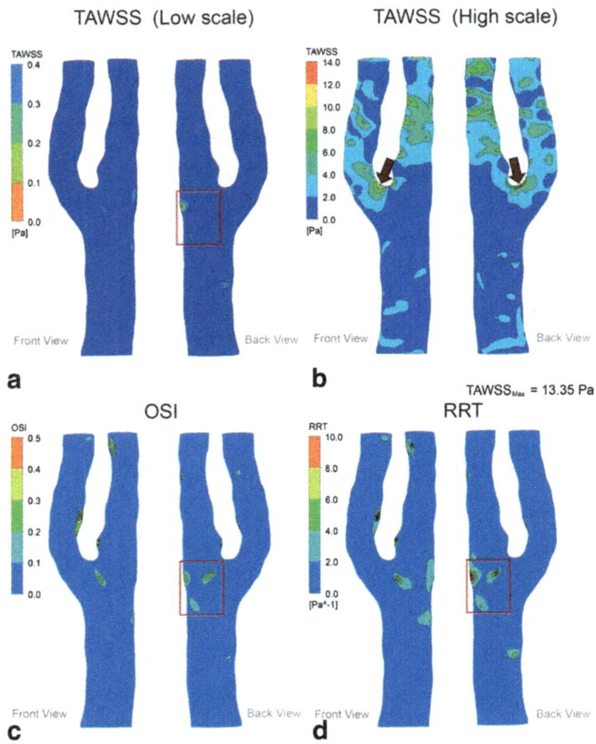

Fig. 7.17 Numerical results of the healthy carotid bifurcation haemodynamics using DPVI modelling method. The plots of the WSS, TAWSS, OSI and RRT flow properties based on carotid bifurcation with downstream peripheral impedance model

The apex of the bifurcation divider-wall experiences maximum TAWSS (13.35 Pa) compared to the remaining portion of the healthy carotid bifurcation. Because peak TAWSS value is much lower than the threshold of 40 Pa, this site remains risk-free from direct endothelial injury from the blood flow.

The comparative difference is shown in Fig. 7.18 where the peak TAWSS magnitude difference (0.46 Pa) is located at the apex of the bifurcation, which only accounts for 3.4 % of the maximum TAWSS value (13.35 Pa) at the same location. This order of percentage difference validates the need for the porous medium impedance model

7.2.8 *Closure*

In this case study, we stepped through each stage of CHD modelling. From medical image reconstruction, we express the arterial geometry that is used as a surface mesh for flow simulation. Then, modelling requirements for steady blood flow were shown. Particular attention was given to the critical issues of:

- the flow being *laminar or turbulent*,
- *steady or unsteady*,

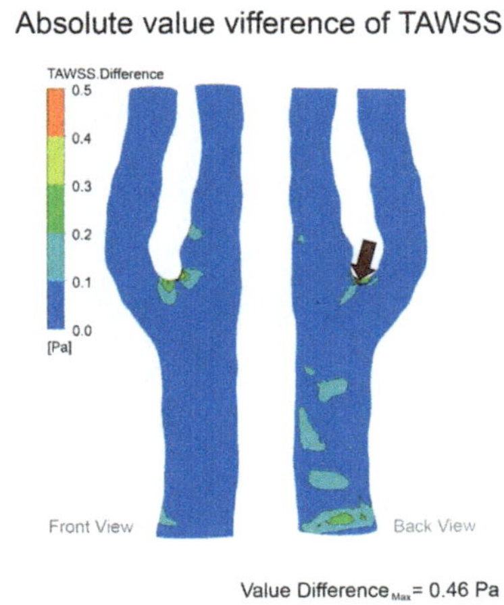

Fig. 7.18 Absolute TAWSS value difference of the healthy carotid bifurcation with prescribed boundary conditions and the DPVI model. It is found that the maximum value difference is located at the apex of the carotid bifurcation with a value of 0.46 Pa

- Newtonian or non-Newtonian,
- the type of *inlet/outlet conditions* for the CCA inlet, and ICA and ECA outlet, and the comparison of a numerical simulation of carotid artery for comparison with available literature to validate the simulation's accuracy,
- corporation of mock vascular geometries that are needed to account for downstream resistance modelling.

In summary, the blood flow could be assumed as laminar. Transitional flows are much more difficult to resolve and care must be taken when analysing the results in terms of a laminar or turbulence model used. The Newtonian assumption was applicable when a higher shear rate (> 100/s) occurs. The type of inlet/outlet condition for this type of artery requires a pressure boundary condition at the inlet and outlet which allows for the flow to be forced through the CCA. Visualisation of the flow field was performed through different post processing techniques such as velocity vector, axial velocity profile, streamline, and WSS contour plots.

7.3 Comparison Analysis of Patient Specific Carotid Bifurcation Models

7.3.1 Medical Image Reconstruction of Patient-Specific Arteries

In this example, imaging of six patients was performed and a total of ten carotid bifurcation models (with different branching and degrees of stenosis) were selected for the simulation Fig. 7.19. The case study subjects are labelled (1) to (6) with the denotation (/L) and (/R) representing the left and right carotid bifurcations respectively. The flow properties of the (/L) and (/R) bifurcations are compared. The

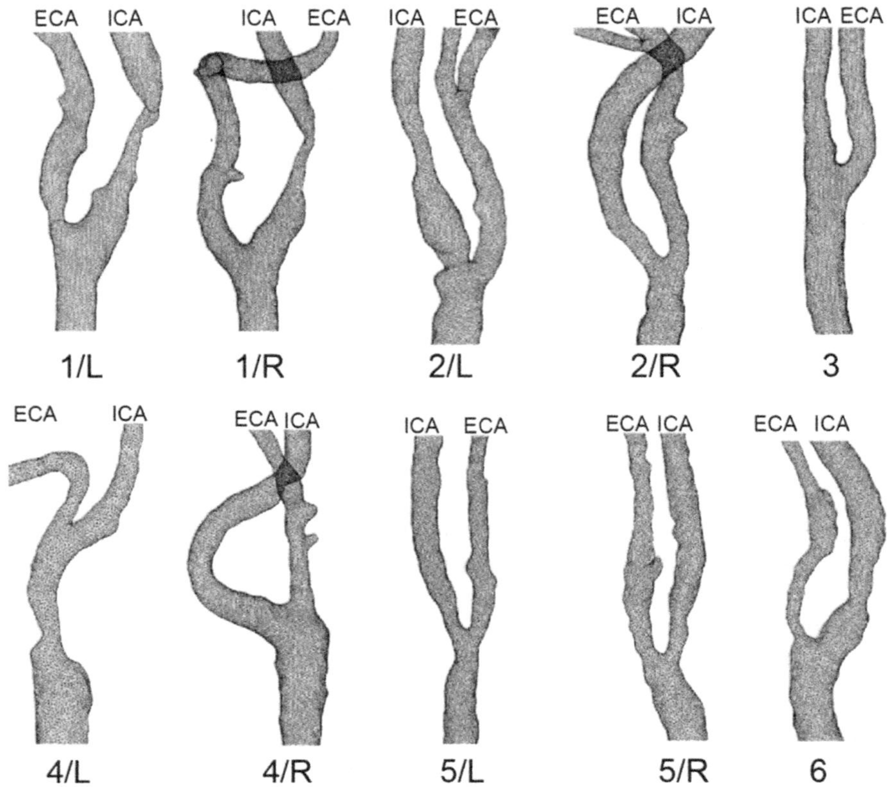

Fig. 7.19 Reconstruction of carotid artery case studies. Carotid bifurcations of 6 patients forming 10 arterial case studies are labelled from 1 to 6, with the symbol "/R" and "/L" representing left and right carotid arteries respectively. The geometry, which is labelled as "3", pertains to the left carotid artery of a healthy normal. ICA—internal carotid artery; ECA—external carotid artery; CCA—common carotid artery

flows of different patients at the same blood vessel location are then analysed for a relationship between the geometry and blood flow property. The flow parameters discussed include geometry, and wall shear stress.

7.3.2 Comparison of Anatomical Geometries

Individual variation in the human arteries may be considerable. The imaged bifurcations exhibited some degree of geometrical non-planarity. Subject (1/L) and (3) being more planar, while (1/R), (2/L) and (2/R) are highly non-planar. The degree of occlusion is defined by a reduction in the cross-sectional area. The ECA for subject (1/L) is approximately 60 % occlusion, whereas ICA of subject (1/R) shows more

Table 7.4 Comparison of stenosis for ICA and ECA with respect to CCA. Carotid bifurcations of patients (1/L), (1/R), (5/L) and (6) are severely stenosed at the location of their ICA. At the ECA, patients (4/L) and (5/R) have the most serious clogs. Note that a higher degree of stenotic severity is indicated by a lower ICA/CCA or ECA/CCA ratio (%)

Comparison of carotid bifurcation stenosis										
Patients	(1/L)	(1/R)	(2/L)	(2/R)	(3)	(4/L)	(4/R)	(5/L)	(5/R)	(6)
CCA (mm^2)	32.1	30.3	20.2	24.5	28.7	41.0	24.1	27.3	31.6	34.6
ICA (mm^2)	2.9	3.0	9.8	12.1	12.7	13.5	8.0	8.7	7.2	8.8
ICA to CCA ratio (%)	9	10	49	49	44	33	33	32	23	25
ECA (mm^2)	17.8	11.7	4.7	6.8	11.2	9.3	8.0	7.1	3.8	14.0
ECA to CCA ratio (%)	55	39	23	28	39	23	33	26	12	40

severe blockage (up to 70 % occlusion). Blood vessel on (2/L) and (2/R) are slightly stenosed (approximately 40 % occlusion).

The ICA and ECA angles with respect to the CCA are described in Table 7.3. Note that the angle between ICA and CCA varies from 10 to 65°. The angles 57° (case (4/L) and 65° (case (4/R) are notably higher than the rest of angles for the other patients. The angle between ECA and CCA ranges from 0 to 35°. The case subjects (1/R) and (6) have angles 33 and 35° that are comparatively higher than the angles pertaining to the other patients.

Detailed comparison for ICA and ECA stenoses based on the ten case studies from the six patients are given in Table 7.4. Patients (1/L), (1/R), (5/R) and (6) have severe stenosis at the ICA, with ICA to CCA ratios of 9, 10, 23 and 25 % respectively, and patients (4/L) and (5/R) are heavily stenosed at the ECA with ratios of 23 and 12 % respectively. Although not displayed in the table, it may be worth noting that the CCA of patient (4/L) has an area reduced to 5.0 mm^2 at its bifurcation. Also, the CCA area for patient (5/R) is reduced until 14 mm^2 at bifurcation.

The geometry of subject (3) was taken from a healthy carotid bifurcation with no signs of stenosis. The CCA, ICA and ECA diameters of the measured sample are 6.00, 3.98 and 3.96 mm respectively, while the diameters of the standard anatomy are 6.2 (at CCA), 4.4 (at ICA) and 3.5 mm (at ECA). The angle between the ICA and ECA is approximately 40°, which is a small deviation compared with the standard geometry at 50°.

7.3.3 Comparison of Wall Shear Stress Computational Models

7.3.3.1 Wall Shear Stress Contours at Peak Systole

It has been indicated by previous studies that wall shear stress promotes luminal thinning and plaque rupture (Berger and Jou 2000; Gertz et al. 1981; Zohdi 2005) and low WSS regions are more prone to athersclerosis (Plank et al. 2007). The

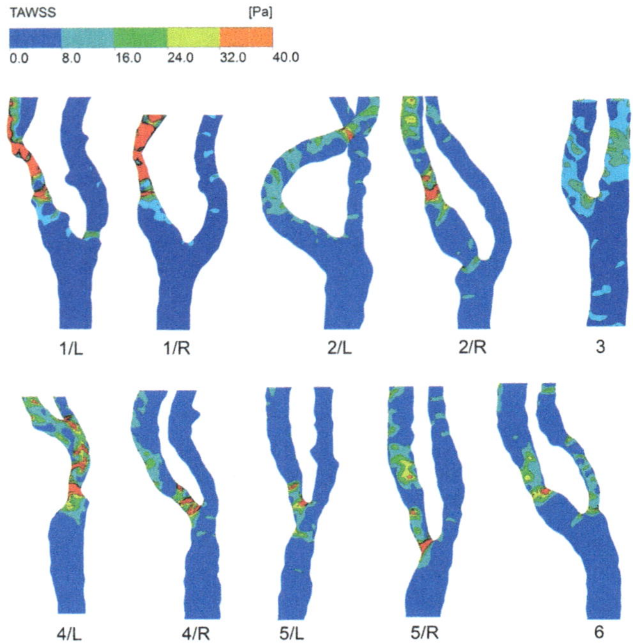

Fig. 7.20 Time-averaged WSS distribution of carotid bifurcations

growth of lesion will lead to stenosis at the artery wall, and further narrowing downstream of the blood vessel will occur. Therefore, this is one of the key parameters that will be used for examination of diseased carotid bifurcations.

Contour maps of the predicted maximum wall shear stress at peak pressure for each of the ten case studies are given in Fig. 7.20. In all models, the maximum wall shear stress (WSS_{max}) shows peak values on the inner ICA and ECA walls near the bifurcation points; and significantly elevated values spiralling around the larger angular branch (either the ICA or ECA) from the inner walls along the superior orientation. The maximum WSS value appeared at the stenosed section due to vessel tapering. For example, the ICA branch on cases (1/L), (1/R), (4/R), (2/L) exhibit maximum wall shear stress.

Low wall shear stress was seen primarily at the roots and along the outer walls of the sinus bulbs. Nevertheless, the shape and location of the wall surface subject to low wall shear stress differ from subject to subject. For all cases, the low WSS zone starts from the CCA and extends asymmetrically to the posterior and anterior aspects, with more pronounced coverage on the ECA walls in subjects (1/L), (1/R), (2/L), (2/R), (4/R) and (3). However, this pattern is reversed in cases (5/L), (5/R) and (6). For subject (4/L), the low WSS zone starts from CCA, and then terminates at the stenosed area, which is located at the CCA just before the bifurcation, and then appears again on the outer wall of ICA branch. WSS_{max} values are presented by Table 7.5 for ten cases.

Table 7.5 Maximum wall shear stress comparison among ten case studies. Comparison among patients is carried out with the maximum wall shear stress (WSS_{max}) values presented in descending order. Subject (6) experiences the maximum WSS at the ECA branch. Based on the geometry comparison, we know that subject (6) has a severe stenosis at the ECA of the carotid bifurcation

Locations and values of maximum WSS		
Patients	Locations	WSS_{max} (Pa)
(6)	ECA branch	370
(4/L)	CCA near bifurcation	239
(1/R)	ICA branch	184
(2/L)	ICA branch	150
(1/L)	ICA branch	144
(5/R)	Bifurcation and ECA branch	123
(4/R)	ICA branch	113
(5/L)	Bifurcation and ECA branch	91
(2/R)	ICA branch	65.5
(3)	ICA branch	53.5

7.3.3.2 Wall Shear Stress at CCA, ICA and ECA Sections

Planes on the CCA, ICA and ECA of five selected subjects were used for analysis. Wall shear stresses are calculated and compared with each other (Fig. 7.21). The ECA and ICA sections are selected at the stenosed location (at place of smallest diameters), while the CCA sections are selected at a common location that is typically located at approximately three times the CCA diameter below its bifurcation. The wall shear stress is chosen at peak systole.

No common trend exists for the patients. Based on the CCA planes of the five subjects, values of WSS for (1/L), (2/L) and (3) vary more than those for (1/R) and (2/R). For example, the maximum WSS for patient (1/L) is 14 Pa and its minimum value is 0 Pa, which results in a difference of 14 Pa. But WSS that pertains to case subject (1/R), which ranges from 2.8 to 9 Pa, has a difference of 6.2 Pa. This may be the result of large geometry change in cross sectional area for patient (1/L).

7.3.4 Comparison of Haemodynamic Properties

Carotid bifurcations of patients (1/L), (1/R), (2/L), (2/R) and (3) were investigated for their geometrical and flow properties. These properties are given in Table 7.6. Note that D_x and D_y are the diameters of the vessel in the *x*- and *y*- orientations. The ECA of patient (1/R) experiences maximum blood flow velocity (denoted as Max Velocity) at 3.34 ms^{-1} as it has the smallest D_x and D_y values. The minimum and maximum wall shear stresses, which are denoted as Min WSS and Max WSS respectively, are given along with the angle that they occur in the radial orientation. For model (1/L), the values of these properties at the ECA section are Min

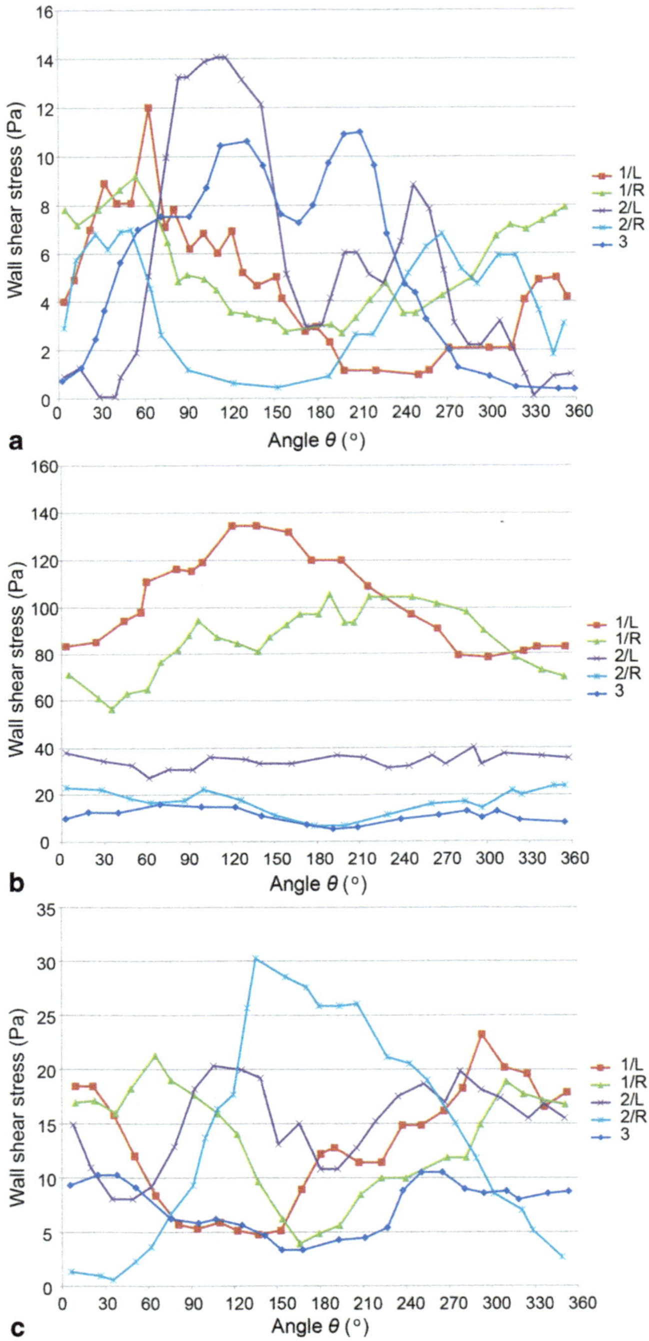

Fig. 7.21 WSS comparison on CCA, ICA and ECA sections. The WSS values are radially distributed based on angle ϕ for sectionings at **a** CCA, **b** ICA, and **c** ECA. The WSS variations are plotted along the angular distribution of the stenosed sections that correspond to the ICA and ECA branches. There is no particular trend that pertains to the WSS values among these 5 cases

Table 7.6 Haemodynamic properties at carotid bifurcation. Patients (1/L), (1/R), (2/L), (2/R) and (3) are investigated based on the diameters of cross section *Dx* and *Dy* in the *x*- and *y*- directions, maximum blood velocity (denoted as Max Velocity), and the minimum and maximum wall shear stresses, which are denoted as Min WSS and Max WSS respectively. The radial location where they occur is also presented

Haemodynamic properties at carotid bifurcation for 5 case arteries

Position	Items	(1/L)	(1/R)	(2/L)	(2/R)	(3)
CCA min Area Section	Dimensions (mm)	D_x=6.34; D_y=6.61	D_x=6.31; D_y=5.88	D_x=6.74; D_y=7.08	D_x=7; D_y=7.3	D_x=6.08; D_y=5.91
	Max velocity (m/s)	0.61	0.73	0.73	0.57	0.69
	CCA Min WSS (Pa)	0.6 at 5°	2.7 at 212°	0.30 at 20°	0.40 at 187°	0.4 at 292°
	CCA Max WSS (Pa)	11.9 at 193°	9 at 93°	15.2 at 140°	7 at 81°	11 at 64°
ICA min Area Section	Dimensions (mm)	D_x=4.41; D_y=5.37	D_x=3.74; D_y=4.09	D_x=2.71; D_y=2.77	D_x=3.31; D_y=3.91	D_x=4.27; D_y=3.69
	Max velocity (m/s)	0.91	0.96	1.49	0.944	0.70
	CCA Min WSS (Pa)	4 at 270°	3.9 at 212°	25.8 at 104°	4.2 at 229°	5 at 148°
	CCA Max WSS (Pa)	23 at 74°	20 at 109°	42.9 at 337°	24 at 38°	14 at 241°
ECA min Area Section	Dimensions (mm)	D_x=3.08; D_y=2.08	D_x=1.51; D_y=2.42	D_x=3.46; D_y=3.27	D_x=1.90; D_y=2.37	D_x=3.78; D_y=4.13
	Max velocity (m/s)	2.20	3.34	0.86	1.08	0.75
	CCA Min WSS (Pa)	76 at 70°	56.3 at 77°	8.0 at 77°	0.9 at 60°	2.7 at 105°
	CCA Max WSS (Pa)	138 at 260°	106 at 271°	21.5 at 159°	40 at 177°	11 at 207°

WSS=76 Pa at 70° and Max WSS=138 Pa at 260° despite having a lower Max Velocity than patient (1/R). Model (1/R) also shows a relatively high maximum WSS while model (2/R) experiences the lowest Max Velocity, Min and Max WSS at the CCA section, at 0.57 ms^{-1}, 0.40 Pa and 7 Pa respectively. This can be related to the relatively larger dimension of the vessel that has D_x=7 and D_y=7.3 mm.

The results produced WSS values that are similar to those by the bifurcation models developed by Lee et al. (2007). However, they can reach a maximum of up to 144 Pa, which is 7 times the typical range of values.

7.3.5 Closure

Different geometry and conditions of the left and right carotid arteries were investigated to evaluate the presence of stenosis. Flow patterns were highly dependent on the anatomical geometry. Based on analytical evaluation and comparison, a relationship between flow properties and geometry variations qualitatively and quantitatively was made. A comparison of the flow property for the left and right carotid vessels of every patient, and evaluation of the flow for different patients at the same location was performed to establish the relationship between the stroke and haemodynamics.

Individual vascular anatomy and pulsatile flow conditions were incorporated into the simulation. It was found that geometries of the carotid bifurcation are highly complex, involving helical curvature and out-of plane branching. These features result in wall shear stresses that are significantly different from those found in the simplified carotid bifurcation models.

This case study presented:

- A comparison of flow characteristics between various stenosed blood vessels simulation that shall act as a basis for evaluation of an appropriate haemodynamics parameter of assessment.
- A CFD prediction of blood flow in arteries based on the suggested haemodynamics parameters will make it possible for support in clinical assessment of atherosclerosis.

7.4 Analysis of Stented Artery Based on Intra-Aneurysmal Flow Simulation

7.4.1 Configuration of Aneurysm Stenting

Based on numerical simulations, the change in intra-aneurysmal flow field for a diseased artery before and after stenting can be studied. Pressure gradient and shear strain mapping can be performed by processing the simulation results and used to analyze the flow in the aneurysm sac. To quantify the complex aneurysmal flow, a monitor point collects flow properties at the center of the sac. Information from these flow characteristics relate to the stress and strain due to swirling of the blood within an aneurysm. This study simulates aneurysmal flow for both non-stented and stented arterial configuration. The information can assist in optimising stent configuration so that aneursym growth can be minimized as much as possible.

Numerical simulations based on untreated and stented aneurysmal arteries are performed to demonstrate their difference in terms of blood flow. The modelling of circular stent struts at the location of the aneurysm is used for the stented artery. Configurations based on 3, 4 and 7 stent struts at the neck of the aneurysm is implemented (Fig. 7.22).

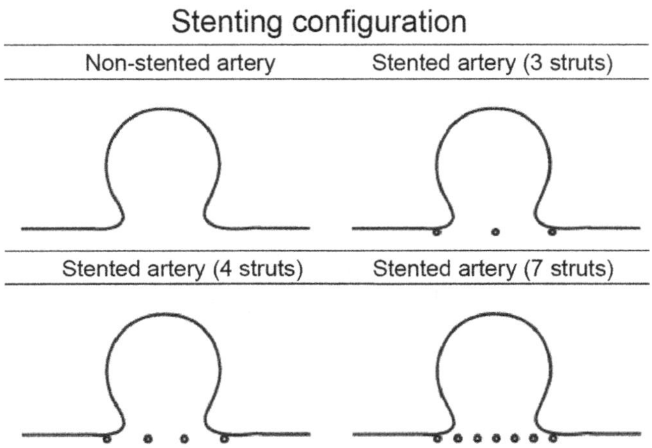

Fig. 7.22 Schematic of non-stented and stented aneurysmal artery

7.4.2 Modes of Aneurysmal Flow

Aneurysmal flows produce a force which propels blood into the aneurysmal sac. This force is proportional to the difference in blood pressure from the vessel to the aneurysm across its neck. The pressure gradient ΔP is influenced by the resistance to that flow. Factors determining this resistance are the dimension of the aneurysmal neck, and in the case of stenting, the effective orifice area of the neck that excludes the stent strut areas.

For a steady, incompressible flow, the shear rate of strain of a fluid element is defined as the rate of decrease of the angle formed by two mutually perpendicular lines on the element. As such, the shear strain rate $\dot{\gamma}$ is defined as a function of τ and is proportional to the velocity gradient in the perpendicular flow direction:

$$\dot{\gamma} = f(\tau) = -\frac{du}{dy}. \tag{7.11}$$

In general, the shear strain rate can be defined as a function of τ, and the linear relationship between the shear stress and strain components with the viscosity as its gradient. Mathematically, the model can be expressed as

$$\tau = \mu\gamma, \tag{7.12}$$

where μ is the viscosity of the fluid.

7.4.3 Computational Modelling and Numerical Details

7.4.3.1 Geometric Domain

The computational domain of the aneurysmal model is shown in Fig. 7.23. The parent vessel and the aneurysmal sac are set at 8 and 5 mm in diameters respectively. The flow enters the vessel with no obstacle and swirl within the aneurysm. For the stented artery, the aneurysmal flow velocity is reduced due to the stent struts obstructing blood movement. For simplicity, the velocity boundary condition that was applied at the arterial entrance assumes a uniform profile and has a direction that is normal to the surface. The artery is modelled long enough so that towards the exit of the vessel, a zero pressure gradient boundary condition along the length of the artery is implemented.

A close-up view of the mesh for this aneurysmal vessel (Fig. 7.24) shows the variation of tetrahedral cells at the aneurysmal wall and around the stent struts. A domain size of more than 6k cells is generated. A mesh independence study was carried out to ensure an optimum mesh for resolving the flow gradients.

Blood flow through the arterial vessel is assumed incompressible, homogeneous, laminar, and Newtonian. The fluid properties are reported in Table 7.7. Typical val-

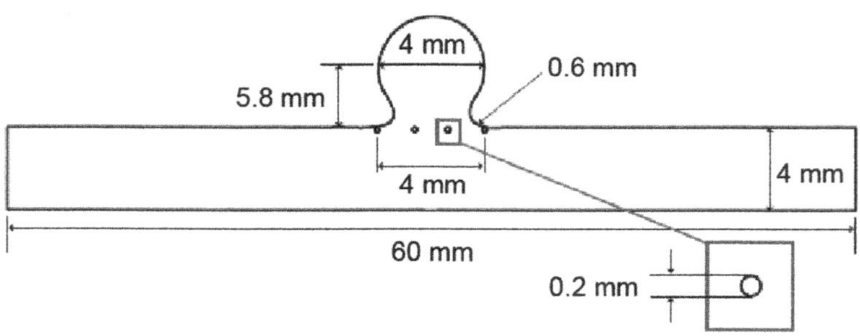

Fig. 7.23 Dimensions of aneurysm model

Fig. 7.24 Computational mesh of aneurysm model

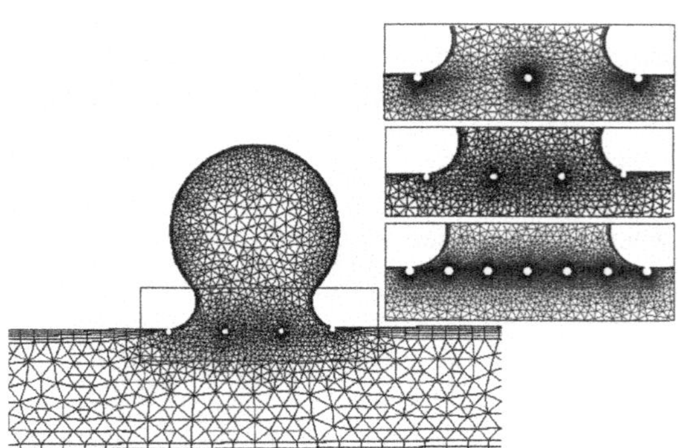

Table 7.7 Fluid properties used in numerical simulation

Parameters	Value	Units
Dynamic viscosity	3.5×10^{-3}	Pa
Density	1050	kgm^{-3}
Entrance velocity	0.04377–8.088	cms^{-1}
Exit velocity	0–8.33	cms^{-1}
Pressure	0–100	Pa

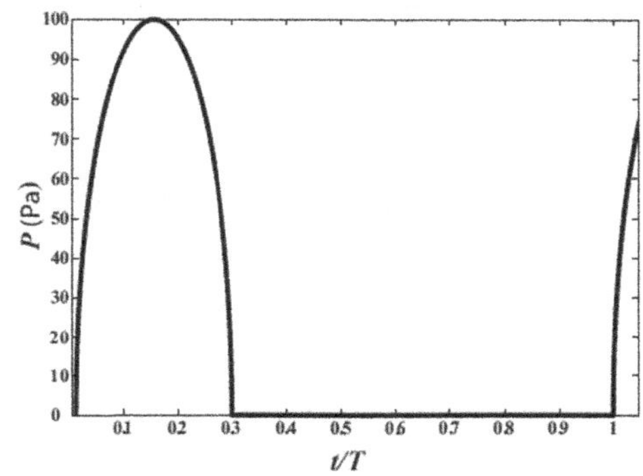

Fig. 7.25 Plot of pressure versus time

ues for Reynolds number in the cerebral artery has a range of 100–300, and is less than 30 within the aneurysm (Kim et al. 2010b). To model the unsteady nature of the flow field, transient CFD using ANSYS CFX was performed. A rigid wall condition is assumed.

Flow in the vessel is simulated from the early-systole state to end diastolic state of the cardiac cycle T based on an idealistic physiological waveform (refer to Fig. 7.25). The model is simulated for one cardiac cycle T from 0 to 1. The pressure starts from zero, and the peak pressure and flow rate occurs at $t/T = 0.15$. The pressure then reduces to zero again at $t/T = 0.3$ and then continues at zero pressure till $t/T = 1$. This pressure variation is repeated in the subsequent cycles.

7.4.4 Aneurysmal Flow Results

Streamline tracing is used to deduce the path of the blood flow. A clearly defined large-scale vortex is present in the aneurysm sac as shown in Fig. 7.26. The results show that an aneurysmal vortex reduces in strength after stenting at time steps $t/T = 0.2$, 0.5 and 0.6. Figure 7.27 presents the counter-clockwise vortices in the aneurysm sac using vorticity and velocity vector fields. In the non-stented artery, a large-scale vortex exists in the aneurysm sac. When stented, the vortex decreases as demonstrated by its lower vorticity values. The pressure gradient ΔP and shear strain rate $\dot{\gamma}$ is characterised by using a monitor point at the centre of the aneurysmal sac (indicated by a crosshair in the flow field plots). The pressure gradient and shear strain rate fields can give a quantifiable indication of the vortex differ-

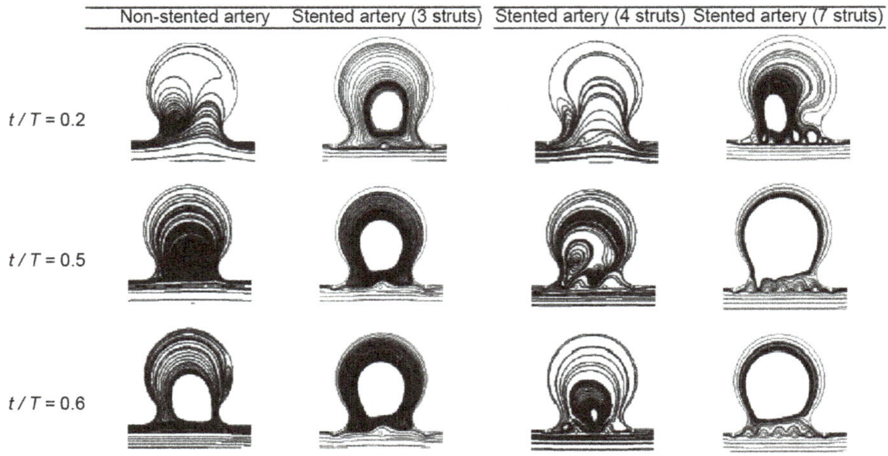

Fig. 7.26 Streamline tracing of aneurysmal flow

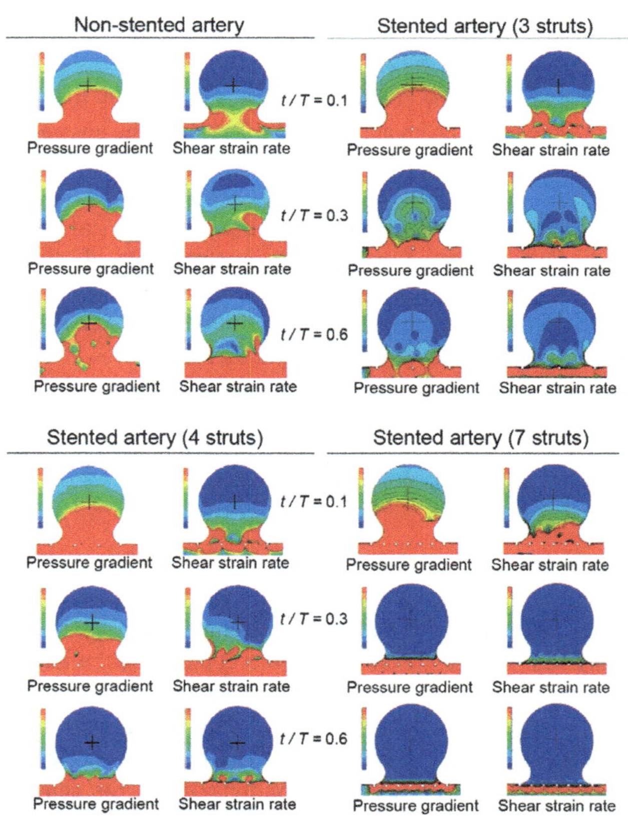

Fig. 7.27 Aneurymal flow analysis based on non-stented and stented arteries

ences for the untreated and stented aneurysmal artery based on time frame indices $t/T=0.1$, 0.3, 0.6 and 1.0.

7.4.5 Parametric Study for Design of Stent in Aneurysm

Pressure gradient and shear strain rate maps were analysed for the start and end of systole as well as the diastolic phase of one cardiac cycle which correspond to t/T=0.1, 0.3 and 0.6 respectively. At the monitor point in the centre of the aneurysmal bulge, these fluid mechanical values were extracted for comparison. Pressure gradients were almost similar at the start of the time cycle ($t/T=0.1$) at $\Delta P=31.61$ kgm^{-2}s^{-2} for the non-stented flow and $\Delta P=31.61$, 37.10, and 39.70 kgm^{-2}s^{-2} for stented aneurysmal flows with 3, 4 and 7 struts present at the aneurysmal neck. However, ΔP decreases with time. After stenting with 3 struts, ΔP decreases. With stenting using 4 struts, ΔP decreases further. Based on stenting with 7 struts, the pressure gradient drops steeply to 0.392 kgm^{-2}s^{-2} at $t/T=0.3$, $\Delta P=0.0346$ kgm^{-2}s^{-2} at $t/T=0.6$, and $\Delta P=0.00843$ kgm^{-2}s^{-2} at $t/T=1.0$, which shows that stent is over-designed such that the drop in pressure gradient becomes too excessive.

The same variation exists for the shear strain rate which is almost the same for the non-stented case as the stented case initially at $t/T=0.1$ (with $\dot{\gamma}=1.53$, 2.11 s^{-1}, 1.72 s^{-1} and 2.14 s^{-1} (for untreated and stented cases with 3, 4 and 7 struts respectively). However, g$^{\cdot}$ decreases to 3.073 (for 3 struts), 1.66 (for 4 struts) and 0.359 s^{-1} (for 7 struts) at $t/T=0.3$, and to 1.69 (for 3 struts), 1.06 (for 4 struts) and 0.0579 s^{-1} (for 7 struts) at $t/T=0.6$, which shows gentle reduction compared to the pressure gradient. After stenting with 7 struts, $\dot{\gamma}$ is reduced drastically. This denotes too much blockage of flow into the aneurysm such that the shear strain decreases excessively.

Based on the numerical simulation and flow analysis, flow fields on untreated and stented aneurysmal arteries were characterized. The streamline tracings enable the visualization of a large-scale vortex in an aneurysmal sac. The results demonstrate that the stenting causes a reduction of pressure, velocity, vorticity shear rate. Reduced pressure exerted by blood on the aneurysmal sac will decrease the risk of rupture. However, lower volume of flow into the sac increases the viscosity of blood in the aneurysm (Kim et al. 2010b). Reduced vorticity in the sac also corresponds to a lower fluid shear stress and shear strain rate. Note that high shear stress is necessary for preventing platelet-dependent thrombosis (Sukavaneshvar et al. 2000a). Moreover, reduced blood into the aneurysm also means flow stagnancy and the induction of thrombosis increases. All these undesirable conditions will aggravate aneurysm rupture (Liou and Liou 2004a). Therefore, the type of stents deployed has to be of sufficient porosity to minimise aneurysmal rupture (Kim et al. 2010b) but prevent platelet aggregation. The study of fluid mechanical properties in non-stented and stented aneurysmal flow can enable medical experts to evaluate the effectiveness of stent designs and their corresponding porosities in prevention of aneurysm dilation leading to rupture.

7.4.6 In-Vitro Flow Measurement of an Aneurysm, Based on PIV

Fluorescent stereo PIV can be used to perform flow measurement of the aneurysm for flow comparison with the CFD simulation (Takahashi et al. 2011). Fig. 7.28, shows a DS20 Laser with a 1 kHz double-pulse, 527 nm wavelength at 5 mJ/pulse and thickness of 80 ~ 100 μm used. The fluid medium is a glycerol and water mixture at 3:2 ratio (by weight). This results in a density of 1.16 kg/m^3, viscosity of 5.3 cP at 37°, and refractive index of 1.409 ± 0.002 (532 nm at 37°). The fluorescent particle, FLUOSTAR ®, which has a density of 1.1 kg/m^3 and diameter of 13 μm, was added as motion tracking particles at a concentration of 1.1 g/L. The measurement used an interrogation size of 16 × 16 pixel^2, with spatial resolution of 110 μm, and laser light of 80 ~ 100 μm thickness, and scanning plane of 80, phase-locked time-binning of 30 ms. Measurements were taken at peak time phase resulting in Re = 450.

The residual neck size, and vessel channel geometry affects the flow field inside the aneurysm, and consequently influences success of the aneurysm treatment. Ideally the aim is to achieve reduced inflow and intra-aneurysmal velocities (Byun and Rhee 2003). Stereo PIV allows 3D measurement and characterization of abnormal flows such as an impinging jet in a near-wall region of the aneurysm (Fig. 7.29). By correlating with pathophysiology, early prevention of aneurysmal rupture using the stenting procedure can be implemented.

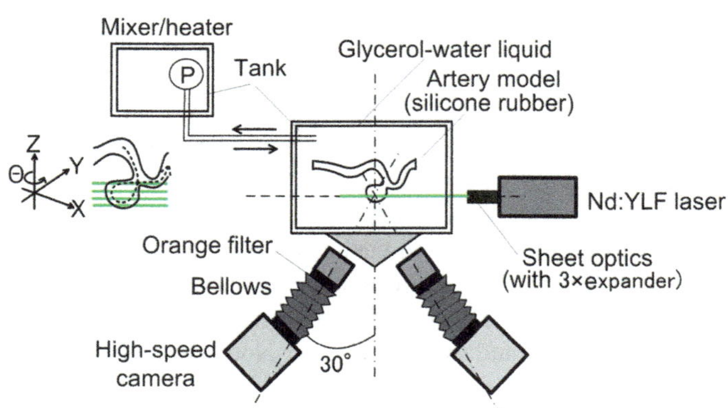

Fig. 7.28 High-magnification index matching stereo PIV setup. In-vitro experimental setup is achieved by immersing a transparent phantom model of the aneurysm into a glycerol-water mixture, which serves as the blood analogue. This setup can enable the measurement of velocity flow field at multiple cross-sections through the aneurysm

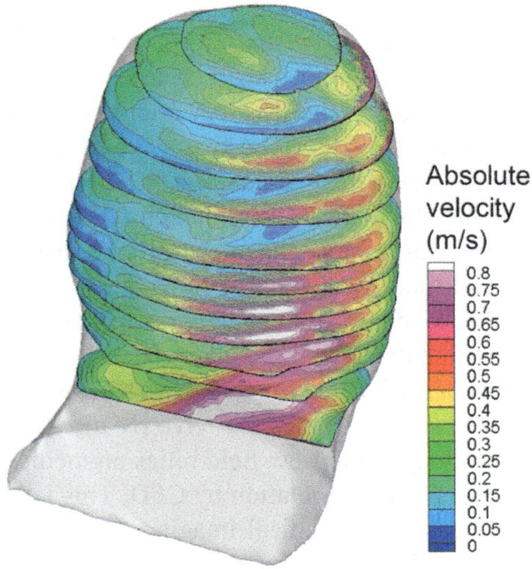

Fig. 7.29 Flow measurement and visualisation in an aneurysm. Counter flow and impinging jet is shown to act on the aneurysm. Impingement of jet onto an aneurysm may result in pre-ruptured cases, and therefore is of importance in investigation. Early detection of such undesirable case can be achieved and stenting can be implemented in advance well before rupture occurs

7.4.7 Closure

In this case study, simulation of stenting inside a diseased artery at the entrance of the aneurysm was performed. The reason for stent insertion inside the aneurysm is to exclude it from systemic blood circulation, so that the aneurysm will shrink gradually (due to reduced pressure) and eventually become smaller. Although this may induce stagnation and possible thrombosis formation after stenting the chance of rupture is low, as long as the aneurysm does not grow. Aneurysms can form on both sides of the artery wall and so for future work, simulations of different types of diseased arterial geometries will be more useful for understanding of the effect of stent treatment. Three dimensional stent models for patient-specific aneurysm would also enhance the clinical value this case study.

7.5 Analysis of Blood Flow in Cardiac Chamber

7.5.1 Introduction to Heart Chamber Flow Visualisation

Computer simulations particularly advantageous for situations where experimental measurements require further detail, and/or when there is a need to prepare a predictive platform. Another potential scenario is the simulation of mechanical structures that are implanted in the human heart.

This case study uses the right atrium of the heart as an illustration of flow quantification and visualization techniques. Specifically, we focus on vortices that are generated within a heart chamber. Cardiac flow analysis based on vorticity mapping can define the flow behaviour within the heart chamber (Wong et al. 2009a; Wong et al. 2010c). The first stage of flow analysis is to generate the blood flow field in the cardiac structure based on phase contrast magnetic resonance imaging and then by numerical simulation. The second stage is to use the flow field for vorticity measurement and visualization. Then, the final stage is to quantify the vorticity field statistically and examine the changes in vorticity over time.

Phase contrast magnetic resonance imaging (MRI) belongs to a class of experimental flow imaging by means of quantifying the phase shifts in proton spins of the imaged fluid, and has been widely used to measure flow fields of blood in the cardiac structures in-vivo (Markl et al. 2007). Another approach to generate a haemodynamic flow field relies on medical imaging, anatomical reconstruction, and numerical simulation by CFD. This involves reconstructing the cardiac vessel geometry from medical imaging whereby blood is transported. The next step is to mesh the flow volume and establish the boundary flow conditions, which can be determined by medical imaging measurement, such as Doppler ultrasound or phase contrast magnetic resonance image data (Marshall et al. 2004b).

Validation of the CFD simulation in human proximal airways with hyperpolarized 3He magnetic resonance phase-contrast velocimetry in-vivo measurements have shown good agreement (de Rochefort et al. 2007). The flow imaging results of the cardiovascular structure in-vivo is often taken as benchmark data since blood velocities are intrinsically measured with good accuracy. The simulation is typically used as a predictive tool for further detailed analysis that are difficult to measure. Velocity measurement and mapping systems by phase-contrast-MRI are often used to validate computational and predictive frameworks with good reliability. The use of phase-contrast-MRI and CFD can quantify the vortical flow for the cardiac chamber and may open up new insights into the flow behavior within this anatomy.

7.5.2 Application of Medical Imaging in Computational Heart Modelling

A realistic right atrium model is reconstructed based on medical imaging (Merrifield et al. 2004a; Saber et al. 2001). Segmentation of the myocardium was performed through a series of time-resolved image slices. Imaging modalities for assessment of cardiac chambers or myocardial function include computed tomography and MRI, but MRI is superior since the tissue and blood regions are well-distinguished in terms of boundary image intensity. The geometry is meshed and a CFD simulation is performed.

7.5.2.1 Region of Interest for Cardiac Flow Analysis [illegible]

Examination of the heart where the atria are positioned [illegible] the optimal cross-sectional area of the right atrium. A sl[illegible] dicular to the axis joining the top of the heart to the apex [illegible] 7.30). The velocity mapping on the right atrium of the heart is used since the scan sections cut through the middle portion of atria.

Inflow from the inferior vena cava (IVC) and superior vena cava (SVC) result in respective flow diversions that generate two vortices in opposite directions. During atrial diastole, there is maximum suction of blood into the right atrium via the vena cava. The chamber dilates due to the inflow of blood and the closure of the tricuspid valves (TV). The maximum inlet flow at the superior and inferior vena cava is set to 155 and 128 cms^{-1}. At this instance, two strong vortices of opposite rotation are developed. During the systolic phases, the atrium compresses and causes the outflow of blood into the right ventricle which causes the valves to open. One of the vortices becomes significantly weaker than the dominant vortex during their evolution over a cardiac cycle.

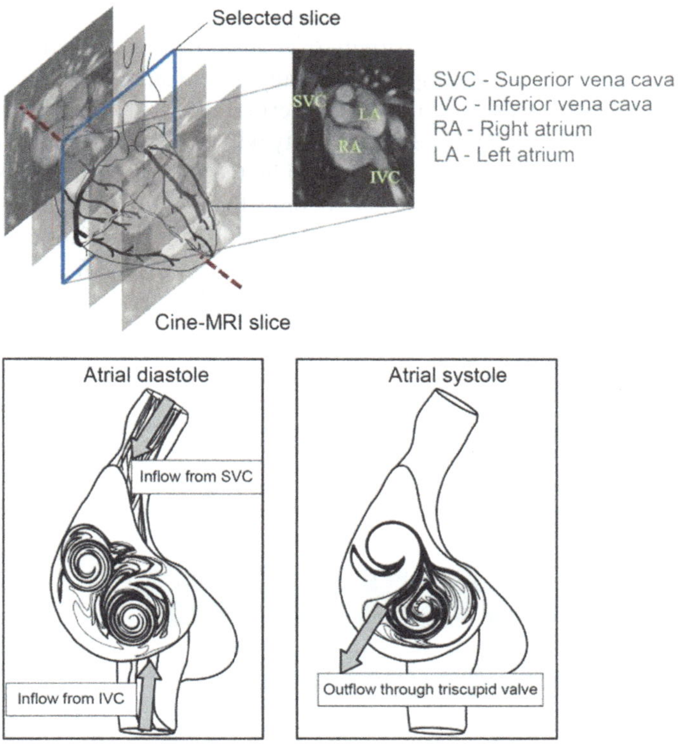

Fig. 7.30 Sectional view of blood flow in the right atrium

7.5.2.2 In-Vivo MRI Flow Measurement

Phase-contrast-MRI was applied onto the short axis orientation through the atria of the heart (Wong et al. 2009a). Scan acquisition of 25 phases or time frames (based on their indices from $n_t = 1$ to 25) was completed based on retrospective gating. The human subject volunteer was a normal male of age 22 years at the time of scanning. The imaging used a two-chamber view of the right and left atria. The main objective was to capture information on the large-scale vortices in the right atrium. These vortices serve as principle features of the flow structures to analyze. The acquisition parameters include: echo time TE = 1.6 ms, repetition time TR = 47.1 ms, field of view FOV = (298 × 340) mm^2 at a (134 × 256) pixel matrix, and a velocity encoding (VENC) of 150 cms^{-1}. The in-plane resolution of 1.54 mm/pixel was determined by the pixel spacing and the through-plane resolution based on a slice interval of 6 mm.

Figure 7.31 describes phase contrast in MRI based on spin phase shifts that are proportional to the blood velocity (Markl et al. 2003a; Yu et al. 2003). Phase contrast MRI signals can be represented using intensity images. The intensity of each pixel corresponds to the blood velocity at the measured location. To quantify a velocity in one spatial dimension, at least two phase images must be taken for subtraction of flow-induced phase shift. These are taken from background phases caused by susceptibility-induced inhomogeneities and coil sensitivity changes (Baltes et al. 2005). The flow sensitivity of a phase contrast signal acquisition is given the term *velocity encoding* and is usually abbreviated as VENC (Svensson 2003). The VENC limits are defined by a set criterion which affects flow measurements (Lagerstrand et al. 2006). Blood velocity will be aliased to an artificially low value if it exceeds the maximum VENC level by the flow sensitisation gradients.

The phase contrast MRI technology is clinically attractive because it is able to provide quantitative information on blood flow without the need for contrast agent to be introduced into the human body. Moreover, ECG-gated phase contrast MRI can provide cine-images for temporal velocity mapping (Thompson and McVeigh

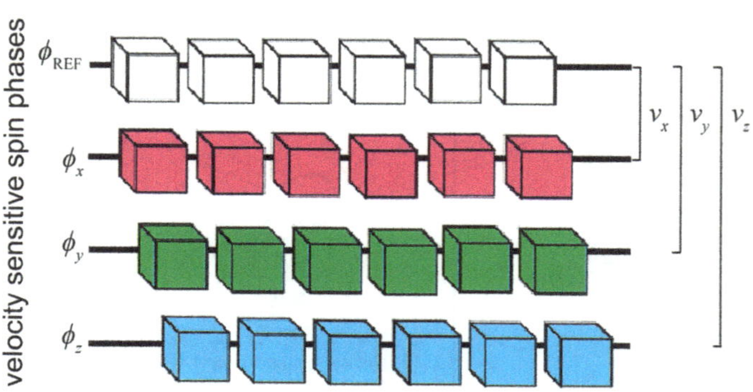

Fig. 7.31 Phase contrast MRI velocimetry. Phase contrast MRI works on the concept that hydrogen nuclei from blood that has been exposed to magnetic fields accumulate a phase shift in spin that is proportional to the blood velocity in the x, y, and z directions. Velocities v_x, v_y and v_z bare functions of the subtractions of spin phases φ_x, φ_y and φ_z bof measured volumes with that of the reference phase φ_{Ref}

2004). The phase contrast images are graphical representations of the velocity components (x- and y- directions) maps.

7.5.2.3 CFD Modelling

Creating a CFD mesh of the right atrium chamber and its related vessels are depicted in Fig. 7.32. Here, 3D surfaces were constructed through segmented slices of the heart chamber. After processing of the geometry, the final output is a tetrahedral mesh with refinements in the near wall consisted of approximately 800,000 cells. The blood flow through the right atrium is assumed incompressible, homogeneous, laminar, and Newtonian. The fluid properties are reported in Table 7.8. Due to the unsteady nature of the flow field, a transient CFD simulation was performed. The range of Reynolds numbers in the atrial flow throughout one cardiac cycle is dominantly laminar while it may be argued that turbulence can be induced through the curvatures within the geometry. The use of a turbulence model therefore does not necessarily imply greater accuracy in capturing the flow field. Instead the incorrect use of a turbulence model, especially models based on fully developed turbulent flows, will produce an overly diffusive flow because of the additional turbulent viscosity μ_t term. At best, low-Reynolds number turbulence models such as the *k-ε* and *k-ω* provide improvements over their standard versions. However, in order to

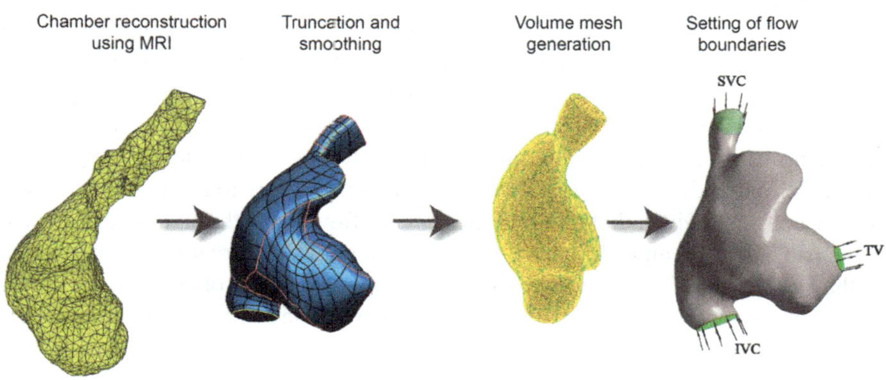

Fig. 7.32 Flow chart of CFD modelling for the right atrium

Table 7.8 Fluid properties used in the CFD simulations

Parameter	Value	Units
Dynamic Viscosity	0.004	Pascal
Density	1176	kgm^{-3}
Range of IVC (Inflow) Velocities	−1.2200 to 40.1450	cms^{-1}
Range of SVC (Inflow) Velocities	−0.5120 to 20.3040	cms^{-1}
Reynolds Number	518–2738	

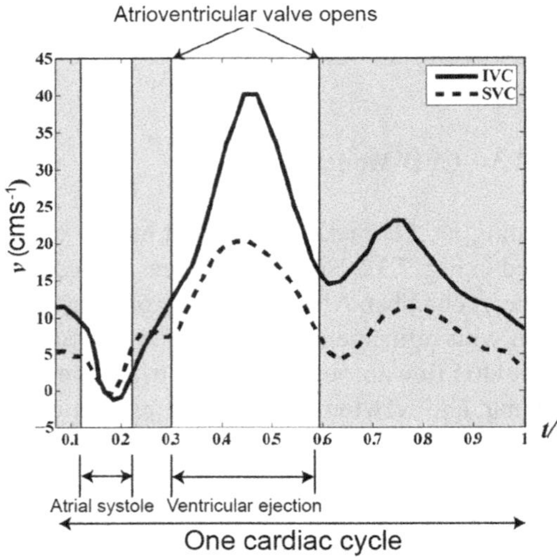

Fig. 7.33 Plot of venae cavae average flow velocities versus time

fully resolve any transitional behaviour, the sophisticated LES and Direct Numerical Simulations models need to be used since they are based on the Navier-Stokes equations. In this case study, a laminar flow model.

Figure 7.32 shows flow entering the right atrium chamber through the inferior-and- superior vena cava (IVC and SVC), and exits through the tricuspid valve (TV). For simplicity, the velocity boundary condition that was applied at the IVC and SVC assumes a uniform spatial profile and has a direction that is normal to the surface. A parabolic profile may not be the correct assumption given that the upstream flow may be highly skewed at high velocities. In addition, different degrees of skewness exist for different velocities. To avoid these complications, a uniformly spaced profile is used. For the TV outlet, a transient pressure profile with zero gradient in the spatial domain is implemented at a virtual outflow to ensure that the flow will be constant. Given that the IVC and SVC in-flows vary throughout the cardiac cycle (see Fig. 7.33 and Table 7.8), the inlet boundaries require a-priori knowledge of the IVC and SVC velocity time histories. Logically, this is obtained from physical experiments (Wexler et al. 1968). For this study, in-vivo measurements were used to generate the time histories.

Flow in the right atrium chamber is simulated from the end-diastolic state to the early-systole state, i.e., corresponding to the normalised time segment of 0.60–1.00 in one cardiac cycle (refer to Fig. 7.33). To avoid start-up effects from unsteady flows, the flow field was simulated over three periodic cycles where the final flow field predictions were taken at the third cycle. Spatial discretisation of the advection terms in the Navier-Stokes equations was performed by using a high resolution scheme, while the transient scheme used a second order backward Euler scheme with a time step size of 0.002 s.

While the MRI scans were taken over a complete cardiac cycle of the heart chamber, the scope of the numerical simulations was limited to examining the flow in the right atrium during the diastole and systole phases. In particular, focusing on particular phases of the cardiac cycle allows the flow analysis to be carried out in the simpler context of an internal flow in a rigid vessel. This is due to the fact that modelling the entire cardiac cycle requires a fluid-structure interaction (FSI) approach and the physical structure of the heart chamber is a dynamic parameter, making complicated to simulate the entire cardiac cycle. The rigid boundary wall assumption instead of elastic ones based on the FSI approach allows the simulation to converge and hence produce a set of predicted results. However this comes at the expense of real physiological cyclic movement of the heart contracting and expanding. In reality, a contraction in the heart will reduce the volume and any cross-sectional area where the blood flows through. Considering mass conservation, the contraction is expected to intensify the velocity and vorticity fields. However, if we implement FSI for this simulation, the blood pressure during atrial systole will stretch the vessel elastically causing the velocity and vorticity fields to attenuate.

7.5.3 Comparison of CFD and PC-MRI Vorticity Fields

Flow results by numerical simulation and phase contrast magnetic resonance imaging are presented in this section. Then, their velocity and vorticity fields are compared in terms of flow pattern development. Figure 7.34 shows three snapshots of the CFD flow visualisation in the right atrium solution domain corresponding to the time instants of $t/T=0.66$, 0.74 and 0.82 per time cycle of 1.0. These were selected to match the MRI time frame indices $n_t=9$, 11 and 13 out of a total of 25 frames. Collectively, they form the basis for comparison between the CFD simulations and phase-contrast-MRI experiments.

The velocity vectors in the CFD plots are superimposed on the vorticity contours. The velocity vectors are used to depict flow direction on the plane itself, thereby implying a tangential projection. A normalized scale is used for the vector length. For the vorticity contours, the clockwise/anti-clockwise direction of the rotation is provided by using the y-axis as a reference axis, while the figure shows the in-plane view of the selected slice. The normal to the slice is almost lying parallel to the y-axis and the plane is chosen is based on the MRI plane.

Cardiac flow analysis that is based on histograms of vorticity field distributions for both modes of flow generation can give a clear overview of their flow differences with some form of quantification (Wong et al. 2009a, b, 2010c).

7.5.4 Computational Haemodynamics Analysis of Heart Chamber

Based on the distribution of CFD and phase-contrast-MRI vorticity fields, the difference in their means and standard deviations are in the order of zero (Difference in means: $\Delta\omega \leq 2.02\ \text{s}^{-1}$, and difference in standard deviations: $\Delta\sigma \leq 4.07\ \text{s}^{-1}$ for a range of vorticity values at approximately -25 to $+25\ \text{s}^{-1}$). The agreement in the CFD simulations and MRI experiments reinforces the evidence

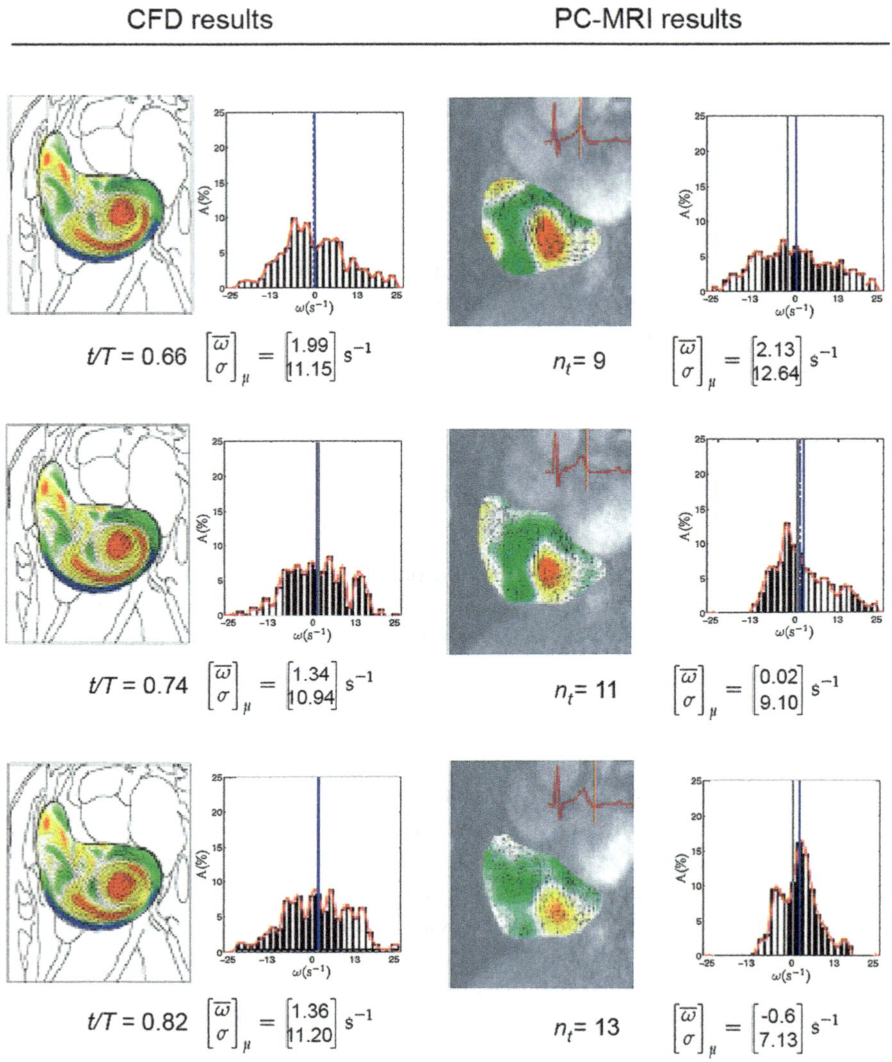

Fig. 7.34 Velocity and vorticity fields based on computational (CFD) and experimental (PC-MRI) results

of swirling flow in the right atrium (RA). Although we have verified phase-contrast-MRI flow mapping with CFD, further comparison of the two flow modes can be performed based on variation of vorticity interrogation size and resolution of the flow fields to understand the effects of flow measurement and visualisation reliability (Wong et al. 2010b).

From a fluid mechanics perspective, blood swirling is caused by the interaction between the two paths of flow from the inlets (IVC and SVC). In Fig. 7.30, the generation and suppression of vortex structures in the right atrium flow is dynamically varying throughout the main cycle. Given that only one dominant vortex has been visualised for the time segment of the cardiac cycle, it appears that the presence of the weaker, clockwise vortex can be attributed to the peaking of IVC in-flow during atrial diastole. The potential extensive datasets from the CFD simulations and phase contrast MRI can allow further insight into the vortex structures present in the right atrium flow.

The Reynolds numbers are typically lower towards end-systole and early diastole due to the slower motion of blood. The blood flow is transitionally turbulent during the onset of systole. The blood inflow velocities at the superior and inferior vena cava tends to increase during the period of systole and causes a collision within the atrium, which generates a turbulent flow that comprise of two large-scale vortices. The range of Reynolds numbers in the atrial flow throughout one cardiac cycle is both laminar and turbulent. The simulation results are largely in agreement with the MRI results. In terms of the flow features, the velocity vectors implied that the flow is moving primarily in the anti-clockwise direction. Partly as a consequence, there is a dominant vortex core-like structure that remained throughout the cardiac cycle. That this dominant vortex showed a positional dependence on the TV outlet.

In terms of clinical applications, medical imaging modalities such as multislice CT and MRI have been increasingly used to diagnose cardiovascular disease owing to rapid technical developments. Despite the improvements in CT, its application in cardiovascular disease is limited for evaluating the morphological changes of cardiac chambers. However, functional MR imaging, especially phase-contrast-MRI allows for both anatomic and functional assessment of cardiovascular disease, and demonstrates huge potential in the clinical practice. Analysis of haemodynamic flow through the heart will assist clinicians to better understand the formation of thrombosis, and the effect of flow on arterial branches arising from the heart, which helps to avoid atherogenicity (Friedman et al. 1993). Therefore, future studies involving CFD analysis of bilateral heart chambers that include both venous inflow and arterial outflow will be clinically meaningful.

7.5.5 Closure

In this case study, the flow pattern development in a chamber of the heart based on phase contrast magnetic resonance imaging and numerical simulation was compared. For the computational study, a static geometry was assumed and dynamic flow boundary conditions in a series of surface geometries pertaining to different times of a cardiac cycle was applied. Although the fluid structure interaction of the blood-wall region is neglected, the generated flow fields did not deviate significantly from the true physiological flow. Experimental study is based on in-vivo magnetic resonance-based flow imaging.

The flow visualization tools using streamline tracing and vorticity field enabled the presentation of our results effectively. Vortices exist in three spatial dimensions, but we have simplified our analysis to present two dimensional results for clarity in flow evaluation. Nevertheless, the visualization of flow patterns remains concise for the two modes of flow field generation.

The comparison of flow results by experimental and computational approaches showed good qualitative and quantitative agreements. The simulated results were validated using the measured flow from velocity-encoded medical imaging successfully. The proposed functional analysis using combined phase contrast MRI and CFD is of interest to research into the pathologies or interventions which involve altered relations between form, flow, mobility, and timing of the heart.

7.6 Intra-Atrial Flow and Mitral Plane Velocity Profile[2]

The left atrium is the upper left chamber of the heart and connected to the left ventricle via the mitral valve. Intraventricular flow has been widely investigated during the last decades both in vivo, in vitro and by computational fluid dynamics (CFD). In this section we present a three-dimensional computational fluid dynamics framework of the left atrium and its pulmonary veins. MRI is used to render subject-specific atrial and venous geometries to investigate: the diastolic flow field in an anatomically representative model of the left atrium and the pulmonary veins; the impact of different venous entry locations on the intra-atrial flow and; the resulting mitral plane velocity distribution. Three 3D models with different venous entry locations were created.

7.6.1 Left Atrium Models

3D geometries were obtained from MRI recordings of a young healthy adult. This included the left atrium and each of the pulmonary vessel stubs were acquired from a 25-year-old healthy male on a Siemens Avanto 1,5 T system. A temporal resolution of 26 ms was achieved, resulting in 40 frames during one cardiac cycle. Images were acquired in the four-chambers (Fig. 7.35) and the short-axis orientation. A total of 14 slices with a slice thickness of 5 mm were needed in both orientations to cover the atrium and the mitral plane. In-vivo measurements of the flow velocity through the mitral orifice and the mass flow rate through each of the venous inlets were performed by velocity phase mapping scans.

The segmentation data was used to build a subject-specific atrial geometry. The first model, denoted C_A has its four pulmonary veins located in the anatomically correct positions adopted from the magnetic resonance recordings. The geometry of C_A is shown in Fig. 7.35a, b. There is close proximity between the ostia of the right pulmonary veins and between the ostia of the left pulmonary. In general, and

[2] This section has been contributed to by **S. K. Dahl**, in Dahl, S., Thomassen, E., Hellevik, L., Skallerud, B. (2012a). Impact of Pulmonary Venous Locations on the Intra-Atrial Flow and the Mitral Valve Plane Velocity Profile. Cardiovascular Engineering and Technology 3:269–281.

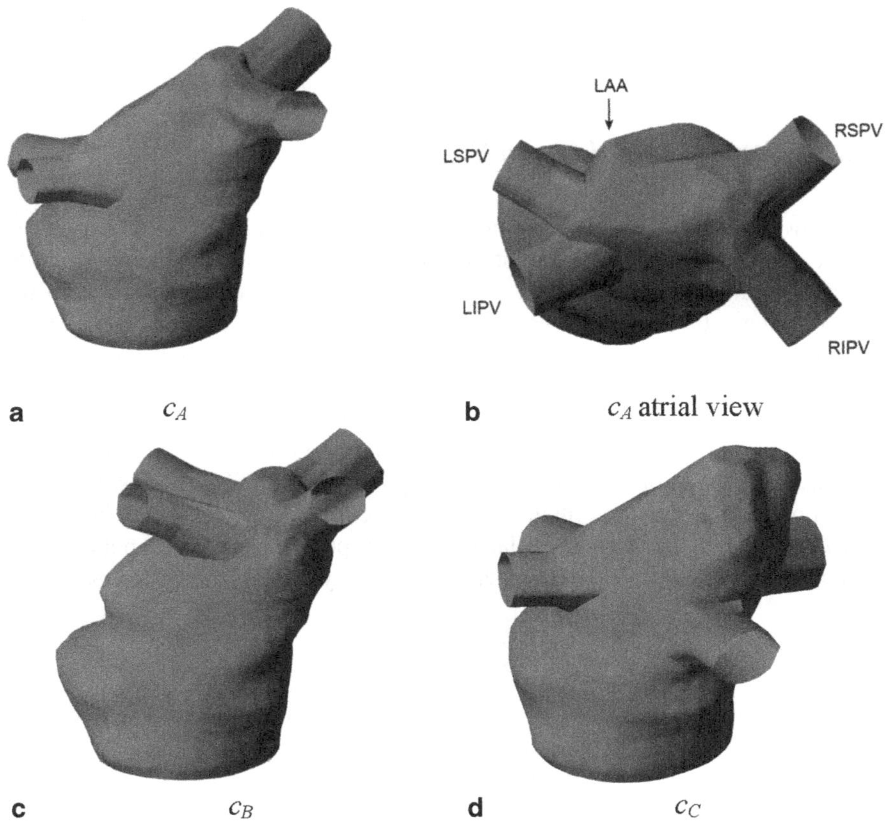

Fig. 7.35 Geometry of model-C_A from **a** an inferior, and **b** atrial view, whereas model-C_B and model-C_C are shown in **c** and **d**, respectively. The name of the four veins and the location of the LAA are indicated in **b**. *LSPV* left superior pulmonary vein; *RSPV* right superior pulmonary vein; *LIPV* left inferior pulmonary vein; *RIPV* right inferior pulmonary vein

also in our subject, the entry locations of the left pulmonary veins are closer to the mitral plane than the right pulmonary veins. The left atrium appendage (LAA) lies adjacent to the ostium of the left superior pulmonary vein and is indicated with an arrow in Fig. 7.35b.

To examine the impact of the venous entry locations, two additional models were constructed; C_B and C_C. There are several possible entry locations for the pulmonary veins. We chose to keep the angle fixed between the pulmonary vein trunks, as seen from an atrial view, while the pulmonary veins vertical distance to the mitral valve plane were modified. The geometries of the atrial chambers were identical with C_A. In C_B the left pulmonary veins are moved up to the same level as the right pulmonary veins, the isthmus line is then in accordance with the maximum distance measured in Schmidt et al. (2006). We assume the interpatient variability also applies for the right pulmonary valves, and thus in C_C the RPVs were moved down and located at the same vertical position as the left pulmonary valves in C_A.

7.6.2 Computational Model Setup

The flow simulations were performed using the finite volume method in Ansys Fluent v13.0 (Ansys Inc.). The diastolic duration was measured at 0.665 s and the simulation time step size was set to 0.5 ms. The models were meshed with tetrahedral elements with an element size determined from grid convergence tests. Laminar flow was assumed and the blood was modelled as an incompressible, Newtonian, homogeneous fluid, with a density of $\rho = 1050$ kg/m^3 and a viscosity of 3.5×10^{-3} kg/ms, which is a reasonably good approximation for blood flow in large cavities (Ku 1997). The mass flow rate for each of the four pulmonary vessels was measured using MR phase mapping scans reported from in vivo studies. For every vein, a power function was fitted to the diastolic mass flow rate as can be seen in Fig. 7.36. The flow pattern is similar for the four pulmonary vessels, but the mass flow rate differs, where the left superior pulmonary vessel has the lowest inflow rate.

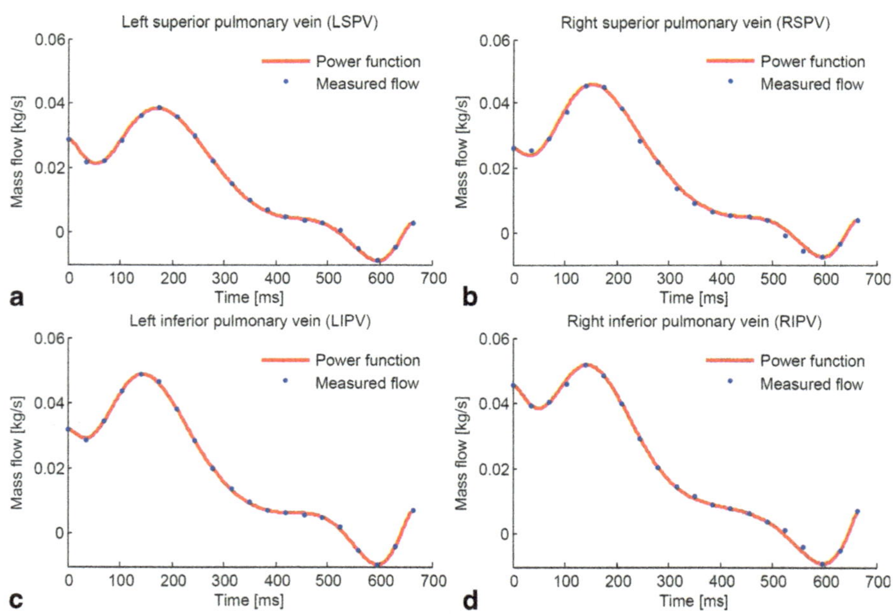

Fig. 7.36 The power function (*solid line*) was fitted to the mass flow rate measured by MR velocity phase mapping (*dotted line*). The graphs represent the flow through the LSPV, RSPV, LIPV and RIPV, respectively

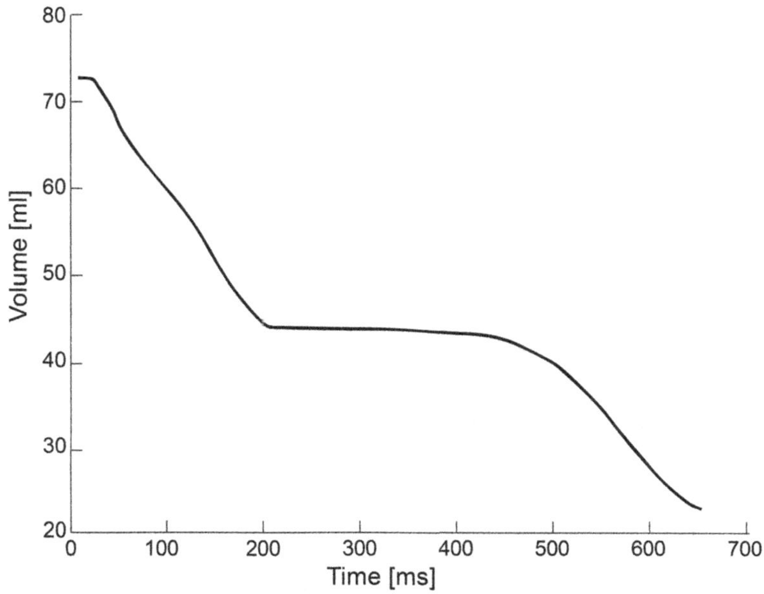

Fig. 7.37 Volume change of the atrial cavity measured from MR recordings throughout diastole

The atrium was kept fixed during the simulation. However, the measured venous mass flow rates do not account for the extra mass flow exiting the atrium due to atrial contraction throughout diastole. The additional flow due to atrial volume change was calculated from the segmented atrial geometry and evenly distributed among the four venous flow inlets. The change in atrial volume during diastole is illustrated in Fig. 7.37. A no-slip condition was imposed at the walls.

7.6.3 Results

7.6.3.1 Intra-Atrial Flow Pattern in Model-C_A

Figure 7.38 shows flow entering the right atrium chamber through streamlines for each pulmonary vessel and the velocity distribution at the mitral orifice at t = 100 ms, relative to the beginning of diastole. The flow field is presented by streamlines to provide an instantaneous image of the flow field by using information from a single time frame. For the (a) right inferior pulmonary vein the streamlines enters the left atrium near its septal side and the streamlines follow the smooth contour of the atrial wall towards the septal/inferior side of the mitral plane. The jet is deflected by the second jet from the left inferior pulmonary vein, but these two jets do not interfere at this stage. The streamlines originating in the (b) right superior pulmonary vein behave in a similar manner but enter the mitral valve at the septal/superior side. The outer, most superior, part of the right superior pulmonary vein inflow, is disturbed

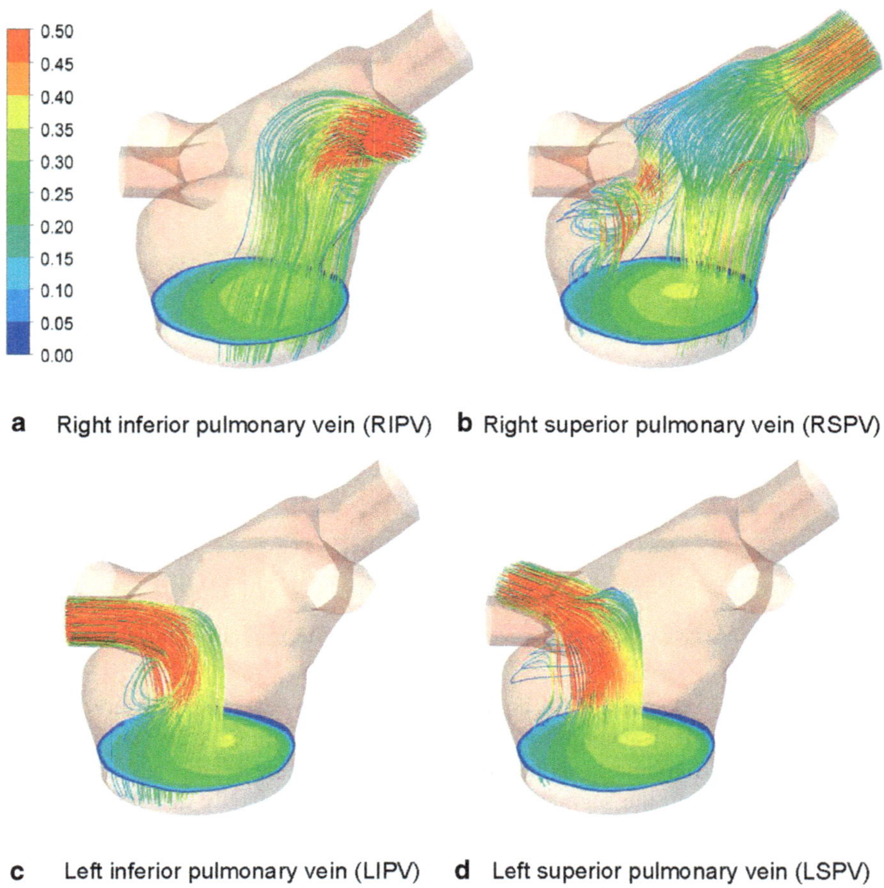

Fig. 7.38 Model-C_A: Velocity contour at the level of the mitral valve and the streamline for each of the four veins during the acceleration of the first filling wave, $t = 100$ ms. The velocity scale is given in [m/s]

by the jet from the lower superior pulmonary vein and turns into a rotational motion towards and into the left atrium appendage before crossing the mitral plane at its lateral side just below the appendix.

In this model the trunks of the left pulmonary veins are oriented almost perpendicular to the atrial chamber. The directions of the inflowing jets, Fig. 7.38c and d are almost parallel to the mitral plane. The flow from the left veins have the highest velocities due to their small orifice areas. As they flow into the atrial chamber, the jets are deflected by the flow descending from the right pulmonary veins. The jets change direction with almost 90 degrees before heading towards the mitral plane. The main part of the flow from the left superior pulmonary vein enters the central/

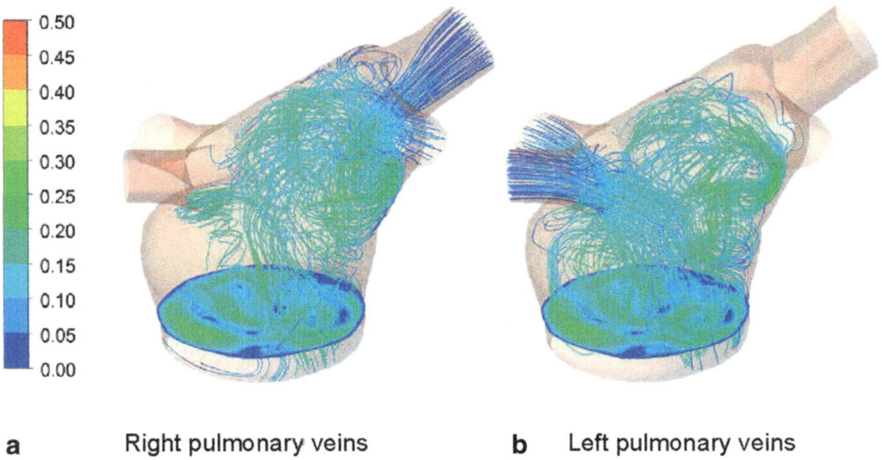

Fig. 7.39 Model-C_A: Velocity velocity contours at the level of the mitral valve with a more complex flow field of mid-diastole compared to early diastole. In **a** and **b** the streamlines of the *right* and *left* PVs are illustrated, respectively. The velocity is given in [*m/s*]

superior part of the mitral orifice, whereas the rest is involved in the swirling motion caused by the interaction with the right superior pulmonary vein jet. The left inferior pulmonary vein's inflow, on the other hand, occupies the centre towards the inferior side of the mitral valve. Some of the streamlines originating from the left inferior pulmonary valve recirculate into a clockwise vortex before entering the valve.

During deceleration the degree of recirculation increases. As the transmitral flow diminishes in mid-diastole and the atrial volume stays constant, the vortices increase in size and number, resulting in an even more complex flow field as illustrated in Fig. 7.39. At the onset of the second filling, the vortices disappear and the main part of the venous inflows is again constrained into more direct paths towards the mitral valve.

7.6.3.2 Intra-Atrial Flow Pattern in Model-C_A

Figure 7.40 illustrates the maximum transmitral velocity for C_A, C_B and C_C during diastole. During the acceleration of the first filling the maximum velocity is the same for the three cases. When approaching peak velocity, the results start to deviate. The maximum is reached at t = 160 ms for C_A with a velocity of 0.50 *m/s*. In C_B and C_C, the maximum velocities are 16 and 33 % higher than in C_A and occur 45 and 30 ms later, respectively. The variation in maximum velocity decreases in mid-diastole, but increase again at atrial contraction. This time the highest velocity occurs in C_A.

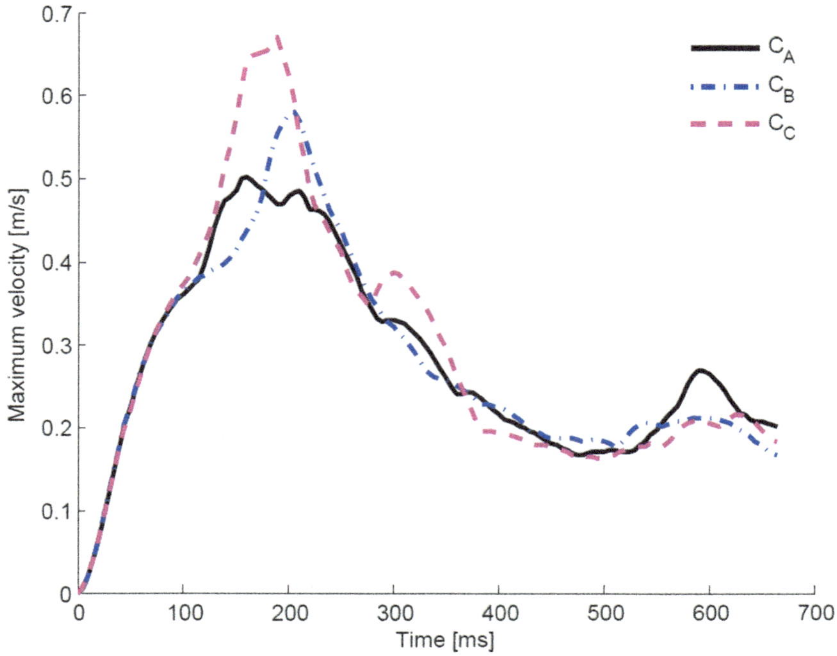

Fig. 7.40 Maximum transmitral velocity of *left* atrium models: C_A, C_B and C_C during diastole

Figure 7.41 shows flow entering the right atrium chamber through the shows the streamlines from the left inferior pulmonary valve and the right superior pulmonary valve for the three models: C_A, C_B, C_C at their respective maximum velocity. The shortest atrial transit time for blood originating in the left inferior pulmonary valve, was found for model-C_A. For model-C_C the distance from the left inferior pulmonary valve to the mitral valve is exactly the same as for C_A, but the average transit time was higher. This is because the left and right jets collide with each other in model-C_C and induce a recirculating motion towards the upper part of the left atrium. In model-C_B, some of the streamlines go directly to the septal side of the mitral opening, while another portion turn into a swirling motion towards and also into the left atrium appendage.

In model-C_C, the right jets collide with the jets from the left pulmonary valves and change direction completely. Whereas a large part of the flow from the left inferior pulmonary valve was forced into a swirling motion, the jet from the right superior pulmonary valve turns directly towards the septal side of the mitral valve. For model-C_B, the streamlines originating in the RSPV are similar to the streamlines origination in the left inferior pulmonary valve, but the number of streamlines having a direct path towards the mitral valve is less.

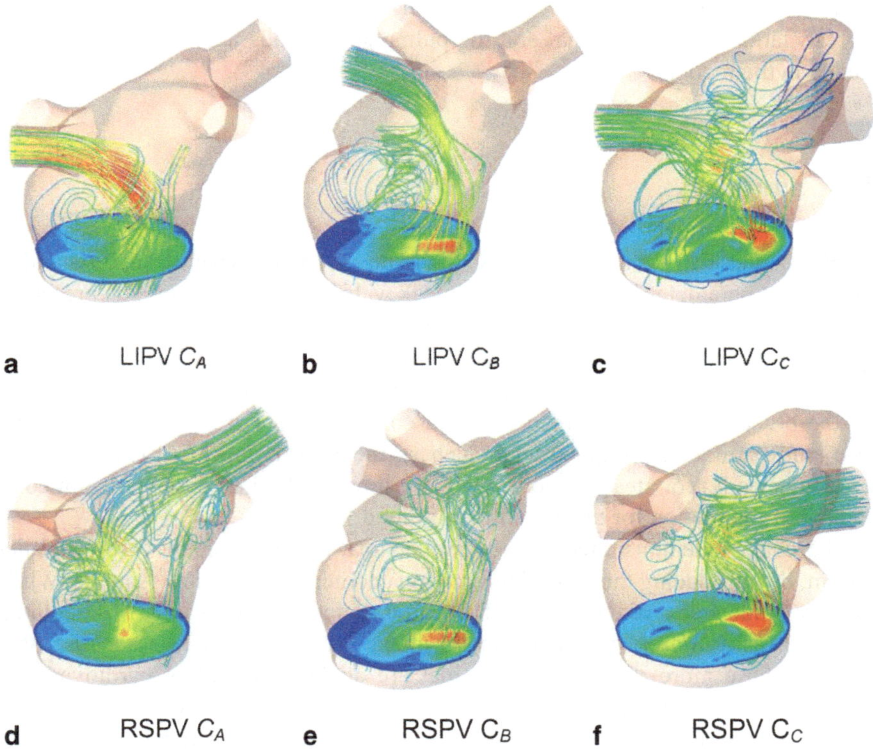

Fig. 7.41 Streamlines from the LIPV (left inferior pulmonary valve) and RSPV (right superior pulmonary valve) are plotted for each configuration at their maximum velocity. The maximum velocity occurs at $t=160$ *ms*, $t=205$ *ms* and $t=190$ *ms* for C_A, C_B and C_C, respectively

7.6.4 Closure

In this section we presented three 3D CFD simulations focusing on the intra-atrial flow and the resulting mitral plane velocity profile during left ventricle diastole. The anatomically based 3D geometries of the left atrium and the pulmonary veins were obtained from MRI recordings of a young healthy adult. The entry locations of the pulmonary veins were different in the three models. Four jets enter the asymmetric atrium, and the resulting flow field is therefore complex. The results clearly illustrate that the locations of the pulmonary valves have a significant impact on the intra-atrial flow.

When comparing the flow field in the three models, the results clearly illustrate that the pulmonary veins have a significant impact on the intra-atrial flow and the final mitral plane velocity profile. Because the interpatient variability in venous number and branching patterns is large, the mitral plane velocity profile should be considered as a subject-specific property. Therefore, we suggest that in order to obtain a physiological correct simulation of ventricular filling, a subject-specific representation of the left atrial and the pulmonary venous anatomy should be included in the model.

7.7 Summary

In this chapter, CFD is used to address the issues in haemodynamic analysis and to identify the parameters of interest for clinical evaluation. The information presented in the chapter highlights the conceptual development of the computational haemodynamics technique and performs proof-of-concept experiments.

Case studies were presented to demonstrate the modelling strategies to account for the blood flow in the human cardiovascular system. For example, in the first case study, modelling requirements such as the flow regime (laminar/turbulent, steady/unsteady) for steady blood flow were shown. The first case study aimed to set a good foundation for the modelling of blood flow.

The chapter demonstrated further applications of CFD for the cardiovascular system, by introducing the carotid bifurcation geometry and its flow features. While the case studies covered a variety of research topics in cardiovascular modelling, there are many more topics that remain unexplored. The next chapter discusses some of these more advanced topics that are at the forefront of cardiovascular modelling and research. For example, some methods are proposed for integrating advanced numerical techniques related to Fluid Structure Interaction (FSI) and later examined based on how this can be used for more realistic physiological arterial wall movements. The future of blood flow modelling in the cardiovascular system will continue to advance and sophisticated models will include more physiology and anatomical features.

7.8 Review Questions

1. What are the main factors that contribute to the blood flow patterns in the atherosclerotic artery?
2. List the physiological functions within the artery and discuss its modelling requirements and whether it is feasible to model.
3. What modelling considerations should be made when simulating oscillatory pumping of blood in artery (10–15 ml/s)?
4. What should you expect from the results when applying a turbulent flow model for: a) laminar flow b) transitional flow
5. What does a quasi-steady flow mean and when can you apply such an assumption?
6. When should Newtonian and non-Newtonian flow behaviour be assumed?
7. Where is the highest wall shear stresses found in the atherosclerotic artery? What are some of the causes for this phenomenon?

8. What flow patterns are found in the stenotic region? Why do you think these flow patterns exist?
9. Discuss the flow distribution in the main artery channel. How does this affect plaque aggravation?
10. What are some of the difficulties that require simplifications or assumptions when dealing with the blood flow simulation through a carotid bifurcation?
11. In the atherosclerotic artery, what flow regime (laminar or turbulent) would you expect the blood flow to be? Discuss in terms of the Re number.
12. In what cases would you need to apply artificial geometrical extensions to a carotid bifurcation model?
13. Which branch of the carctid bifurcation is the blood flow biased towards and why? Investigate further and determine if this is consistent among the population.
14. What haemodynamics parameters can be used to analyse flow in aneurysms?
15. What is the range of Reynolds numbers of flow in aneurysms?
16. What haemodynamics parameters can be used to analyse flow in heart chambers?

Chapter 8
Applications of FSI for Cardiovascular Haemodynamics

8.1 Introduction

Computational simulations increase in complexity when it considers blood flow with elastic wall structures of the cardiovascular system. In this chapter we present applications of fluid-structure interaction in cardiovascular haemodynamics to provide the reader with an overview of some of the latest developments in this emerging field.

8.2 Flow in an Idealised Stenotic Artery Bifurcation

It is widely accepted that severe stenosis contributes to some cardiovascular diseases such as heart attack, ischemia, and atherosclerosis. In this example we present a comparative between an idealized healthy and stenosed carotid artery bifurcation by studying the interaction of the artery wall with vascular blood flow.

8.2.1 Computational Considerations

An idealised geometry of a healthy carotid bifurcation model is used by Tada and Tarbell (2005) shown in Fig. 8.1. Detailed dimensions of the geometry and modelling parameters are given in Table 8.1. For a diseased carotid bifurcation model, a stenosis is prescribed on the idealised model along the outer ICA sinus wall through varying degrees of occlusion. The partitioned FSI approach was used in the ANSYS software package. Fluid structure interaction is performed between ANSYS-Mechanical for the solid domain and ANSYS-CFX for the fluid domain. Coupling of the two solvers was performed until convergence of interface variables (such as displacements and pressure) was reached per time step. At each coupling loop, the calculation of blood flow was initiated. The calculated pressure field from the fluid domain was transferred to the solid domain as an applied force

J. Tu et al., *Computational Hemodynamics—Theory, Modelling and Applications*,
Biological and Medical Physics, Biomedical Engineering,
DOI 10.1007/978-94-017-9594-4_8

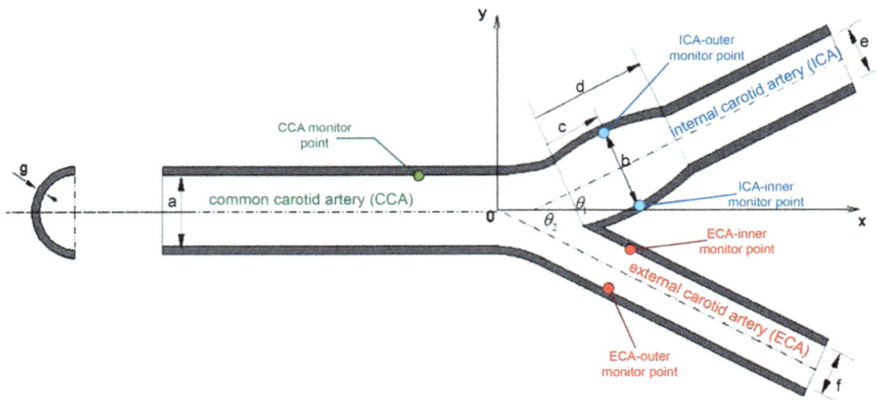

Fig. 8.1 Geometry of carotid bifurcation. The degree of stenosis occurs at the maximum ICA sinus cross-section denoted by measurement variable b

Table 8.1 Artery vessel structure, blood properties, dimensions of carotid bifurcation, and parameters for the fluid-structure interaction. These values are used to define the artery geometry and structural behaviour, as well as the physical properties of the blood flow model

	Symbols	Value	Units
Material properties Young's modulus of artery	E	5.00E+05	Pa
Poisson ratio of artery	v	0.5	
Density of blood	ρ	1050	kg m^3
Viscosity of blood	μ	3.50E−03	Pa s
Dimension			
CCA internal diameter	a	0.006	m
Maximum sinus internal diameter	b	0.00666	m
Distance from bifurcation point to location of maximum sinus internal diameter	c	0.00546	m
Total length of sinus	d	0.0138	m
ICA internal diameter	e	0.00444	m
ECA internal diameter	f	0.0039	m
Artery wall thickness	g	0.0007	m
ICA bifurcation angle	θ_1	25	degree
ECA bifurcation angle	θ_2	25	degree
Parameters for fluid-structure interaction			
Relaxation factor for coupling	R	0.75	–
Convergence criteria for coupling	ε_0	1.00E−05	–
Time step size	t_s	0.01	s
Time	t	–	s
Pressure pulse period	T	0.92	s

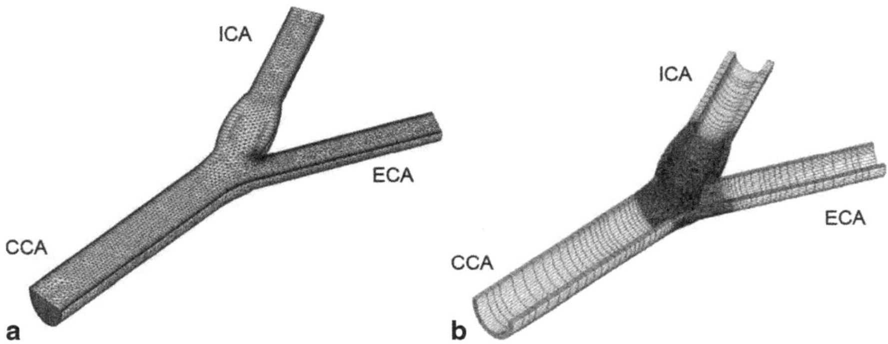

Fig. 8.2 Computational model showing the (**a**) fluid and (**b**) solid domain meshes. Grid independence analysis was performed at 3 different mesh refinement levels: coarse (*20,000 elements*), medium (*500,000 elements*) and fine (*1,000,000 elements*). Only 2 % of the dissimilitude between the fine and medium mesh is observed

in the solid solver to calculate the artery deformation. The coupling performance is enhanced by the "artificial compressibility" method (Raback et al. 2001). The blood flow is modelled as laminar where the peak Reynolds number, which occurs at the stenosis, is approximately 1000. The time step size is set to 0.015s and the results are obtained at the 4th cardiac cycle to avoid any start-up effects from initial conditions.

Figure 8.2 shows the computational mesh of the stenosed model with approximately 1,000,000 cells made up of tetrahedral cells in the inner fluid domain and prism cells at the near wall region along the wall boundaries

8.2.2 Simulation Details

In the solid domain, each end of the artery (CCA, ICA, ECA) are modelled as fixed supports while a symmetrical condition is assumed at the plane of the bifurcation. In the fluid domain, a pressure inlet boundary condition is imposed using a time-varying pressure waveform as shown in Fig. 8.3a referred from Tada (2005), while outlet boundaries are governed by transient mass flow rate profiles as given in Fig. 8.3b. For simplicity, the arterial wall is assumed, the material elasticity properties are treated as isotropic based on the Hooke's Law (Salzar et al. 1995; Thubrikar and Robicsek 1995), and the blood flow is assumed to be Newtonian with a constant dynamic viscosity of 0.0035 Pa-s. An alternative approach to represent the material hyper-plastic elasticity property and the non-Newtonian flow behaviour is the Mooney-Rivlin elastic model (Torii et al. 2009b) and Casson fluid model, which require more computational sources and longer simulation time.

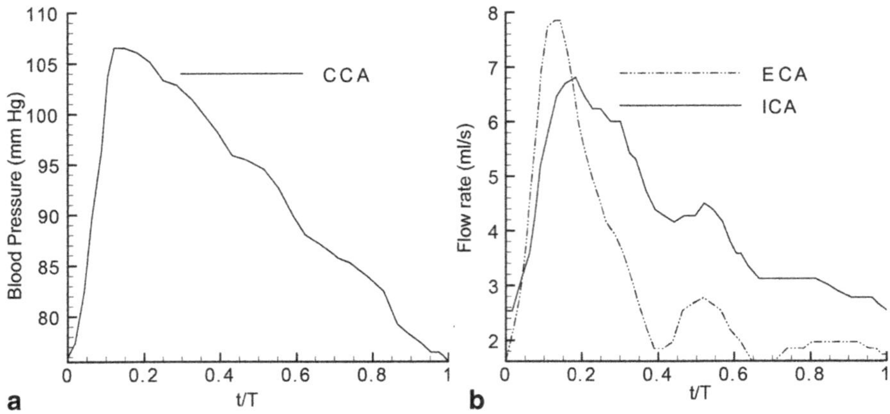

Fig. 8.3 Flow conditions imposed at the ends of common, internal and external carotid arteries (*CCA, ICA, and ECA respectively*). A pressure waveform is applied at the CCA and flow rates are prescribed at the ICA and ECA. The period of pressure pulse and flow rate cycle is denoted by *T* (*in seconds, s*)

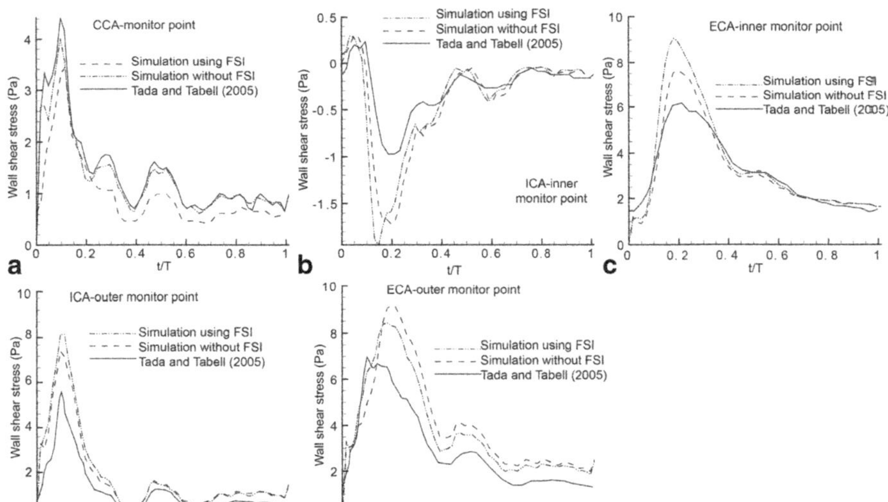

Fig. 8.4 Wall shear stress profiles based on FSI and no FSI simulations at monitoring points as (**a**) CCA-monitor point; (**b**) ICA-inner monitor point; (**c**) ECA-inner monitor point; (**d**) ICA-outer monitor point; (**e**) ECA-outer monitor point

8.2.3 FSI Analysis of Diseased Carotid Bifurcation

Figure 8.4 demonstrates the transient variation of wall shear stress at various location of the artery. Five monitoring points are shown in Fig. 8.1 labelled as the CCA-monitor point; ICA-inner monitor point; ECA-inner monitor point; ICA-outer

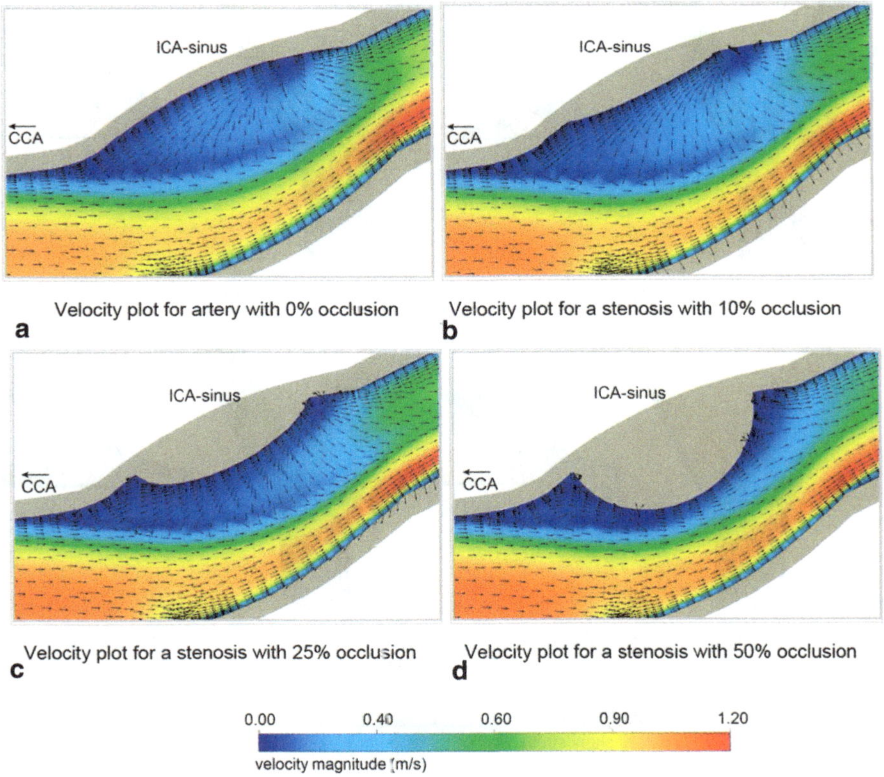

Fig. 8.5 Vector plot and velocity contour plot of carotid bifurcation with different degrees of stenosis. The simulation pertains to an atherosclerotic artery are presented for the (**a**) 0 % or healthy case; (**b**) 10 % occlusion; (**c**) 25 % occlusion and; (**d**) 50 % occlusion

monitor point; and ECA-outer monitor point. Wall shear stress profiles derived from simulations based on a rigid vessel wall assumption (no FSI), or a flexible vessel wall (with FSI) are compared with the work by Tada and Tarbell (2005). In general, the comparison of no-FSI with FSI shows a 5 % discrepancy in the peak wall shear stress. Figure 8.4a–e show that the wall shear stress profiles have similar shapes to the blood flow pressure waveform. The effect of blood vessel compliance reduces the wall shear stress peak slightly at each monitoring point.

Figure 8.5 illustrates blood flow patterns in the healthy and varying stenosed carotid bifurcations. The occlusion in the arterial lumen have been deliberately exaggerated to highlight its effects on the blood flow, stresses and deformation. In Fig. 8.5a, representing a healthy carotid bifurcation, the blood flow separates at the terminal end of the CCA, and forms a recirculation zone inside the ICA sinus. This recirculation zone is located near the outer wall of the sinus, and is relatively large compared to the diameter of the sinus. The recirculation zone ends with a reattachment point around the distal end of the sinus. Blood motion inside the recirculation zone is slow and reverses. Consequently, wall shear stress in this region is also very

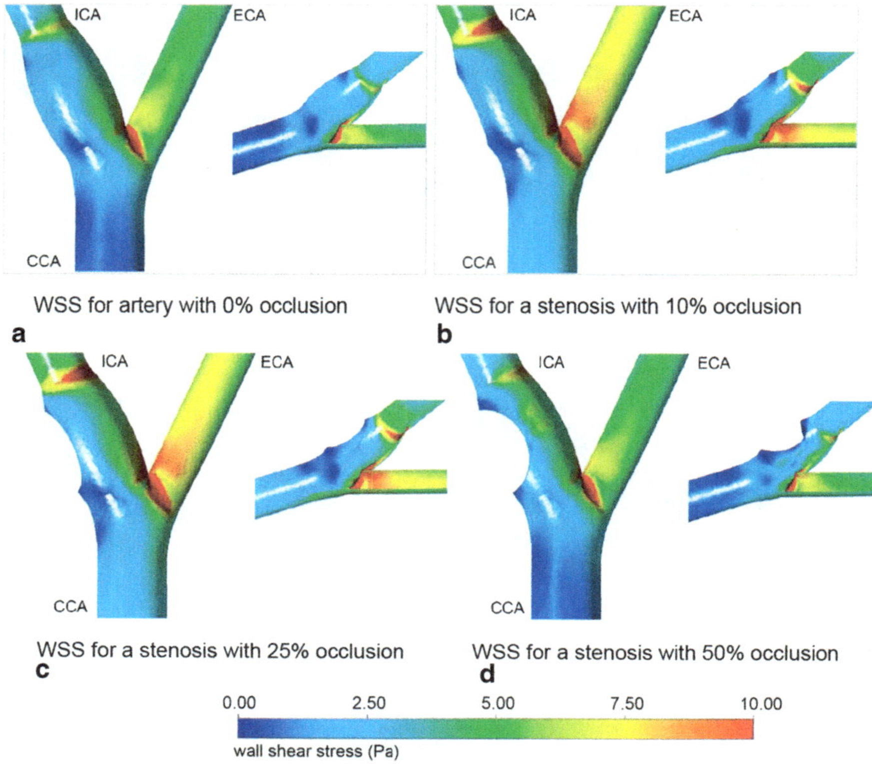

Fig. 8.6 Wall shear stress contour plots of carotid bifurcation with different degrees of stenosis. The simulation pertains to an atherosclerotic artery are presented for the 0% or healthy case (**a**); 10% blockage (**b**); 25% blockage (**c**) and 50% blockage (**d**). The wall shear stress scale is in Pa

small and has negative value for the large part of the heart cycle as demonstrated earlier in Fig. 8.4b. The results also show that at the beginning of each cycle, the recirculation zone disappears for a short period of time (at $t/T = 0$ to 0.1) since the sign is positive in this period. Figure. 8.5b and c show that as the stenosis becomes more aggravated, two recirculation zones at the proximal and distal ends of the plaque are formed.

Figure 8.6 presents the wall shear stress spatial variation on the arterial surface of the carotid bifurcation. Based on the different degrees of stenosis, the wall shear stress variation of the healthy artery can be used for comparisons with reference Fig. 8.4 and the flow patterns in Fig. 8.5.

The FSI simulations allow examination of vessel deformation, and maximum principal stress distribution of the artery. Figure 8.7 are contour plots of vessel deformation to highlight high deformation regions in the carotid bifurcation. For the healthy carotid (Fig. 8.7a) maximum deformation occurs just upstream of the bifurcation point. For the stenosed arteries, high deformation occurs at the stenosis and upstream of the bifurcation (Figs. 8.7b–d). The deformation pattern does not change with the degree of stenosis. However, the deformation magnitude changes

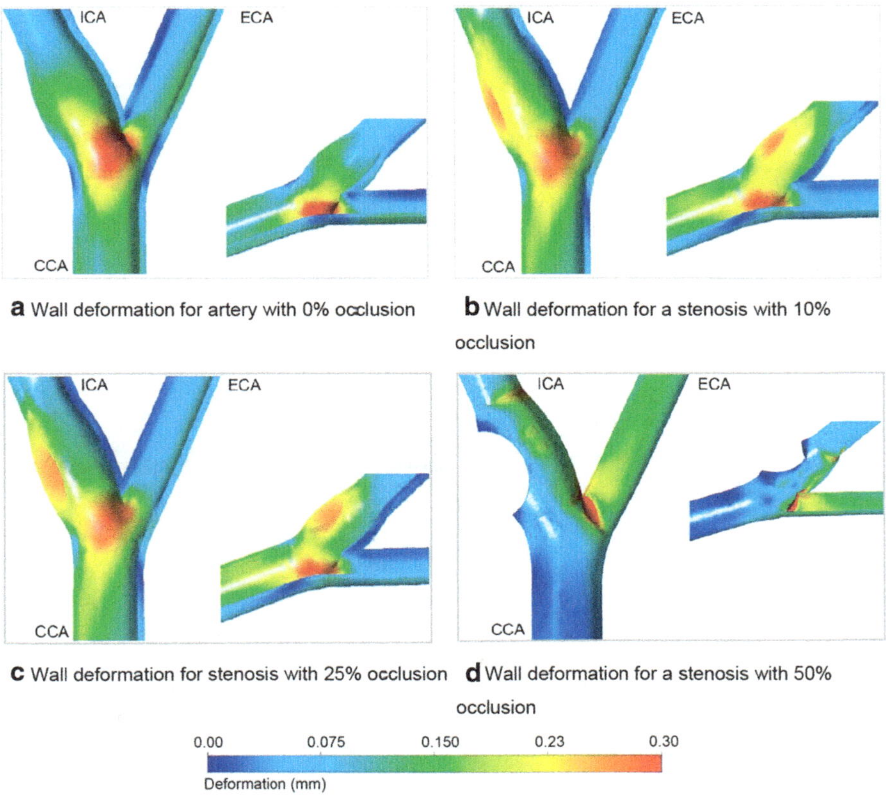

Fig. 8.7 Deformation contour plots of carotid bifurcation with different degrees of stenosis. The simulation pertains to an atherosclerotic artery are presented for (**a**) 0 % or healthy case; (**b**) 10 % blockage; (**c**) 25 % blockage and; (**d**) 50 % blockage. The deformation scale is in metres

with time, corresponding to the input pressure waveform (this is discussed in the next section of this Chapter). Furthermore as the degree of stenosis increases, the wall deformation in the stenosis region reduces due to the increase in its structural integrity. In contrast, deformation in the region just downstream of the bifurcation is independent of the stenosis occlusion.

Figure 8.8 shows maximum principal stress contours along the artery stenosis plaque and also a reference centreline used later for analysis. Principal stresses may be defined as the minimum and maximum values of the normal stresses possible in a material. Its physical meaning can be visualised by considering a material under stress and looking at the stresses in from different directions. At a certain orientation all shear stresses are zero and the state of stresses is defined by three normal stress components only. These three normal stress components are referred to as principal stresses. Positive stresses are tensile, and negative stresses are compressive and they exhibit a direction component. A region of high positive maximum principal stress is found along the border of the stenosis while a region of negative maximum principal stress resides near the stenosis apex. For the 10 % blockage case, the maxi-

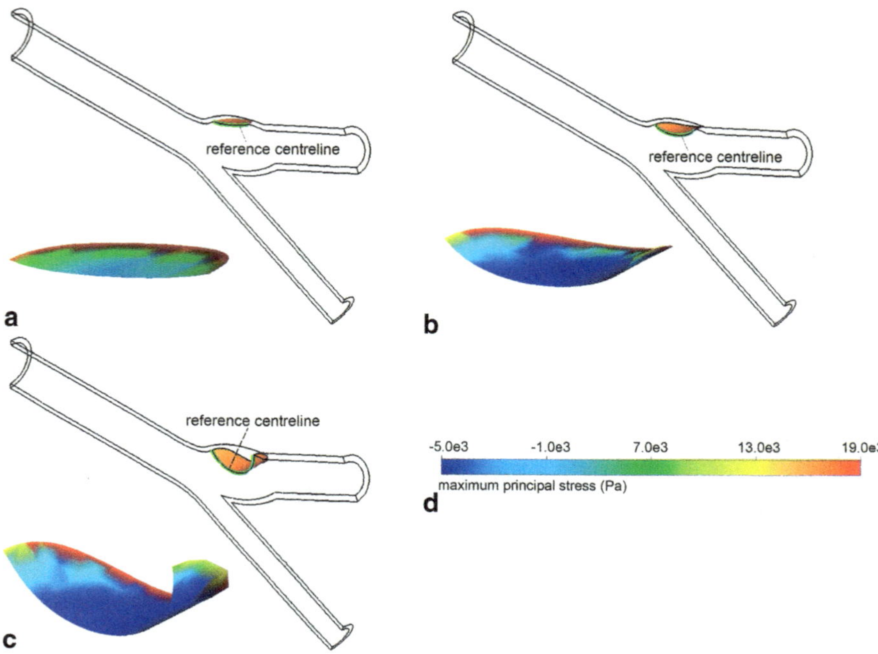

Fig. 8.8 Maximum principal stress contours of the carotid bifurcation with different degrees of stenosis (**a**) 10 % occlusion; (**b**) 25 % occlusion and (**c**) 50 % occlusion. The maximum principal stress scale is in Pa

Fig. 8.9 Plots of the maximum principal stress on stenosis in diseased carotid bifurcations. At peak flow rate, the maximum principal stress decreases from approximately to −400 to −4000 Pa when the degree of stenosis increases from 10 to 25 %

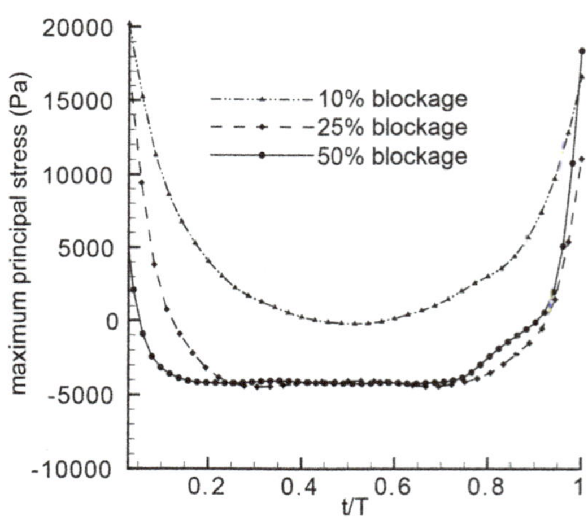

mum principal stress pattern seems highly symmetrical. However, as we increase the degree of stenosis, the distribution starts to become non-symmetrical.

The maximum principal stress distribution is plotted in Fig. 8.9. For the 25, 50 % blockage cases, the maximum principal stresses at the proximal end of the stenosis are greater than those at the distal end for the peak and minimum flow rates. The pattern becomes non-symmetrical when increasing the degree of stenosis, and also the minimum maximum principal stress becomes more negative.

8.2.4 Comparison Between FSI and Non-FSI Models

The study by Valencia and Villanueva (2006) is examined here which looked at the unsteady non-Newtonian blood flow and mass transfer in symmetric and non-symmetric stenosis arteries (Fig. 8.10). This was compared for rigid and elastic models. The compliant arterial wall uses the Mooney–Rivlin model. It is shown that by using FSI, the arteries undergo significant dilation and compression due to the stenosis.

Figure 8.11 presents wall displacement and wall shear stress for symmetrical and non-symmetrical stenotic artery due to 70 % occlusion and at systolic phase of 0.932 s. The maximum and minimum artery diameters at the inlet were at 3.4 mm and 2.3 mm at systole and diastole respectively. The maximal effective stress at systolic time is around 0.25 MPa. However, at the stenotic throat, the effective stress is about one order of magnitude smaller at approximately 0.05 MPa.

Valencia and Villanueva (2006) also evaluated the instantaneous recirculation lengths at systolic time for the rigid and elastic wall arteries, and compared the symmetrical and non-symmetrical stenosis for each. Based on different degrees of stenosis, their comparison is shown in Table 8.2. For the velocity inlet condition, the recirculation length after the stenosis shows a positive correlation with the degree of stenosis, while for the pressure inlet condition, this correlation becomes negative. For a stenosis with 80 % occlusion, the flow does not show any recirculation zone based on the pressure inlet boundary condition.

The study presented demonstrates that the degree of the stenosis influences its recirculation length, wall displacement, and effective wall stress. Furthermore using rigid walls produced less dilation and compression during the cardiac cycle in com-

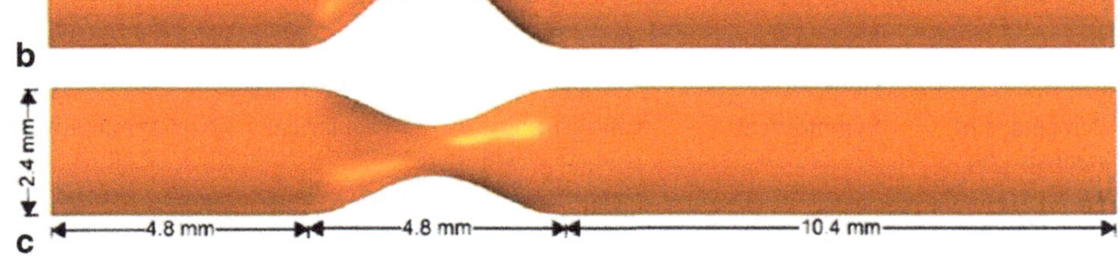

Fig. 8.10 Geometrical models of stenosed arteries. Arterial channels based on (**a**) symmetric stenosis and rigid wall; (**b**) non-symmetric stenosis with elastic wall; and (**c**) symmetric stenosis artery with elastic wall. (Image from Valencia and Villanueva 2006)

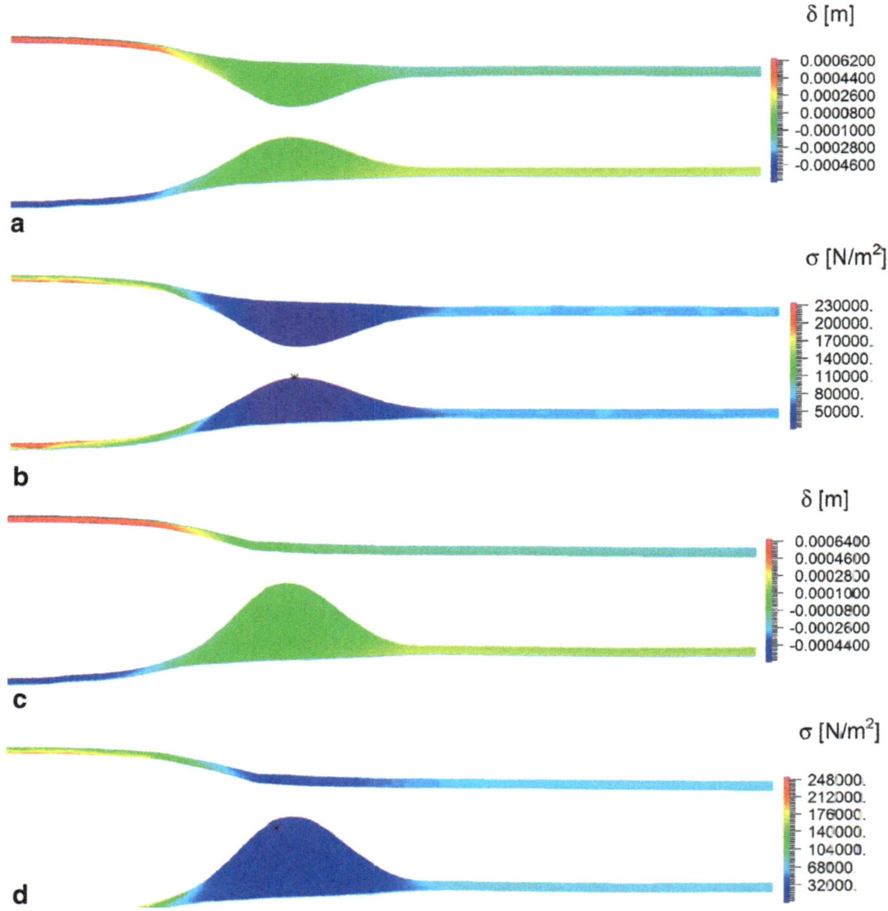

Fig. 8.11 Analysis of wall displacement and effective wall shear stress for stenotic arteries. (**a**) Wall displacement for the symmetric stenotic artery; (**b**) wall shear stress for the symmetric stenotic artery; (**c**) wall displacement for the non-symmetric stenotic artery, (**d**) wall shear stress for the non-symmetric stenotic artery. (Image from Valencia and Villanueva 2006)

Table 8.2 Instantaneous recirculation lengths at systolic time 0.932s. The rigid and elastic arteries are compared based on recirculation lengths at different percentage of stenosis (%), and the symmetry of the stenosis. (Adapted from Valencia and Villanueva 2006)

	Recirculation length (mm)			
	Rigid artery		Elastic artery	
Percentage of stenosis (%)	Symmetrical	Non-symmetrical	Symmetrical	Non-symmetrical
50	0	4.91	2.60	4.59
60	7.52	7.04	3.43	4.69
70	13.6	8.67	2.83	3.64
80	24.8	10.2	0	0

parison with a FSI model. Furthermore, using pressure as the prescribed inlet boundary condition allows a more realistic haemodynamic simulation of stenotic arteries.

8.2.5 Closure

Different degrees of stenosis have a significant influence on three basic characteristics of atherosclerosis. Firstly, the presence of severe stenosis can dramatically alter the blood flow pattern. Flow reversal is smaller at advanced stages of the disease. This leads to diminishing low wall shear stress in the stenotic region. Secondly, the stenosis apex experiences compressive stresses which intensify as it grows. This explains why plaque can still grow at advanced stages when the wall shear stress is higher. Finally, the relationship among pulse pressure, maximum displacement, and maximum principal stress suggests that elevated heart rates can lead to a higher risk of stroke due to longer exposure to high vessel wall displacement and stresses. Furthermore, the relationship between stenosis with recirculation length, wall displacement, and effective wall shear stress can be established. These results can be used as a platform for future work in atherosclerotic research aiming for improved understanding of the complicated process of the disease.

8.3 Flow in a Realistic Carotid Artery Bifurcation

In this example we present an FSI simulation of a realistic carotid artery bifurcation model which is patient-specific and realistic unlike the ideal case presented in the previous section. Studies on the pathophysiology of vascular disease have shown that atherosclerotic plaque is closely related to wall shear stress (Wong et al. 2013b) due to blood flow inside the blood vessel (Cunningham and Gotlieb 2005; Malek et al. 1999). For regular vascular functioning, the endothelial cells located in the blood vessels must experience the appropriate wall shear stress. Under a low shear stress condition, generation and growth of atherosclerotic plaque can unofrunately progress easily.

Medical imaging diagnosis using CT or MR mainly focus on the geometrical occlusion of the arterial vessel, and lacks the detailed blood flow information such as a local wall shear stress distribution. This has limitations in the use for the clinical diagnosis and treatment of cerebral vascular diseases. To predict the evolution of atherosclerotic plaques related to vascular disease, the haemodynamic blood flow and arterial deformation inside patient-specific carotid artery bifurcation models have to be investigated.

The arterial model was reconstructed from the CT images and a fully coupled FSI simulation performed based on this model by imposing physiologically reasonable flow boundary conditions. The differences between rigid wall and elastic wall approaches were addressed, and detailed blood flow information inside a compliant

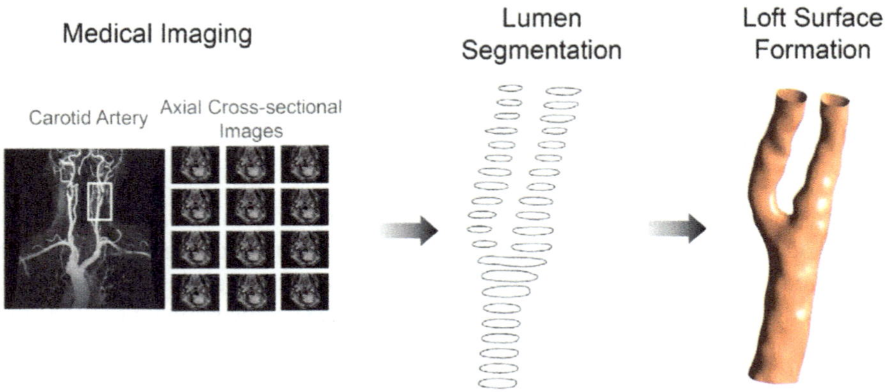

Fig. 8.12 Reconstruction of an abdominal aortic aneurysm based on multi-slice CT angiography

carotid artery were found for the prediction of wall shear stress closely related with the atherosclerotic plaque progression.

8.3.1 Geometric Models and Material Properties

A healthy left carotid artery model was generated from CT images (voxel size of $0.6 \times 0.6 \times 0.6$ mm^3) using Blender version 2.48 (Blender Institute, Netherlands). For simplicity, only images within the vicinity of the bifurcation were used, e.g. all daughter branches were removed. The reconstruction procedure is shown in Fig. 8.12 where a pixel based segmentation approach is used to detect the inner lumen profiles of the arterial vessel (see Chap. 3). These lumen profiles were lofted and integrated into surface data to create a water-tight CAD model.

With the reconstructed arterial geometry, the computational meshes in the fluid and solid domains for the FSI analysis were generated (Fig. 8.13). A structured mesh was used with a mesh refinement at the near wall regions.

Although blood flow profiles can be obtained from clinical examinations using phase-contrast MRI or Doppler ultrasound this is time consuming and costly compared with computational studies. In this example physiological flow conditions, shown in Fig. 8.14, were applied as pressure inlet and velocity outlets (Fig. 8.15).

The arterial wall was assumed a hyperelastic material exhibiting a non-linear stress-strain behaviour like rubbers and polymers. A widely accepted strain energy density function, the Mooney-Rivlin model, was used to express its elasticity as

$$\begin{aligned} W = {} & C_{10}(\bar{I}_1 - 3) + C_{01}(\bar{I}_2 - 3) + C_{20}(\bar{I}_1 - 3)^2 + C_{11}(\bar{I}_1 - 3)(\bar{I}_2 - 3) \\ & + C_{02}(\bar{I}_2 - 3)^2 + C_{30}(\bar{I}_1 - 3)^3 + C_{21}(\bar{I}_1 - 3)^2(\bar{I}_2 - 3) \\ & + C_{12}(\bar{I}_1 - 3)(\bar{I}_2 - 3)^2 + C_{03}(\bar{I}_2 - 3)^3 + \frac{1}{d}(J - 1)^2 \end{aligned} \tag{8.1}$$

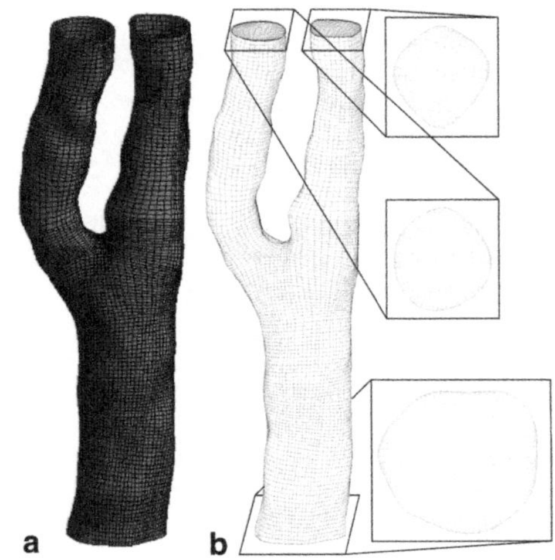

Fig. 8.13 Computational mesh results of the used carotid artery model (**a**) structure domain, (**b**) fluid domain

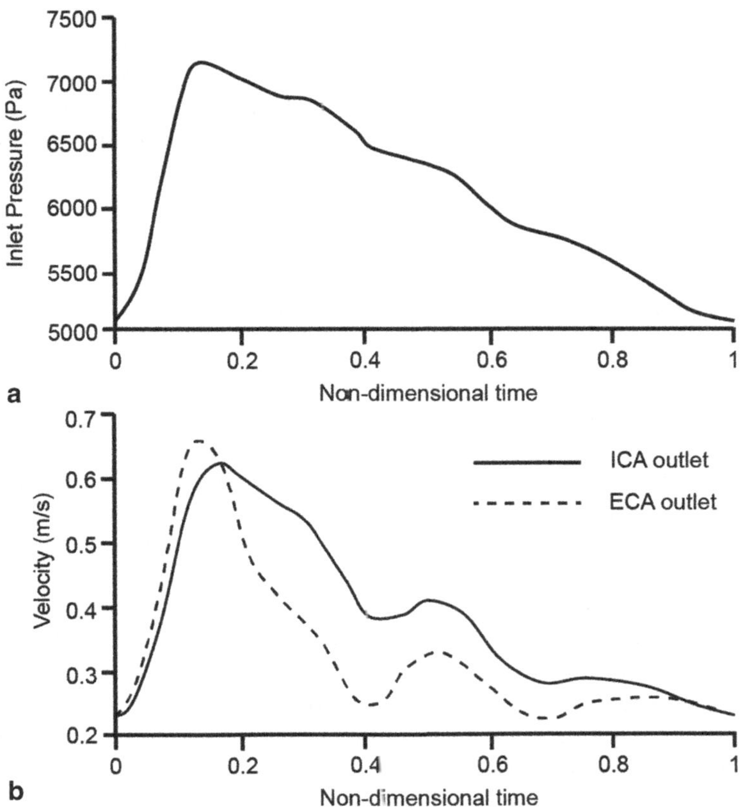

Fig. 8.14 Physiologically reasonable flow waveforms (**a**) inlet pressure waveform, (**b**) outlets velocity waveforms. ICA is the internal carotid artery; ECA is the external carotid artery

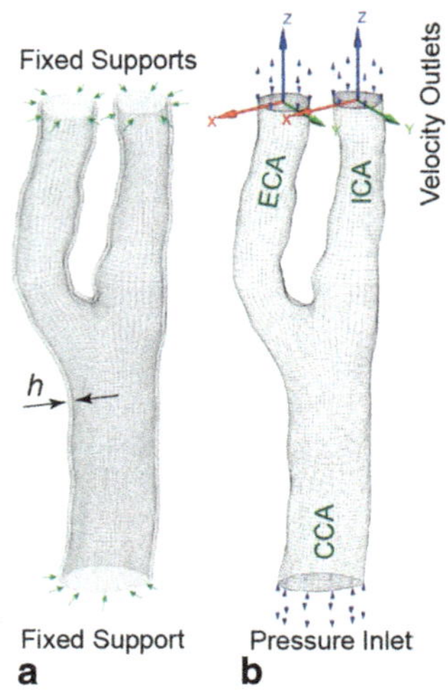

Fig. 8.15 Boundary conditions setup (**a**) fixed supports were placed at all branches ends for the structure domain, (**b**) pressure inlet and velocity outlets were prescribed at the inlet and two outlets of the fluid domain. ICA is the internal carotid artery; ECA is the external carotid artery

where. I_1 and. I_2 are the first and second deviatoric strain invariants and J is the determinant of the elastic deformation gradient tensor. The 0.5 mm thickness is assigned to the arterial vessel, and all model parameters being used are listed below: $C_{10}=0.070$ MPa, $C_{20}=3.2$ MPa, $C_{21}=0.0716$ MPa, $D_1=0.1$ MPa, and $C_{ij}=0$ MPa for the remaining tensors, following the method by Koshiba et al. (2007). The time step size was set as 0.002 s, and 500 time steps were implemented for a cardiac cycle. 24 hours of CPU time was used for a simulation of one cardiac cycle by using two Xeon CPUs with a speed of 2.83 GHz.

8.3.2 Haemodynamics Inside the Healthy Carotid Artery

Qualitative comparison of flow velocity between a CFD simulation (rigid wall assumption) and a fully coupled FSI simulation (elastic wall assumption) is illustrated in Fig. 8.16. The time-averaged blood flow velocity, averaged over one cardiac cycle, is used for comparison. The CFD simulation predicts the maximum averaged velocity occurring at the moderate stenosed site on the ICA (internal carotid artery) branch, and the maximum velocity value is 0.63 m/s (Fig. 8.16a). Compared with the FSI approach (Fig. 8.16b), the CFD model results show an over prediction of 14.5 %. By allowing deformation of the arterial vessel, the carotid artery vessel deforms in response to the mechanical loading induced by the pulsatile blood pressure – especially at the moderate stenosis site located at the downstream ICA branch. Therefore, more blood flow can be transported via ICA branch. Due to the fixed arterial geometry for the CFD simulation, the flow resistance of the ICA branch is greater than the FSI simulation, and more blood flow is distributed to the ECA branch.

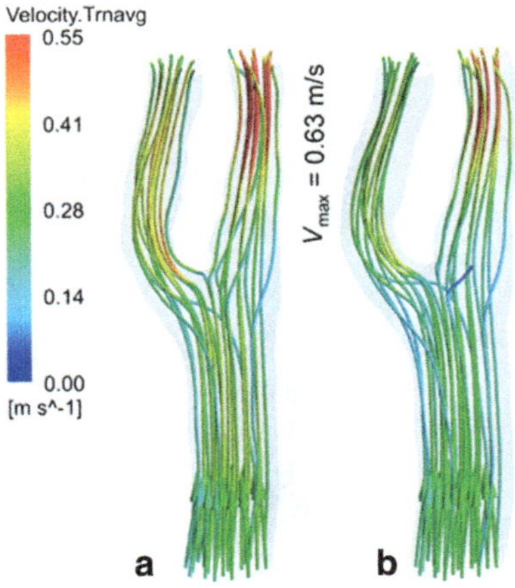

Fig. 8.16 Time averaged velocity comparison (**a**) simulation based on rigid wall approach (*CFD-only*), (**b**) simulation based on the compliant wall approach (*FSI approach*)

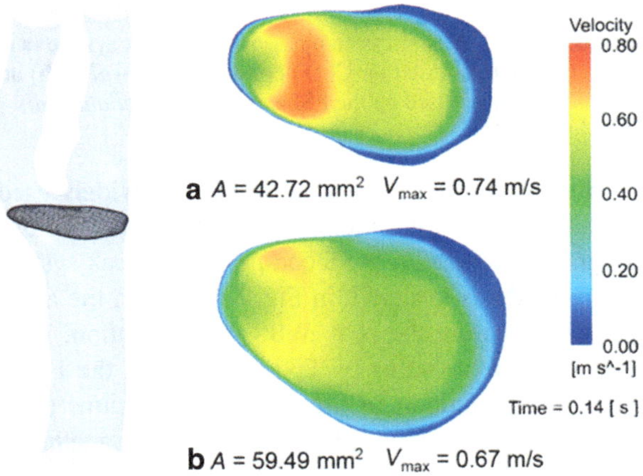

Fig. 8.17 Plane velocity comparison at the peak systole (**a**) velocity distribution based on rigid wall approach, (**b**) velocity distribution based on the compliant wall approach

Figure 8.17 depicts the velocity at one selected cross-section within the carotid bifurcation region at peak systole. Its area increases from 42.72 mm^2 to 59.49 mm^2, when switching from the CFD modelling approach to a fully coupled FSI modelling approach. The FSI simulation contributes an area expansion of 39.3 % at this location, and this expansion significantly mitigates the peak velocity from 0.8 m/s to less than 0.6 m/s. Due to the increased flow resistance of the ICA branch in the CFD-model, more blood flow occurs in the ECA branch, leaving a relatively large area of low velocity at the near wall region in front of the ICA branch. The low velocity region for the FSI simulation is considerably reduced since the downstream flow resistance is not as great as the CFD simulation, and the blood flow is distributed more evenly.

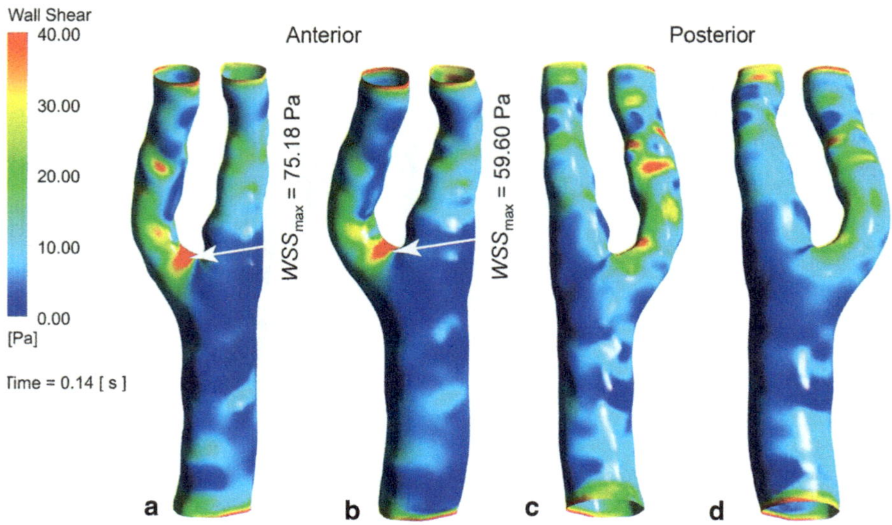

Fig. 8.18 Wall shear stress comparison at the peak systole (**a**) and (**c**) front and back views of the WSS distribution based on CFD approach (*rigid wall*), (**b**) and (**d**) front and back views of the WSS distritution based on the FSI approach (*compliant wall*)

Since the wall shear stress (WSS) is a widely used bio-marker for vascular disease, its results give more meaningful data than blood flow velocity for clinical diagnosis purpose. A WSS comparison at peak systole between CFD and FSI modelling methods is shown in Fig. 8.18. From the anterior view, both modelling approaches capture the peak WSS stress location. The CFD simulation predicts the maximum WSS value by 26 % more than the FSI approach. From the posterior view, the prediction of high risk WSS regions (where the WSS exceeds 40 Pa) shows more differences between these two methods. The CFD simulation shows more regions located on the ECA branch of high risk WSS area, while the FSI simulation does not.

8.3.3 Closure

The haemodynamics in a healthy carotid artery subject was numerically investigated with a fully coupled FSI approach, and the simulation results were compared with a rigid wall CFD approach. With the mechanical modelling of arterial vessel, the influence of vessel deformation was taken into account, and a decrease of blood flow velocity and wall shear stress were observed. The numerical studies shows vessel compliance needs to be modelled using the FSI simulation approach, and the haemodynamic characteristics of the carotid artery can be used to predict the lumen area at risk for vascular disease diagnosis purposes.

8.4 Flow in the Left Coronary Artery

This example presents an FSI study of flow in a coronary artery with different branch angulation, using anatomically accurate and idealized human coronary artery models. This study aims to elucidate the link between coronary artery angulation, coronary haemodynamics and local mechanical forces (tensile stress) to enable a better understanding of the role of haemodynamics in atherosclerotic disease initiation and progression in the vicinity of bifurcations. Using a coupled FSI modelling approach, five idealized left coronary artery models with various angles ranging from 70° to 110° were developed to investigate the influence of branch angulations. In addition, one CT image-based model was reconstructed to further demonstrate the medical application potential to patient specific models.

8.4.1 Geometric Models and Material Properties

The anatomical replica model was reconstructed from multi-slice CT angiography of a left coronary segment (Chaichana et al. 2011). Figure 8.19 shows the CT image-based model which exhibits an angle of 90° generated from CT images (voxel size of $0.6 \times 0.6 \times 0.6\,\text{mm}^3$) using the software Blender version 2.48 (Blender Institute, Amsterdam, Netherlands). In addition, an idealized model was developed with averaged anatomical data for vessel diameter, length and curvature based on 19 post-mortem casts of normal human coronary artery trees (Nerem and Seed 1983). An overview of the idealized model is shown in Fig. 8.20 and its basic dimensions are listed in Table 8.3. To analyse the influence of bifurcation angulation (θ), five idealized models were constructed with angles of 70°, 80°, 90°, 100° and 110° respectively, which are in the physiological range reported by Girasis et al. (2010). To isolate the effect of a single geometric factor, the angle formed by the left main stem (LM) and left anterior descending (LAD) was kept constant (Johnston and Kilpatrick 1997). The angle variation was achieved by changing the orientation of left circumflex (LCx).

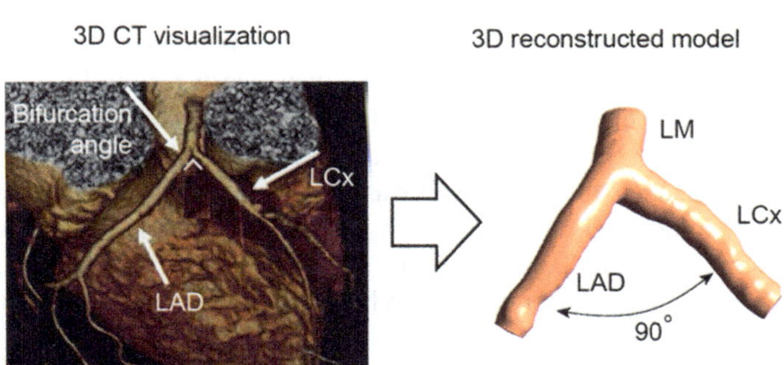

Fig. 8.19 3D CT visualization and reconstruction of the image-based model. *LM*, left main stem; *LAD*, left anterior descending; *LCx*, left circumflex

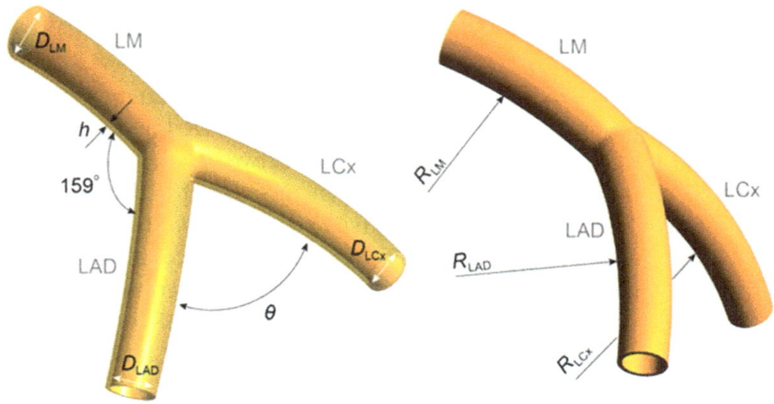

Fig. 8.20 Geometry configuration of the idealized artery model. *h* indicates vessel wall thickness, *D* stands for vessel diameter, and *R* for branch curvature. *LM*, left main stem; *LAD*, left anterior descending; *LCx*, left circumflex

Table 8.3 Anatomical dimensions of the idealized model

Length of LM	11.0 mm
Dia. of LM	4.0 mm
Dia. of LAD	3.4 mm
Dia. of LCx	3.0 mm
Rad. of curvature of LAD	42.8 mm
Rad. of curvature of LCx	39.3 mm
Angulation between LM and LAD	159°
Vessel wall thickness	0.4 mm

LM left main stem, *LAD* left anterior descending, *LCx* Left circumflex

8.4.2 Mesh Generation and Physiological Boundary Conditions

For each coronary artery model, both fluid and structural domains were meshed with hexahedral cells. A near wall grid refinement was imposed on each model to provide better resolution for near wall quantities. Mesh results for the fluid and structural domain of the image-based model and the idealized model ($\theta = 90°$) are depicted in Fig. 8.21. Since it is difficult to obtain the outer wall boundary of the artery from CT images, the vessel wall was artificially constructed with a constant thickness (Colombo et al. 2010) $h = 0.4$ mm. Although arterial wall is a composite tissue comprised of collagen fibers, its structural properties were simplified by adopting a nine parameter Mooney-Rivlin hyperelastic model (Koshiba et al. 2007) due to lack of *in vivo* data.

The inlet and outlet boundary conditions shown in Fig. 8.22a are based on a physiological pulsatile flow rate and pressure at the aorta (Nichols et al. 2011), reconstructed using a Fourier series in Matlab (Math Works Inc., Natick, MA,

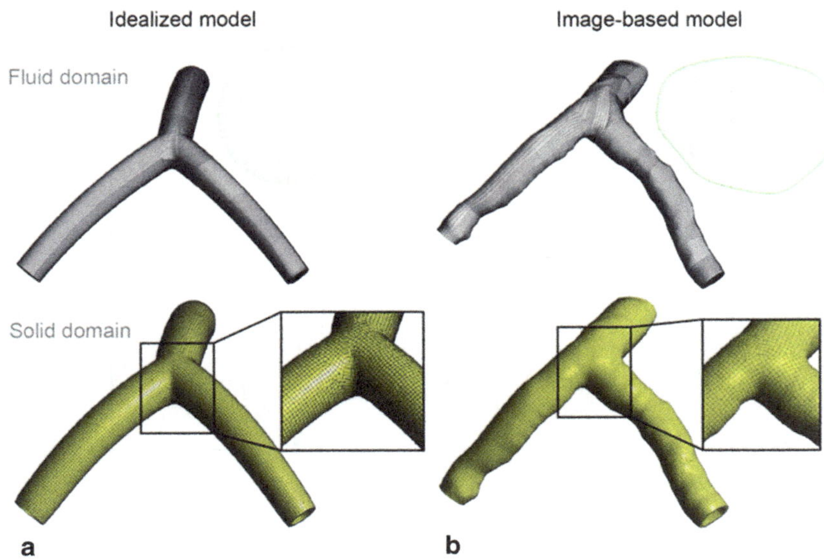

Fig. 8.21 Structure mesh results of idealized (**a**) and image-based (**b**) models

USA). The blood flow distribution in the bifurcation adopts the method by Boutsianis et al. (2004), where 71 % is directed through LAD and 29 % through LCx, and this is unchanged through the entire cardiac cycle. Pulsatile aortic pressure was applied as an inlet boundary condition at the entrance of the main stem, and pulsatile velocity conditions were imposed on both the LAD and the LCx outlet boundaries (Fig. 8.22b). As this study focuses on the local haemodynamic changes under different branch angulations, global coronary wall motion due to its attachment to the moving myocardium is neglected to isolate the effects of wall compliance (Malve et al. 2012; Torii et al. 2009a).

The blood was assumed Newtonian since the shear rate is large enough in coronary arteries (larger than 100 s^{-1}) to maintain a flow regime with nearly constant viscosity (Gijsen et al. 2007; Joshi et al. 2004). The density and viscosity of the blood are 1060 kg/m^3 and 0.0035 Pa-s (Chaichana T et al. 2012) respectively. The blood was treated as laminar and a no-slip condition was applied at the arterial walls. To eliminate local fluid dynamic effects on the reconstructed fluid domain and ensure fully developed outlet flow conditions, a 10-diameter length inlet extension and 15-diameter length outlet extensions were added (Joshi et al. 2004).

The fully coupled FSI models were solved in commercial software packages ANSYS CFX and ANSYS Mechanical (ANSYS Inc., Canonsburg, USA). FSI models are coupled and solved iteratively by these two packages within each time step by applying appropriate kinematic and dynamic conditions at the fluid-structure interface until the coupling system residual is less than a specified tolerance. For each model, transient flow simulations over three cardiac cycles were performed, and results at the last cycle were used for mechanical and haemodynamic analysis.

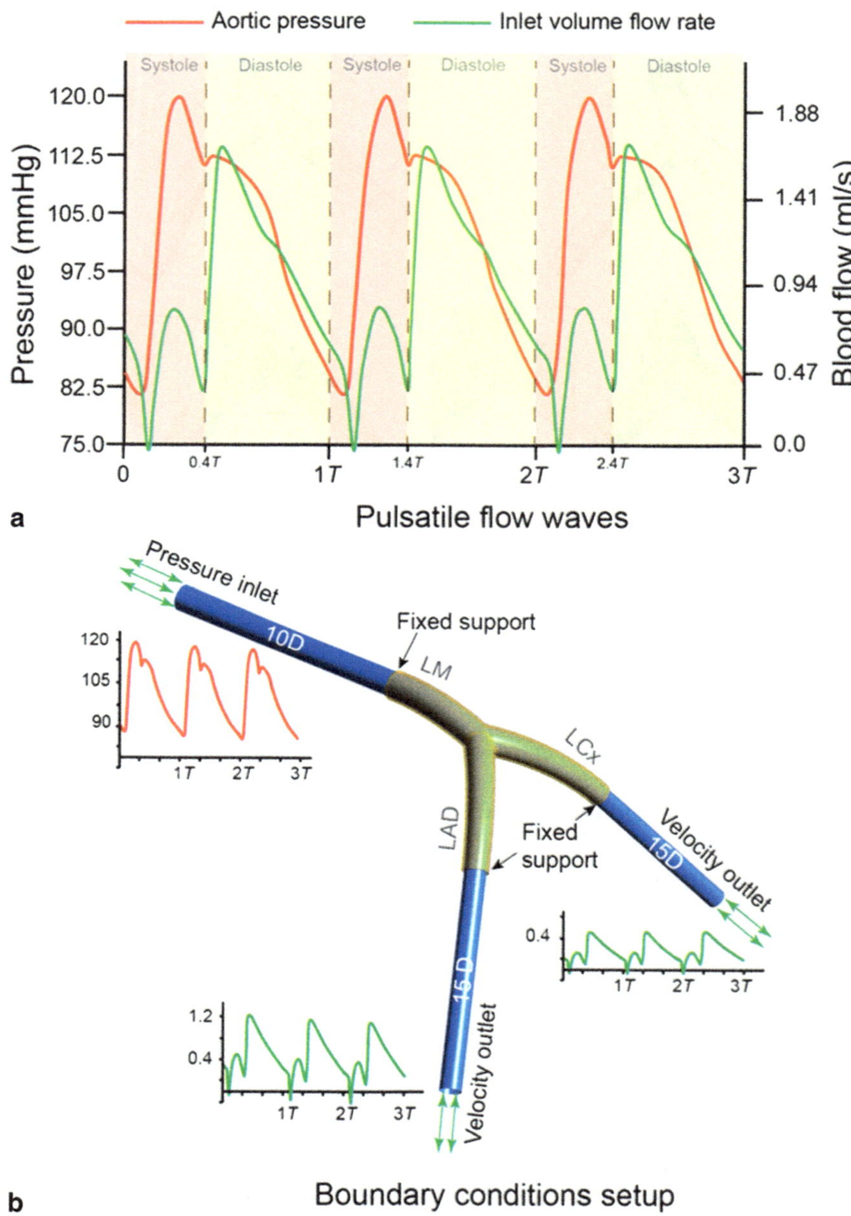

Fig. 8.22 Pulsatile blood flow waves used in this study. *T* represents one cardiac cycle

8.4.3 Mechanical Results Analysis

The first principal stress is used for stress distribution analysis, as it represents the maximum tensile stress included in the vessel wall due to the pulsatile loading of blood flow. Figure 8.23 illustrates the first principal stress distribution for

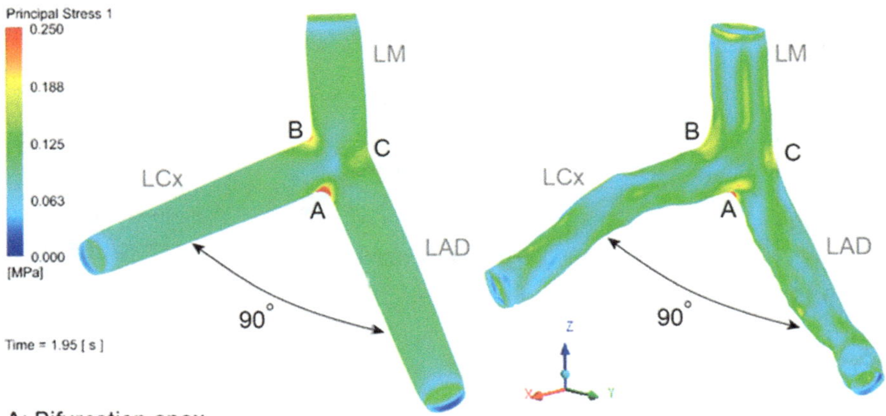

Fig. 8.23 Comparison of first principal stress at peak systole ($t = 1.95$ s). *LM*, left main stem; *LAD*, left anterior descending; *LCx*, left circumflex

the idealized ($\theta = 90°$) and the image-based models at peak systole phase. For the idealized model, high first principal stress concentrates at the bifurcation area. A maximum value of 0.28 MPa occurs at the bifurcation apex, followed by the bifurcation shoulders on LCx side (0.22 MPa) and LAD side (0.15 MPa) respectively. Similarly, the first principal stress distribution of the image-based model shows a similar result when compared with the idealized model ($\theta = 90°$). Due to its irregular vascular luminal shape, a fraction of the luminal left main stem (LM) region experiences stronger stress values than their corresponding locations on the idealized model.

The influence of bifurcation angulation on stress distribution is shown through the variations of first principal stress at the LAD side bifurcation shoulder (Fig. 8.24) the bifurcation apex (Fig. 8.25), and the LCx side bifurcation shoulder (Fig. 8.25). In general, the first principal stress shows a similar profile with aortic pressure, which demonstrates the stress variation mainly driven by the pulsatile aortic pressure. Since the angle formed by LM and LAD is kept constant (159°), the stress variation profiles at the LAD side bifurcation shoulder are similar for all models (Fig. 8.24), and hence the bifurcation angle variation does not affect the stress distribution in this region.

Figure 8.25 depicts the stress variation at the bifurcation apex, where the bifurcation angle correlates negatively with the first principal stress value. For the narrowest idealized model ($\theta = 70°$), the maximum stress $\sigma_{max} = 0.41$ MPa occurs at peak systole, while the widest idealized model ($\theta = 110°$) experiences $\sigma_{max} = 0.24$ MPa at the same time. This represents a value reduction of 41.5 % due to bifurcation angle increase.

In contrast, a positive correlation is found between the first principal stress at the LCx bifurcation shoulder and the bifurcation angle (Fig. 8.26). As the angle increases, the maximum stress also increases 50 % from 0.18 MPa ($\theta = 70°$) to

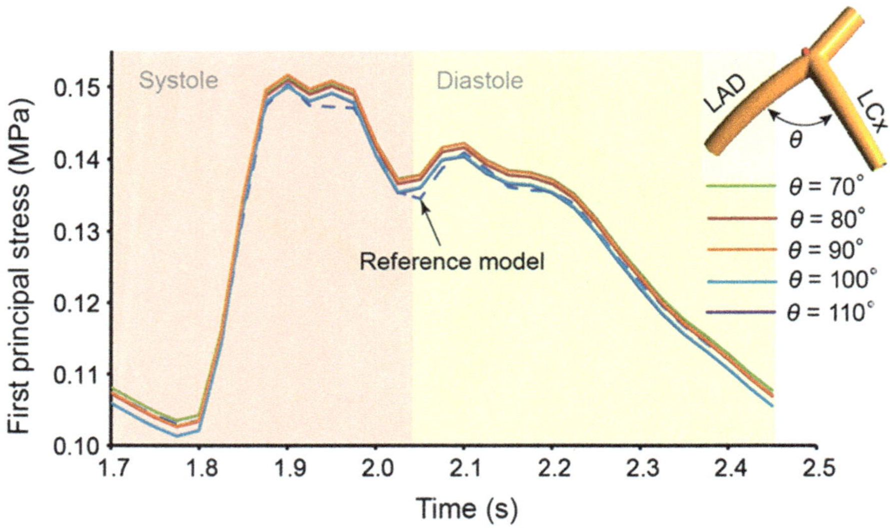

Fig. 8.24 First principal stress profiles at the LAD side bifurcation shoulder denoted by a *red dot* on the bifurcation

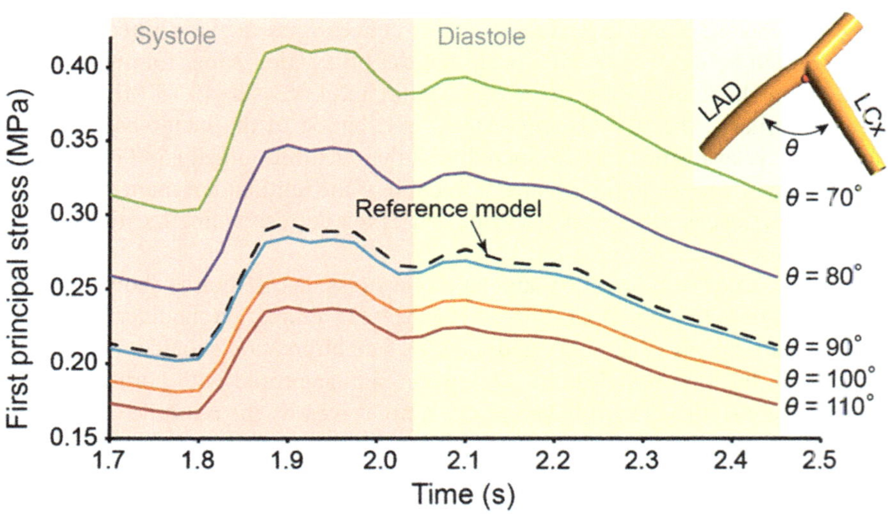

Fig. 8.25 First principal stress profiles at the bifurcation apex denoted by a *red dot* on the bifurcation

0.27 MPa ($\theta = 110°$) at peak systole. Furthermore, the overall stress value at the bifurcation apex (Fig. 8.24) is stronger than that of the LCx shoulder (Fig. 8.25) when $\theta < 100°$. When $\theta = 100°$, the stress difference between these two locations is significantly reduced. For the idealized model with $\theta = 110°$, the maximum stress

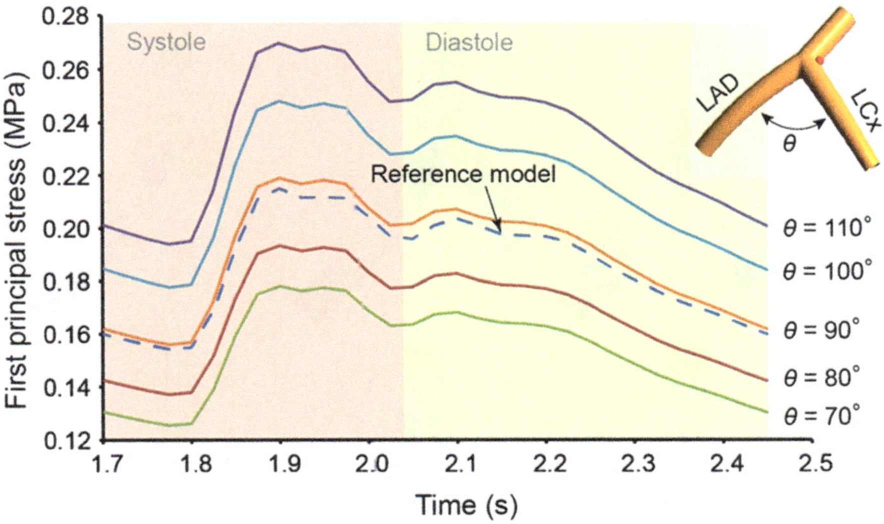

Fig. 8.26 First principal stress at the LCx side bifurcation shoulder denoted by a *red dot* on the bifurcation

occurs at the LCx side bifurcation shoulder (σ_{max}=0.27 MPa) rather than the bifurcation apex (σ_{max}=0.24 MPa).

The image-based model also displays a similar stress variation profile with the idealized model (θ=90°) both on the bifurcation apex and the LCx side bifurcation shoulder. Along with the results shown in Fig. 8.24, it can be demonstrated that the idealized models are capable of representing key FSI results for further analysis.

8.4.4 Haemodynamic Results Analysis

The WSS-based oscillatory shear index (OSI) is a flow indicator used to evaluate the total shear stress exerted on the arterial wall (Fig. 8.27). It can be regarded as the fraction of angle and magnitude change between the instantaneous WSS and the time-averaged WSS ranging from 0 to 0.5. High OSI indicates unsteady and oscillatory flow with low WSS which leads to a predisposition of endothelial dysfunction and atherogenesis (Dong et al. 2013b; Ku et al. 1985b). Generally, high OSI regions are concentrated at the origins and proximal LCx branches with no significant differences among idealized models. This indicates the affected regions are susceptible to progress atherosclerotic changes due to the presence of disturbed flow. As a reference, the image-based model shows a similar OSI distribution except in the distal LAD branch, where a locally high OSI is found caused by a moderate bulge section. Due to this luminal expansion, local flow separation and disturbance appear as a result.

Wall shear stress predicted by FSI and rigid model are compared in Fig. 8.28a. The rigid model predicts a greater distribution of higher WSS values at peak diastole

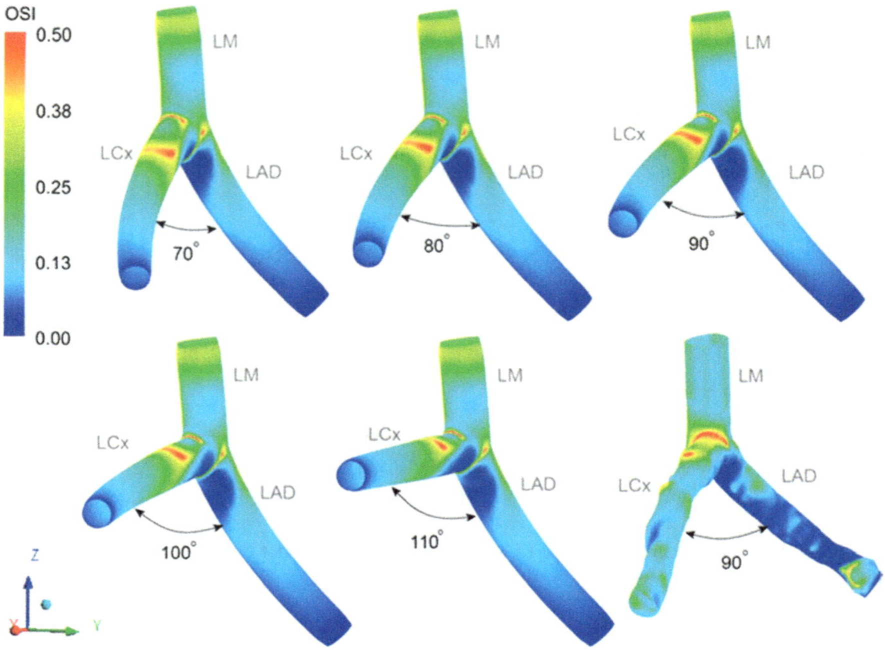

Fig. 8.27 Comparison of OSI distribution for different idealized models

(t=2.05 s) at the bifurcation apex and the narrowed lumen site downstream of the LAD branch. In contrast, the FSI model produces lower WSS values at these locations due to considerable vessel expansion driven by the pulsatile blood flow, which is in agreement with previous studies (Torii et al. 2009a). Instantaneous WSS variation at the bifurcation apex is shown in Fig. 8.28b. At the systole phase, the WSS predicted by the rigid model is slightly larger than the FSI model. However, the WSS magnitude difference becomes greater in the diastole phase. The averaged WSS predicted by the FSI model is smaller than the rigid model by 32 %. This significant difference is mainly caused by the increased mass flow rate occurring at the diastole phase than systole. Since WSS is proportional to the velocity gradient in the near wall region, and if the arterial wall is assumed to be rigid, then the blood flow will be further accelerated leading to a greater velocity gradient in the near wall region than in an FSI modelling approach.

8.4.5 *Closure*

This study provides insight into the connection between bifurcation angle and the development of coronary atherosclerosis from mechanical and haemodynamics

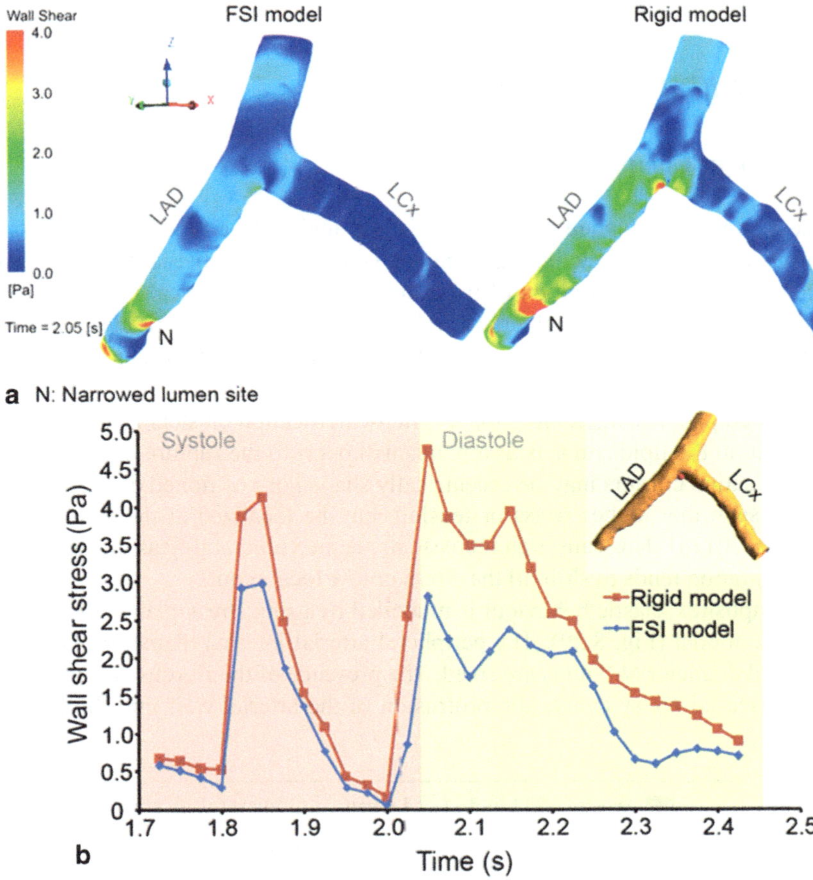

Fig. 8.28 Comparison of FSI and Rigid models for the image-based model (**a**) WSS distribution at peak diastole ($t = 2.05\ s$); (**b**) WSS variation over the last cardiac cycle at the bifurcation apex

point of view, which differs from angiography assessment conventionally used by clinical study. The clinical study by Sun et al. (2011) reports that the mean diameter of LCx in patients with a bifurcation angle $\theta > 80°$ was significantly larger than that measured in patients with bifurcation angle $\theta < 80°$ due to the presence of atherosclerotic plaques, and wider bifurcation angles are closely related to the development of atherosclerosis, thus leading to coronary artery disease. Results from this study are consistent with their reports as high tensile stress and low oscillatory wall shear stress simultaneously occur at the LCx side bifurcation shoulder in wider-angled models, high tendency of inducing atherosclerotic changes are indicated.

8.5 Analysis of Calcified Plaque

8.5.1 Calcified Plaque Models

8.5.1.1 Plaque Composite Model

We assess stress on a plaque that comprises four main tissue types: the lipid (*lp*), the fibrous cap (*fc*), the calcium agglomerate (*cag*), the non-diseased wall (*ndw*). The morphological configuration of these components is of critical importance in the quantification of plaque vulnerability. The properties of these tissues are variable and integration of these various components into a plaque structure produces different stress effects.

In calcified plaques, agglomeration of microcalcification clusters is aligned in a crescent within the lipid and acts as a buoyant support to the rupture of the fibrous cap. Calcification clusters may be eccentrically shaped or positioned distantly from the lumen such that higher stress or tension may be localized at the fibrous cap (Wong et al. 2012a). This causes an increase in plaque vulnerability as the calcification configuration tends to shift all the stress onto a focal point.

The composites' elastic behaviour is modelled by using stress-plane analysis on an idealistic model (Fig. 8.29). The peripheral arterial internal diameter (3.6 mm) and external diameter (4.0 mm) are fixed. The presence of the fibrous cap and lipid within a diseased artery causes the protrusion of the arterial wall into the lumen.

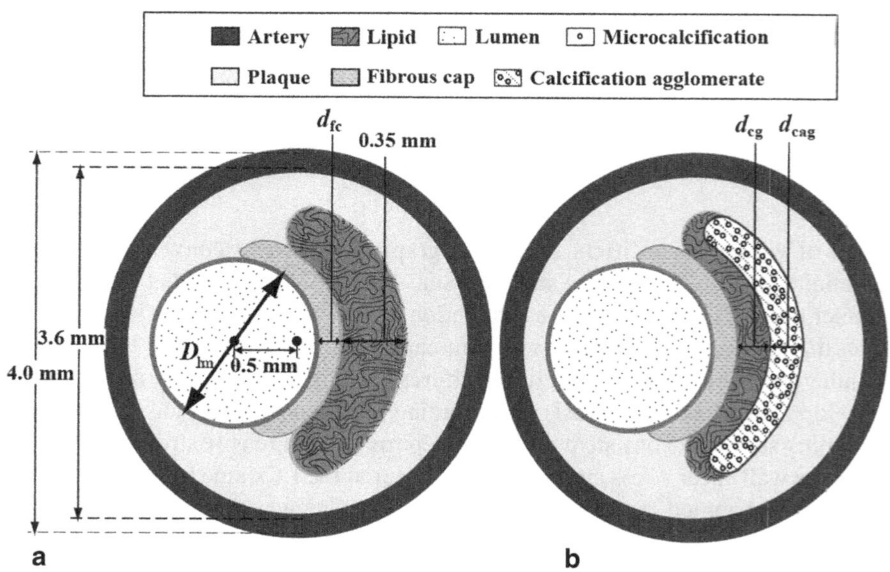

Fig. 8.29 Plaque configuration in atherosclerotic artery. **a** The morphological configuration of these plaque composites is idealized schematically to facilitate geometrical modelling for validation. **b** The definition of a calcification gap is based on distance between the calcification agglomerate and the fibrous cap

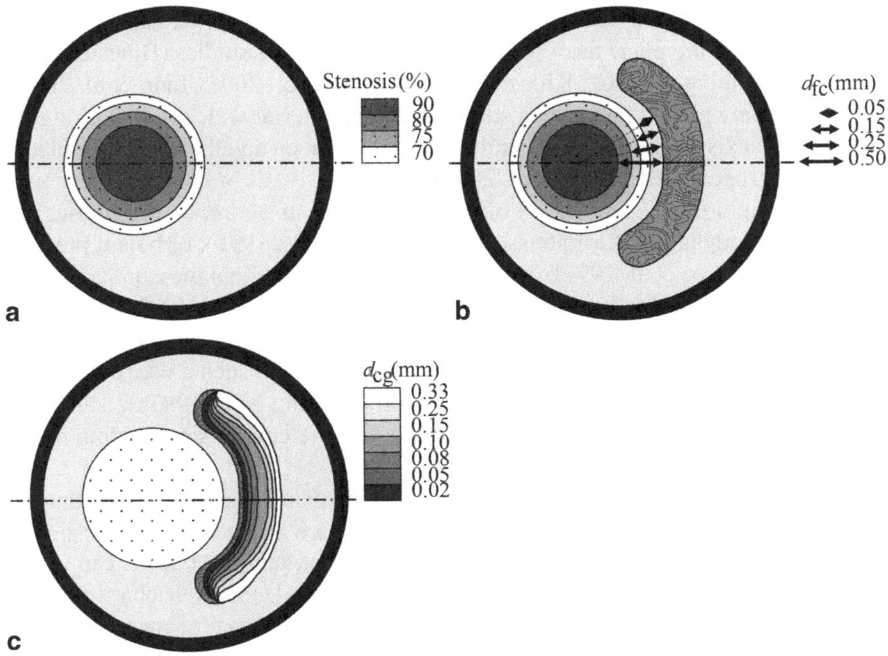

Fig. 8.30 Design of models based on varying geometrical configurations. (**a**) fibrous plaque geometry uses a constant material (**b**) fibrous plaque geometry uses a composite matrix and a constant lipid core (**c**) a set of calcified placues which comprises of a varying calcification gap d_{cg} (*from 0.02 to 0.33 mm*)

The fibrous material from the plaque occupies the interior wall of an artery such that a lumen of varying size is formed. The lumen is modelled with an eccentricity of 0.5 mm with respect to arterial centre, and with varying lumen diameter *L* that corresponds to percentage of stenosis.

The fibrous cap is assumed with the same material make-up as the plaque. For the lipids, a subintimal substance is constructed by extending a 140° crescent with thickness of 0.35 mm (Fig. 8.30a). The calcification gap refers to a lipid gap between fibrous cap and calcification agglomerate (Fig. 8.30b). The plaque morphology is based on fibrous cap width d_{fc}, thickness of calcification agglomerate d_{cag}, and calcification gap d_{cg}. A range of stenosis at 70, 75, 80 and 90 % are modelled. As the position of lipid core (of constant thickness of 0.35 mm) remains consistent, the stenotic reduction results in thinning of fibrous cap.

8.5.1.2 Plaque Rupture Mechanics

Anisotropic modelling of atherosclerotic vessel can be implemented to probe plaque vulnerability (Cheng et al. 1993b; Loree et al. 1992) where a two-dimensional modelling platform for calibrating the extent of plaque rupture is based on mechanical parameters governing the atherosclerotic configuration.

Three-dimensional analyses have also been prepared to justify the accuracy of the results based on the plane analyses of patient-specific case studies (Bluestein et al. 2008; Holzapfel et al. 2002; Kiousis et al. 2009; Li et al. 2006; Tang et al. 2005a, b, 2008). Some plaque mechanics studies examine arterial wall bending along the longitudinal axis since repetitive bending causes strain on an atherosclerotic plaque resulting in rupture (Beaussier et al. 2008).

Plaque rupture is dependent on biomechanical events acting on the fibrous cap such as haemodynamic shear stresses (Gertz and Roberts 1990), turbulent pressure fluctuations (Loree et al. 1991), cyclic variation of intraluminal pressure and maximum principal stress by the pulsatile blood pressure (Loree et al. 1992; Richardson et al. 1989a). In particular, large eccentric lipid cores are of mechanical disadvantage since circumferential tensile stresses are configured in such a way that fibrous caps have a tendency to rupture most of the time (Cheng et al. 1993b). This gives rise to the relationship between plaque rupture and the critical stress acting on the fibrous cap.

Autopsies of patients that are diagnosed with cardiac ischemia showed that the level of macrophages is high, smooth muscle cells are reduced, the proportion of crescentic acellular mass for a lipid core is significant, and the fibrous cap is thin (Davies et al. 1993; Fayad and Fuster 2001; Moreno et al. 1994a; Richardson et al. 1989a). For plaque rupture, 65 μm thickness with an infiltrate of macrophages is defined as the threshold after histological analysis (Burke et al. 1997). This can guide critical risk analysis of plaque condition.

8.5.1.3 Design of Plaque Models

Idealized plane models of the longitudinal atherosclerotic arteries were implemented to study the effects of stenotic severity on plaque circumferential stress. One set pertains to stenosis based on a homogenous wall material while the other is based on plaque with a lipid core where the constitutive model is assumed non-homogenous, anisotropic, and elastic. To numerically simulate this type of plaque-vessel, all plaque constituents were assigned with physiological mechanical properties.

For validation, a non-calcified plaque structural configuration was implemented. Two subset models of plaques with and without the lipid core are shown in Fig. 8.30a and b respectively. The effects of fibrous cap thickness d_{fc} and width of calcification gap d_{cg} on the stress levels of plaque were examined by varying fibrous cap thickness d_{fc} from 0.05 to 0.5 mm. We hypothesize that calcification plays an important role in plaque vulnerability assessment, and therefore the calcification agglomerate is modeled as a 140° crescent of variable thickness d_{cag} and positioned within the lipid. Idealistic models for analysis of calcification structural variation were designed with calcification gap d_{cg}, ranging from 0.05 to 0.33 mm (Fig. 8.30c).

The following parameters were used in a plane-stress model: Young's modulus (E) in circumferential (θ) and radial (r) directions, $v_{r\theta}$ and v_{rz} that are the Poisson ratios in r-θ and θ-z planes respectively, as well as $G_{r\theta}$ that is the shear modulus in r-θ plane.

Table 8.4 Material properties for plaque constituents. The Young's modulus (E_r, E_θ), Poison ratios ($G_{r\theta}$) and shear modulus ($v_{r\theta,}v_{rz}$) are presented for the plaque, non-diseased wall of artery and lipid. The data are revised from Loree et al. (1992)and Holzapfel et al.(2002) Fibrous tissue at $\alpha = 5\,\%$, lipid at $\beta = 20\,\%$ and calcium at $\gamma = 75\,\%$ pertain to the homogenous calcification agglomerate

Parameter	Artery (ndw)	Fibrous tissue (ft)	Lipid (lp)	Micro-calcium (Ca)	Calcification agglomerate (cag)
E_r (kPa)	10	50	1	12,600	9452.7
E_θ (kPa)	100	1000	1	12,600	9500.2
$G_{r\theta}$ (kPa)	50	500	1	12,600	9475.2
$v_{r\theta}$	0.01	0.01	0.01	0.01	0.01
v_{rz}	0.27	0.27	0.27	0.27	0.27

The elastic mechanical property of the calcification agglomerate is established based on defined percentages of fibrous plaque tissue (*ft*), lipid core (*lc*) and calcium (*Ca*):

$$E_i^{cag} = \alpha E_i^{ft} + \beta E_i^{lc} + \gamma E_i^{Ca},$$
$$G_{r\theta}^{cag} = \alpha G_{r\theta}^{ft} + \beta G_{r\theta}^{lc} + \gamma G_{r\theta}^{Ca}, \tag{8.2}$$

where i denotes r and θ represents radial and circumferential orientations respectively. The percentage of compositions α, β, and γ corresponds to fibrous tissue, lipid and calcium, respectively. Young's modulus E_i^{cag} and shear modulus $G_{r\theta}^{cag}$ are based on a linear combination of E_i^j and $G_{r\theta}^j$ with component $j = ft$, lc, and Ca.

The material properties of the plaque constituents are consolidated from Loree et al. (1992) and Holzapfel et al. (2002) in Table 8.4. In this study, we assumed a combination of fibrous tissue ($\alpha = 5\,\%$), lipid ($\beta = 20\,\%$) and calcium ($\gamma = 75\,\%$) as components for a homogenous calcification agglomerate.

The ANSYS finite element program was used to relate the stress distribution within plaque with a luminal pressure P of 14.6 kPa. The finite element mesh were tetrahedral elements with minimal skewness. The axial strain is of the order of vessel dimension and negligible with respect to the circumferential and principal strain, and therefore the analysis is based on plane-stress models. Due to the symmetry of the vessel, a half model was implemented to reduce computational costs of modelling stress. The symmetry condition was applied at the half-vessel that lies on a symmetry line. Adaptive meshing for each component of the plaque was performed to increase the mesh resolution at regions where high strain energies are localized. Such non-homogenous distribution of mesh elements improves accuracy of the numerical solution. Different grid densities were applied for the artery, fibrous plaque, lipid and calcification agglomerate as a variation of strain energies pertain to these elastic materials (Fig. 8.31).

As atherosclerosis is a complex process, multiple parameters are required to accurately model plaque vulnerability. As a prerequisite, preliminary analysis was performed on a simplified version of the model to identify the correlations between maximum principal stress, maximum deformation, fibrous cap thickness

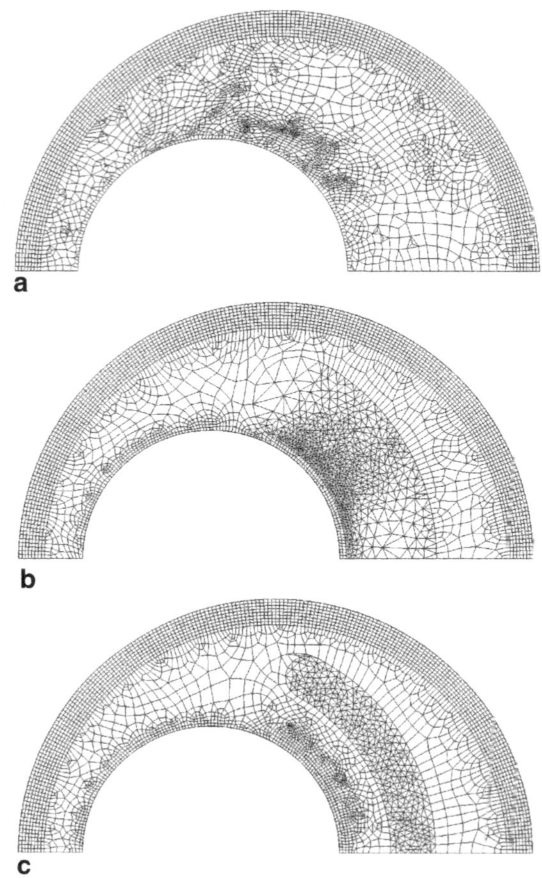

Fig. 8.31 Computer modelling of plaque models based on different configuration of lipid pool and calcification agglomerate at 70% stenosis. (**a**) plaque with no lipid pool (*70% stenosis*), $N = 12{,}928$ plane-strain elements. (**b**) plaque with constant lipid pool (*of thickness 0.35 mm*), $N = 12{,}712$ elements (**c**) plaque with constant lipid pool and calcification agglomerate ($d_{cg} = 0.175$ *mm*), $N = 12{,}123$ elements

and calcification gap. A validation case was also performed against research study by Loree et al.(1992) based on idealised atherosclerotic plaque configuration using planar stress analysis.

Figure 8.32 presents the effect of fibrous cap thickness on peak circumferential stress for cases with constant lipid, fibrous cap thickness reduces with decreasing stenosis. When there is a constant lipid inside the plaque, the level of stress tends to be strongly influenced by the thickness of fibrous cap. We deduce that since the fibrous cap thickness correlates to plaque stability, it is an important parameter when determining plaque vulnerability.

8.5.2 *Boundary Conditions and Material Properties*

To analyze the structure of the plaque components, a simulation is performed involving mechanical property variations due to changes in the lipid core and its

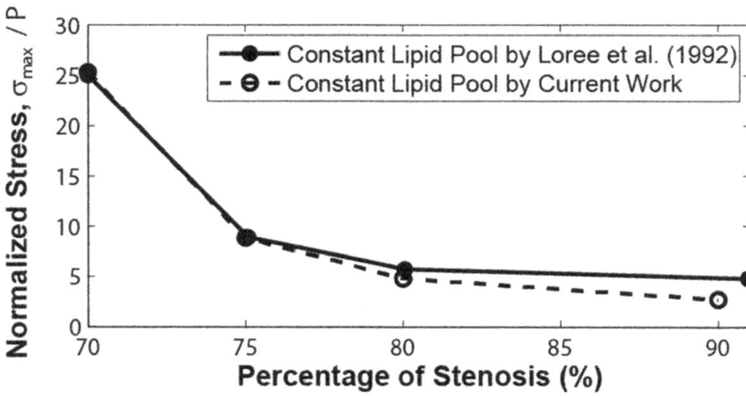

Fig. 8.32 Relationship between circumferential stress and stenosis. The graph of the circumferential stress versus degree of stenosis is presented for plaque with or without lipid pool. The peak circumferential stress is normalized with respect to the luminal pressure (σ_{max}/P). This can give an indication of the accuracy of simulation analysis using results by Loree et al. (1992) as a form of validation

agglomerate of microcalcification. Modelling of the human atherosclerotic artery with varying degrees of lipid core elasticity, fibrous cap thickness and calcification gap (distance between the fibrous cap and calcification agglomerate) form the basis of the rupture analysis.

8.5.2.1 Geometry Reconstruction and Meshing

Tada and Tabell (2005) modelled a healthy carotid bifurcation based on an ideal geometry. Key dimensions of this artery are presented in Table 8.5. For their geometry, a sinus region is included as it is a common feature found in carotid bifurcation. In this case study our model was based on a size scale of 1:2.775 when compared with the two-dimensional verification model. However, this will not affect the structural analysis if all our structural parameters are varied at the same specific ratios to achieve physiological similitude.

Table 8.5 Geometrical properties for carotid bifurcation. The dimensions for a generic carotid bifurcation are presented based on the design by Tada and Tabell (2005)

Location of carotid bifurcation	Dimensions
CCA internal diameter	0.01 m
Maximum sinus internal diameter	0.23 m
ICA internal diameter	0.007 m
ECA internal diameter	0.0065 m
ICA bifurcation angle	25°
ECA bifurcation angle	25°

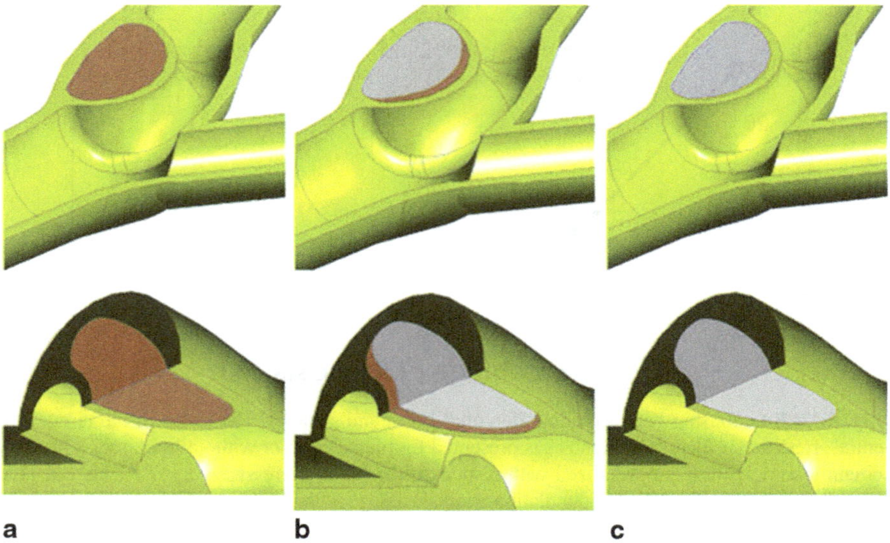

Fig. 8.33 Three-dimensional isometric view of the plaque in a carotid bifurcation. The plaque-arterial configuration can be (**a**) non-calcified plaque, (**b**) mixed type of plaque and (**c**) calcified plaque. Here, the lipid height and width is specified as 6.714 mm and 3.298 mm respectively, with the fibrous cap at 0.500 mm thick

For a diseased carotid bifurcation, the plaque location is located on the outer wall of the internal carotid artery (ICA) sinus in order to study the stenosis effect on the flow as well as the mechanical stress occurs. A three-dimensional crescent structure was incorporated into the sinus of the carotid bifurcation to simulate the presence of plaque. The plaques can be characterized into three types based on the component with it: non-calcified plaque, partially and fully calcified plaque (see Figs. 8.33a–c respectively). Fibrous tissue (α), lipid (β) and calcium (γ) pertain to the homogenous calcification agglomerate at $\alpha = 5\,\%$, $\beta = 20\,\%$ and $\gamma = 75\,\%$ as components for a homogenous calcification agglomerate.

An anisotropic modelling of the atherosclerotic vessel was used to investigate plaque vulnerability. A three-dimensional modelling platform for calibrating the extent of plaque rupture was used based on mechanical parameters governing the atherosclerotic configuration. Studies of plaque mechanics of arterial wall bending along the longitudinal axis has shown that repetitive bending causes strain on an atherosclerotic plaque resulting in rupture (Beaussier et al. 2008).

Figure 8.34 shows the computational mesh for a calcified carotid bifurcation where a refined mesh is required for the calcified region. The mesh consisted of three-dimensional tetrahedral and prism cells. Grid independence was performed at 3 different mesh refinement levels for the structure domain: coarse (130,000 elements), medium (340,000 elements) and fine (1,500,000 elements). For the fluid domain, the mesh refinement levels were: coarse (110,000 elements), medium (300,000 elements) and fine (1,000,000 elements). For both the structure and fluid domain, only 2 % of dissimilitude between the fine and medium mesh was observed.

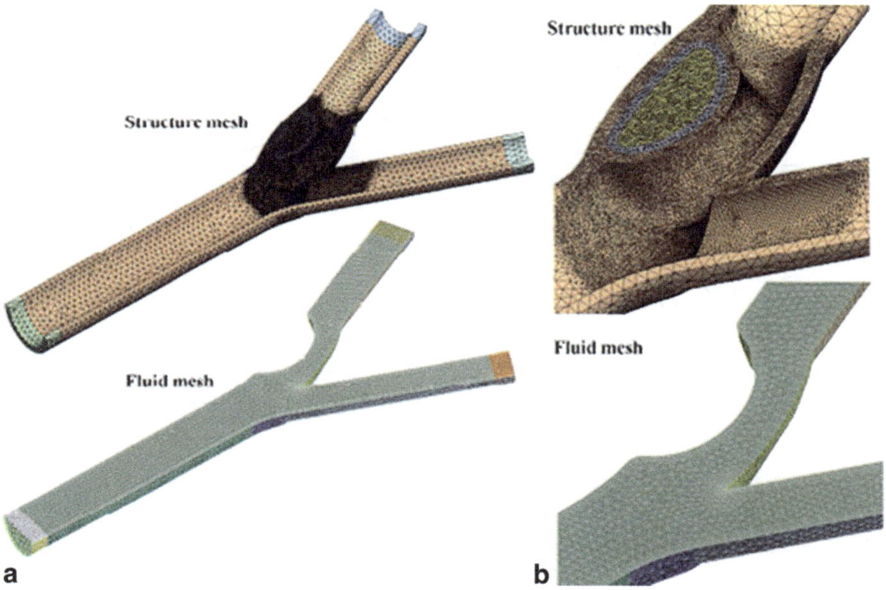

Fig. 8.34 Meshing of calcified carotid bifurcation. The isometric view of the mesh for entire length of carotid bifurcation and a *zoom-in view* of the calcified plaque gives an indication of the mesh resolution required for blood-plaque interaction simulation

8.5.2.2 Blood-Vessel-Plaque Model Setup

The partitioned approach was used to implement the FSI where coupling was performed until convergence of interface variables (displacements and pressure) is reached per time step. At each coupling loop, calculation of blood flow was initiated. The calculated pressure field was then transferred and used as an applied force in the structural domain to calculate the artery deformation. The tolerance for the interface variables was 1E-4. The blood flow was modelled as laminar where the highest Reynolds number reaches approximately 1000 in the stenosis region. The time step size was set to 0.015 s and the results were obtained at the 4th cycle to avoid any startup effects from a transient simulation.

In the solid domain, each artery end (CCA, ICA, ECA) were modelled as fixed supports while a symmetry condition was assumed at the plane of the bifurcation. In the fluid domain, the inlet boundary condition was a time-varying waveform (Tada and Tarbell 2005) while the outlet boundary conditions at the ICA and ECA were time-varying mass flow rates (Fig. 8.35). A smooth, no slip condition was specified at the artery wall. In this work, the arterial wall was modelled as a Hookean and isotropic material for computational simplicity(Salzar et al. 1995; Thubrikar and Robicsek 1995). The blood properties were also simplified as Newtonian fluid. Both blood and artery properties are shown in Table 8.6.

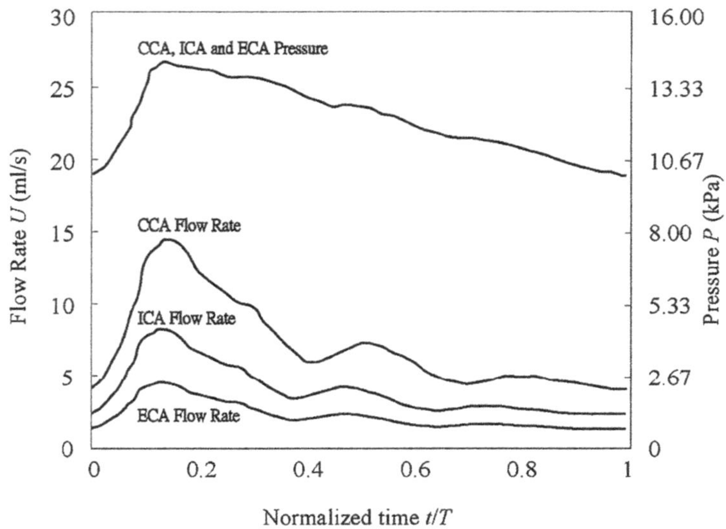

Fig. 8.35 Patient-specific pressure waveform used in simulation. Patient-specific pressure waveform for the fluid simulation within the carotid bifurcation shows a maximum of 107 mmHg used. This waveform is imposed at the entrance of the artery

Table 8.6 Material properties for artery and blood constituents. The Young's modulus and Poison ratio for artery as well as density and viscosity for blood are presented for the blood-vessel interaction simulation

Artery	
Young's modulus (Pa)	5.00E+05
Poisson ratio	0.5
Blood	
Density (kg/m^3)	1050
Viscosity (Pa s)	3.50E−03

8.5.3 Two-Dimensional Structural Modelling

Sub-intimal plaque structures such as fibrous cap thickness play an important role in plaque stress distribution. We analyze the pathological fracture caused by the increases of stress on plaque. In addition the calcification gap (which is defined as the width of the lipid layer sandwiched between the calcification agglomerate and the fibrous cap) is another variable. Due to different elastic materials in the composition, stress concentrations vary throughout the structure (Beattie et al. 1998; Loree et al. 1994). Therefore, it is of interest to simulate how the morphological configurations affect the stress levels on the plaque which lead to fracture. Sensitivity studies on the effects of lipid elasticity and fibrous cap thickness is then explored for the case of a constant lipid core.

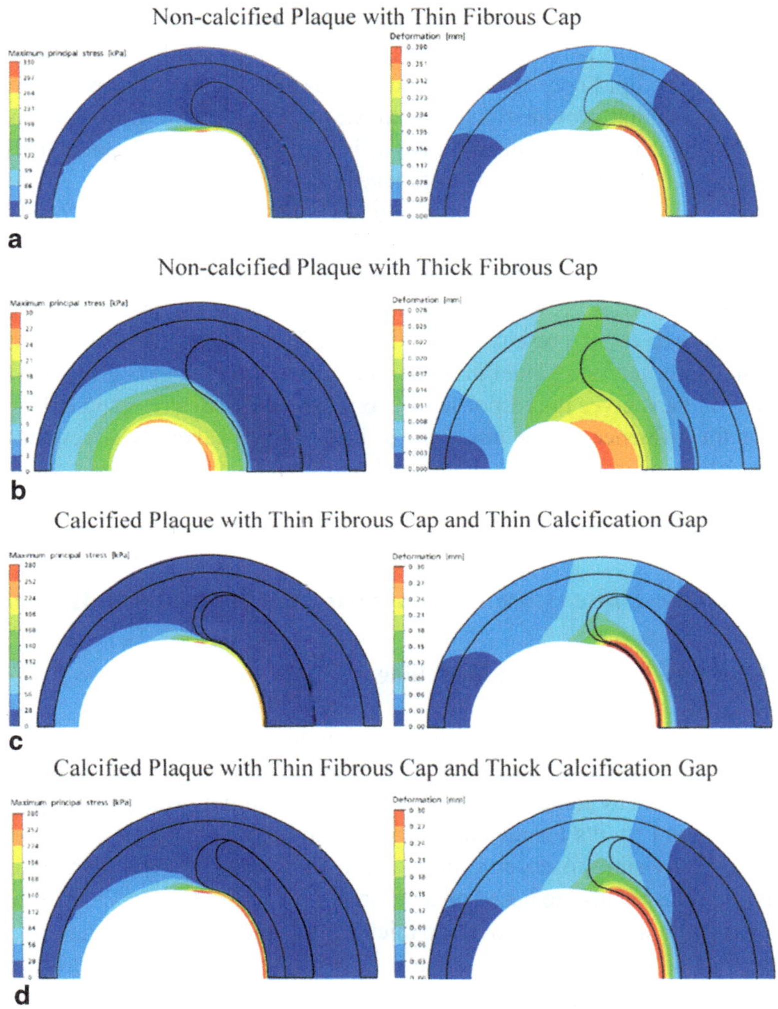

Fig. 8.36 Finite element analysis of plaque models at two-dimensional cross-sections and with different fibrous cap thickness and calcification gap. Peak principal stress and deformation plots pertain to a constant lipid pool ($E_{lp}=1\ kPa$) of fixed thickness ($0.35\ mm$), and with varying fibrous cap thickness d_{fc} and calcification gap d_{cg}. **a**: For $d_{fc}=0.05$ mm, the critical stress σ_{cr} and maximum deformation D_{max} are 331 kPa and 0.390 mm respectively. **b**: For $d_{fc}=0.48$ mm, $\sigma_{cr}=28.7$ kPa and $D_{max}=0.0281$ mm. The geometrical outlines of the plaque composites for **a** and **b** show the structural difference in the deformed plaque due to applied stress. **c**: For $d_{cg}=0.02$ mm and $d_{fc}=0.05$ mm, $\sigma_{cr}=282$ kPa and $D_{max}=0.290$ mm. **d**: For $d_{cg}=0.175$ mm and $d_{fc}=0.05$ mm, $\sigma_{cr}=326$ kPa and $D_{max}=0.370$ mm

Results for plaque models at 70% and 90% stenosis and with a constant lipid pool ($E_{lp}=1$ kPa) are illustrated by Fig. 8.36a and b. Analysis of the different plaque models with lipid cores of fixed size (0.35 mm) shows the effect of fibrous cap thickness d_{fc} on maximum principle stress and deformation. Multiple numerical models based on the variation of d_{fc} and E_{lp} was performed to characterize critical stress and maximum deformation levels. The sensitivity of the mechanical stress properties to the lipid core elasticity and fibrous cap thickness are presented with response curves that provide the interaction between different mechanical properties of the plaque material. This can give us insight into the morphological effect of plaque constituents on maximum stress levels.

Figures 8.36c and d are simulated models with a constant lipid cores whose Young's modulus is set as $E_{\mathrm{lp}}=1$ kPa and a calcification agglomerate that has Young's modulus E_{cag} based on $\alpha=5\,\%$, $\beta=20\,\%$ and $\gamma=75\,\%$. The changes in these mechanical properties are shown when calcium clusters are present. Variation of calcification gap d_{cg} is presented to show its effect on peak principal stress and maximum deformation. Modelling calcified plaque with agglomerate at varying calcification gaps gives the response of maximum principal stress and deformation based on the influence of calcium clusters. This mechanical entity affects structural integrity of the overall plaque content, and plays a major role in plaque vulnerability.

8.5.4 Three-Dimensional Fluid-Structure Interaction Modelling

8.5.4.1 Blood-Vessel-Plaque Modelling

Figure 8.37 are simulated three-dimensional models with constant lipid cores at $E_{\mathrm{lp}}=1$ kPa and a calcification agglomerate where E_{cag} is based on $\alpha=5\,\%$, $\beta=20\,\%$ and $\gamma=75\,\%$ (refer to Table 8.4). We extract the maximum principal stress and deformation contour plots for the carotid bifurcation along its longitudinal axis as it is more easily visible to observe these mechanical property variations along the fibrous cap. The results are based on a constant lipid pool ($E_{\mathrm{lp}}=1$ kPa) of fixed size (0.35 mm), with a specific fibrous cap thickness d_{fc} and calcification gap d_{cg}.

The three-dimensional plaque models at 90 % stenosis under the effect of different fibrous cap configurations are illustrated by Figs. 8.37a and b. The different plaque models with lipid cores of fixed size is effected and the influence of fibrous cap thickness d_{fc} on maximum principle stress and deformation is demonstrated to be similar to the trend shown by the two-dimensional structural analysis, whereby increment in the fibrous cap thickness d_{fc} results in a reduction of critical stress and maximum deformation.

Figures 8.37c and d are simulated blood-plaque-vessel models in which variation of calcification gap d_{cg} is presented to show its effect on peak principal stress and maximum deformation. Here, increment of d_{cg} results in an increase of these two mechanical properties.

8.5.4.2 Plaque Elasticity and Structural Variation

Response curves for stress and deformation versus plaque composite elasticity and fibrous cap thickness are plotted for the 2D (Fig. 8.38) and 3D models (Fig. 8.39).

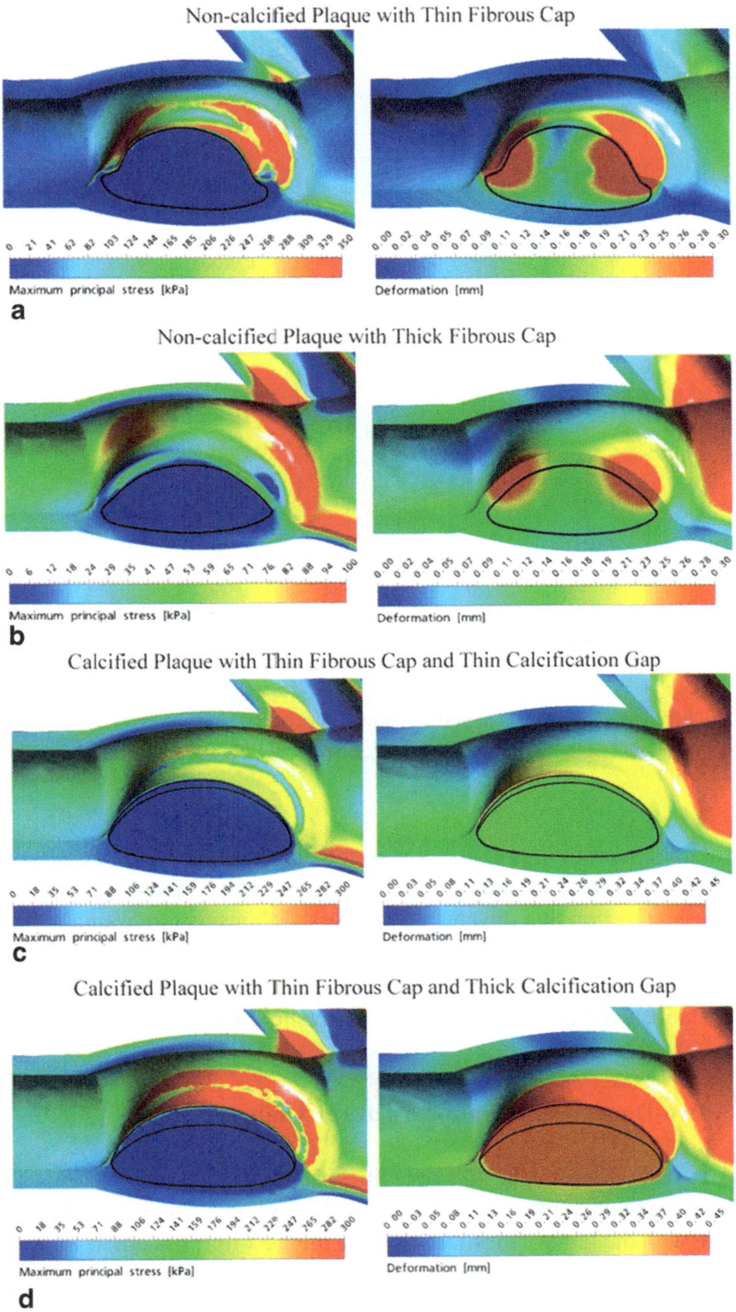

Fig. 8.37 Fluid-structural interaction analysis of three-dimensional plaque models based on a longitudinal orientation and with different fibrous cap thickness and calcification gap. (**a**) $d_{fc}=0.05$ mm, critical stress σ_{cr} and maximum deformation D_{max} are 350 kPa and 0.421 mm respectively. (**b**) $d_{fc}=0.05$ mm, $\sigma_{cr}=95.6$ kPa and $D_{max}=0.293$ mm. (**c**) For $d_{cg}=0.1$ mm and $d_{fc}=0.05$ mm, $\sigma_{cr}=258$ kPa and $D_{max}=0.352$ mm. (**d**) $d_{cg}=0.3$ mm and $d_{fc}=0.05$ mm, $\sigma_{cr}=314$ kPa and $D_{max}=0.467$ mm

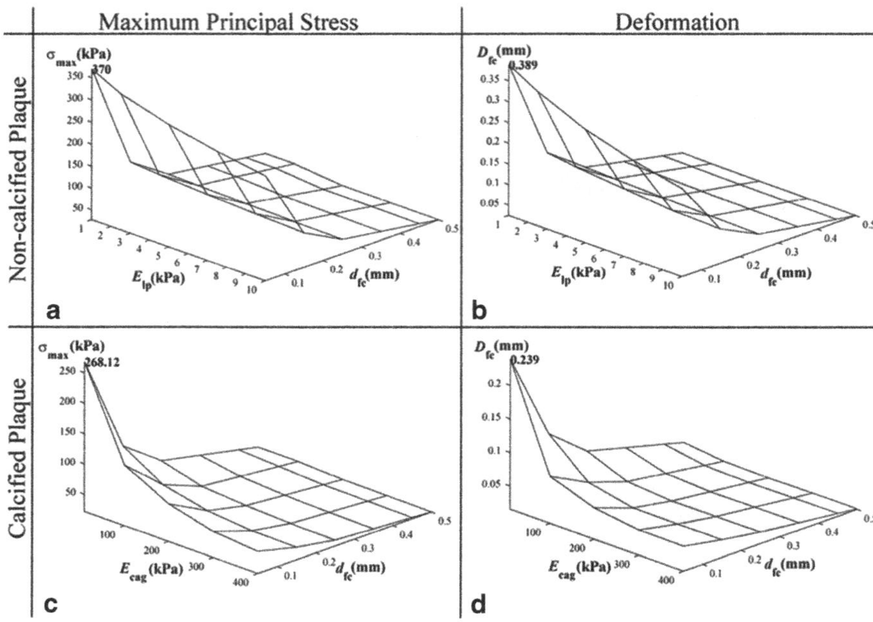

Fig. 8.38 Maximum principal stress and deformation based on elasticity of two-dimensional plaque composite and fibrous cap thickness for non-calcified and calcified plaque. (**a**): maximum principal stress σ_{max} versus Young's modulus E_{lp} and fibrous cap thickness d_{fc} shows that critical stress is 370 kPa. (**b**) σ_{max} versus E_{cag} and d_{fc} shows that critical stress is 268.12 kPa. (**c**) D_{fc} versus E_{lp} and d_{fc} shows that maximum deformation D_{max} is 0.389 mm. (**d**) D_{fc} versus E_{cag} and d_{fc} gives D_{max} = 0.239 mm

In general both maximum principal stress and deformation have a negative correlation with the fibrous cap thickness and Young's modulus of plaque composites. This leads to the suggestion that the change of stress with respect to Young's modulus of lipid core or calcification agglomerate and fibrous cap thickness tends to follow the same variation as deformation. Calcification gap and maximum deformation thresholds are established based on critical stress threshold for plaque rupture.

8.5.4.3 Two-Dimensional Fluid-Structural Analysis

The peak maximum principal stress or critical stress σ_{cr} is 370 kPa in Fig. 8.38a corresponds to the highest plaque vulnerability. This is achieved where the plaque has a lipid core with the highest elasticity and the thinnest fibrous cap. At d_{cg} = 0.02 mm, the fibrous tissue, lipid and calcium structures are present at various concentrations in the calcification agglomerate such that its Young's modulus E_{cag} varies from 10 to 400 kPa (Fig. 8.38b).

Calcified plaque stress values (where E_{lp} = 1 kPa, d_{cg} = 0.02 mm and E_{cag} ranges from 10 to 400 kPa) demonstrate the same correlation with plaque composite

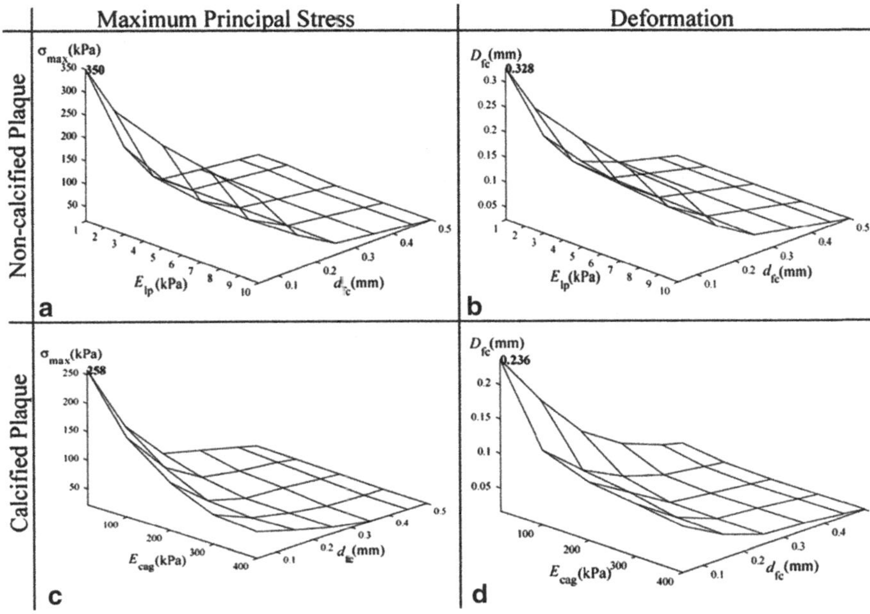

Fig. 8.39 Maximum principal stress and deformation based on elasticity of three-dimensional plaque composite and fibrous cap thickness based on non-calcified and calcified plaque. (**a**) maximum principal stress σ_{max} versus Young's modulus E_{lp} and fibrous cap thickness d_{fc} shows that critical stress is 350 kPa (**b**) at d_{cg}=0.02 mm, the plot of σ_{max} versus E_{cag} and d_{fc} shows that critical stress is 258 kPa (**c**) peak deformation D_{fc} versus E_{lp} and d_{fc} shows that maximum deformation D_{max} is 0.328 mm. (**d**) D_{fc} versus E_{cag} and d_{fc} gives D_{max}=0.236 mm

elasticity and fibrous cap thickness (Fig. 8.38c). Critical stress for a calcified plaque (σ_{cr}=268.12 kPa) is lower than that of a non-calcified one. In general, the stress levels of the calcified plaque are lower than a non-calcified one.

Peak deformation D_{max} at 0.389 mm or 389 µm corresponds to the lower limit of the range that pertains to lipid core Young's modulus and fibrous cap thickness. With calcification, D_{max} is reduced to 0.239 mm (Fig. 8.38d). The overall deformation is generally lower than that for the non-calcified plaque. The deformations are an order of magnitude higher than the fibrous cap for plaque rupture.

8.5.4.4 Three-Dimensional Fluid-Structural Analysis

An improvement in smoothness of the surface curve variation was found based on three-dimensional fluid-plaque simulation in the atherosclerotic carotid bifurcation. The critical stress for a non-calcified plaque σ_{cr}=350 kPa (in Fig. 8.39a) is higher than that of a calcified one with d_{cg}=0.1 mm at σ_{cr}=258 kPa (shown in Fig. 8.39c). D_{max} at 0.328 mm corresponds to the maximum deformation for non-calcified plaque (Fig. 8.39b). With calcification, D_{max} is reduced to 0.236 mm (Fig. 8.39d).

Typically, the simulation results follow the same trend as that of the two-dimensional plaque structural analysis. A drop in value of the blood-vessel interaction model is found when compared based on the two-dimensional structural analysis. However, the critical stress and maximum deformation follows a more accurate trend due to the realism of the blood-plaque configuration being modeled. The two-dimensional analysis can serve as a preliminary verification of the three-dimensional results.

The relationship between calcification gap and maximum principal stress is based on the stress distribution on the fibrous cap having d_{cg} varied from 0 to 0.25 mm and with $E_{lp} = 1$ kPa and $E_{cag} = 100$ kPa. Plaque rupture occurs when stress levels exceed a 300 kPa threshold (Lendon et al. 1991; Vengrenyuk et al. 2006). This stress threshold determines whether plaque fracture will occur and is based on the morphological conditions assumed in our model and for a threshold calcification gap. However it should not be assumed that all plaque fracture will occur at this value (Cheng et al. 1993b).

8.5.4.5 Two-Dimensional Structural Analysis

For the non-calcified plaque with the same fibrous cap thickness, the stress can reach as high as 370 kPa. However, the presence of calcification agglomerate at sufficiently low calcification gap can lower stress levels to below 370 kPa and prevent plaque rupture which may occur at 300 kPa. Since fibrous cap thinness threshold for rupture is 0.065 mm, we implement the case of a fibrous cap as thin as 0.05 mm as a limiting example. The calcification gap is specified as 0.02 mm consistent with Fig. 8.38.

Based on calcified plaque with fibrous cap thickness d_{fc} at 0.05 mm, the relationship between calcification gap d_{cg} and peak maximum principal stress or critical stress σ_{cr} is shown in Fig. 8.40a.). For d_{fc} at 0.05 mm as a conservative setting, a calcification gap value of >0.04 mm causes stress levels to exceed 300 kPa and cause plaque rupture. As calcification gap increases, the critical stress tends to converge to a peak maximum principal stress. The plaque is stabilized when the calcification gap is less than 0.04 mm based on the assumed plaque configuration. The critical stress σ_{cr} has a positive correlation with maximum fibrous cap deformation D_{max} (Fig. 8.40b). This is due to the correlation that exists for the calcification gap with the maximum deformation. Here, D_{max}>0.165 mm causes plaque rupture. For D_{max}> 165 μm, which is 3.3 times the fibrous cap thickness (0.05 mm), stress levels exceed 300 kPa.

8.5.4.6 Three-Dimensional Fluid-Structural Analysis

The critical stress profiles for a three-dimensional fluid-structure analysis are shown in Fig. 8.41. A similar trend to the two-dimensional analysis is found for the three-dimensional model with calcified plaque where the calcification gap is specified at

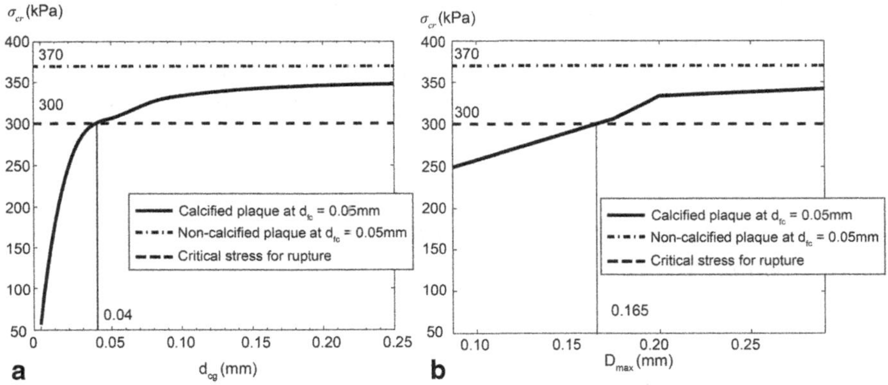

Fig. 8.40 Critical stress on fibrous cap with respect to width of calcification gap and maximum deformation on fibrous cap for two-dimensional plaque-rupture analysis. (**a**) of critical stress σ_{cr} versus width of calcification gap d_{cg} reflects the decrease in plaque vulnerability for increasing occupancy of the calcification agglomerate (*which is inversely correlated to* d_{cg}). (**b**) Critical stress σ_{cr} versus maximum deformation D_{max} of fibrous cap for d_{fc} at 0.05 mm demonstrates that σ_{cr} becomes lower as d_{cg} minimizes the deformation

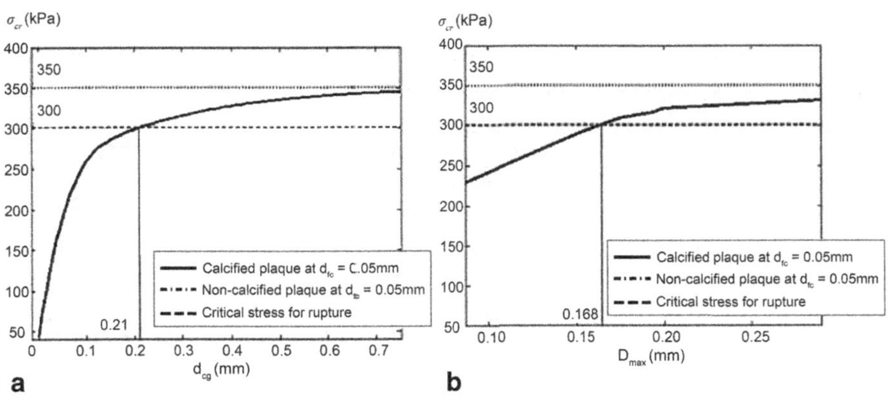

Fig. 8.41 Critical stress on fibrous cap with respect to width of calcification gap and maximum deformation on fibrous cap for three-dimensional plaque rupture analysis. (**a**) critical stress σ_{cr} versus width of calcification gap d_{cg} shows that calcification gap value of >0.21 mm causes stress levels to exceed 300 kPa and cause plaque rupture. (**b**) Critical stress σ_{cr} versus maximum deformation D_{max} of fibrous cap for d_{fc} at 0.05 mm demonstrates that D_{max} >0.168 mm causes plaque rupture

0.1 mm. Figure 8.41a illustrates the relationship between calcification gap d_{cg} and critical stress σ_{cr}, while Fig. 8.41b correlates the critical stress σ_{cr} with maximum fibrous cap deformation D_{max}. We note a slight reduction in critical stress below the 350 kPa threshold. The limiting calcification gap occurs at $d_{cg}=0.21$ mm before plaque rupture takes place for critical stress at 300 kPa. A larger calcification gap is found (compared to the results of the two-dimensional model) occurring due to the artery-plaque structure that is 2.775 times larger in size.

The maximum deformation also assumes the same trend as the two-dimensional structural stress analysis. The larger artery in the carotid bifurcation model, produces a reduction in terms of value for critical stress. The maximum deformation is observed to be approximately the same at $D_{max} > 168$ µm for the stress levels to exceed 300 kPa.

8.5.5 Correlations Between Plaque Progression and Wall Shear Stress

In this section we review a selection of work in the literature regarding carotid plaque progression using FSI. It is well accepted that atherosclerosis initiation and progression correlate positively with low and oscillating flow wall shear stresses.

Tang et al. (2008) investigated advanced plaque growth under elevated oscillating flow wall shear stresses conditions. Correlations were quantified between plaque progression measured by wall thickness increase and plaque wall (structure) stress conditions. For the 3D FSI model, the flow was set as laminar, Newtonian, viscous, and incompressible. The arbitrary Lagrangian-Eulerian (ALE) formulation was used. Inlet and outlet boundaries were fixed in longitudinal (axial) direction, but allowed to expand/contract with the flow. No-slip conditions, natural traction equilibrium boundary conditions, and continuity of displacement were assumed on all interfaces between all components and the interface between solid and fluid. Material stress–stretch curves for fibrous tissue (vessel), lipid pool, and calcifications, prescribed inlet and outlet pressure conditions and the corresponding flow rates obtained from the FSI model are shown in Fig. 8.42.

Baseline results are presented in Fig. 8.43 which shows typical stress distributions (Stress-P1, which is the maximum principal stress) including the flow behaviors obtained from the 3D FSI plaque model. Elevated flow shear stress is observed in the stenotic region where vessel wall is thick.

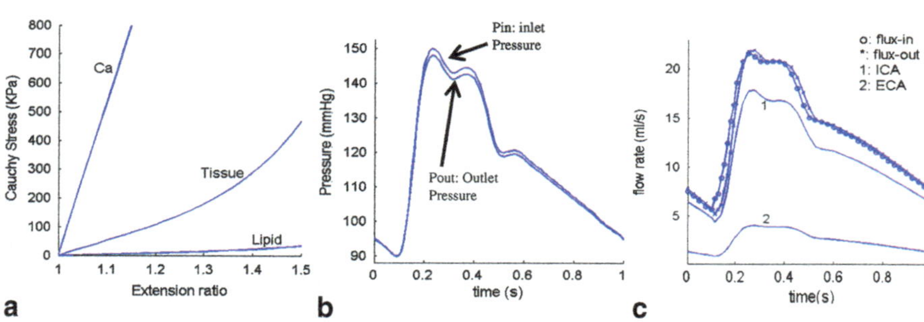

Fig. 8.42 Material curves and pressure conditions for the multi-component plaque model. (**a**) Stress–stretch curves derived from the modified Mooney–Rivlin model. (**b**) pressure conditions specified at the inlet (*CCA*) and outlet (*ICA and ECA*); (**c**) flow rates obtained from the 3D plaque model solutions corresponding to the pressure conditions. (*Image from Tang et al. 2008*)

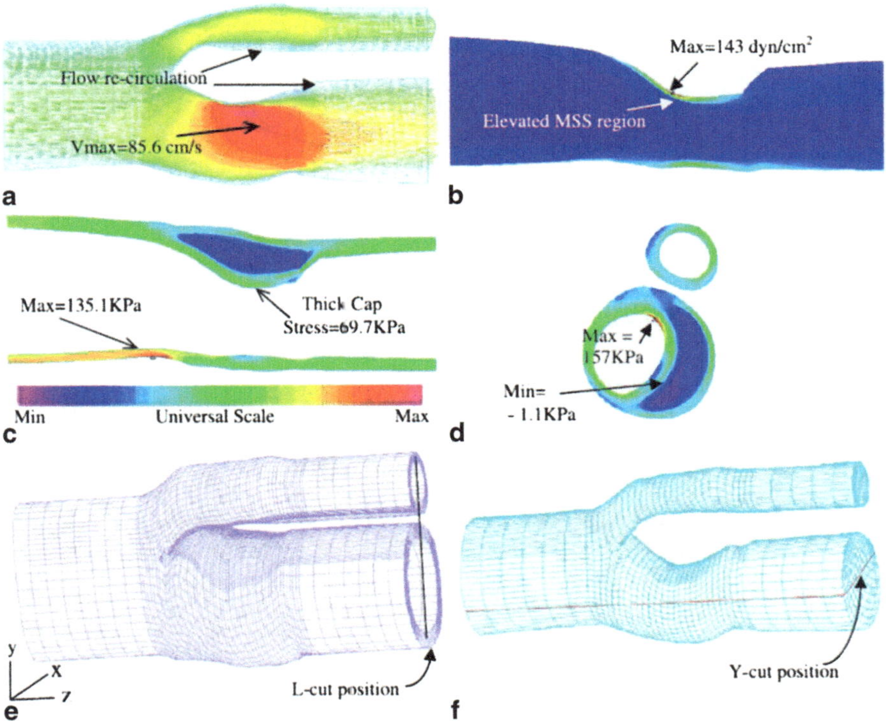

Fig. 8.43 3D FSI model baseline solution plots showing both structure stress and flow velocity and flow maximum shear stress behaviors. (**a**) Velocity on a L-cut surface; (**b**) Flow Max-Shear-Stress on a Y-cut surface; (**c**) Stress-P1 on a Y-cut surface, Pin = 110 mmHg; (**d**) Stress-P1 on a cross section cut; (**e**) L-cut position; (**f**) Y-cut position. (Image from Tang et al. 2008)

Figure 8.44 gives six matching slices (out of eleven) of a plaque sample. WTI (wall thickness increase) vs. Stress-P1 at Time 2 using results from 3D FSI modelling show that the Pearson's Correlation value was −0.528 ($p<0.0001$) which is highly statistically significant. To compare 2D and 3D FSI models, WTI vs. Stress-P1 (at Time 2) using results from 2D models gave a Pearson's Correlation of −0.497, $p=0.0001$. Tang et al. (2008) notes that stress values from 2D and 3D plaque models can be different because 3D model includes additional stretch and slices. While 2D and 3D models give different Stress-P1 values, the results show similar distribution patterns. They suggest that plaque thickening tends to occur where the wall is thicker due to the *negative correlation between WTI and Stress-P1*. This work showed that both lower plaque wall stress and lower flow shear stress may contribute to continued plaque progression and it is suggested that this should be taken into consideration in future investigations of diseases related to atherosclerosis.

The previous example showed that atherosclerosis initiation and early progression can be negatively correlated with flow wall shear stresses. Furthermore low and oscillating blood flow shear stresses have shown to correlate positively with

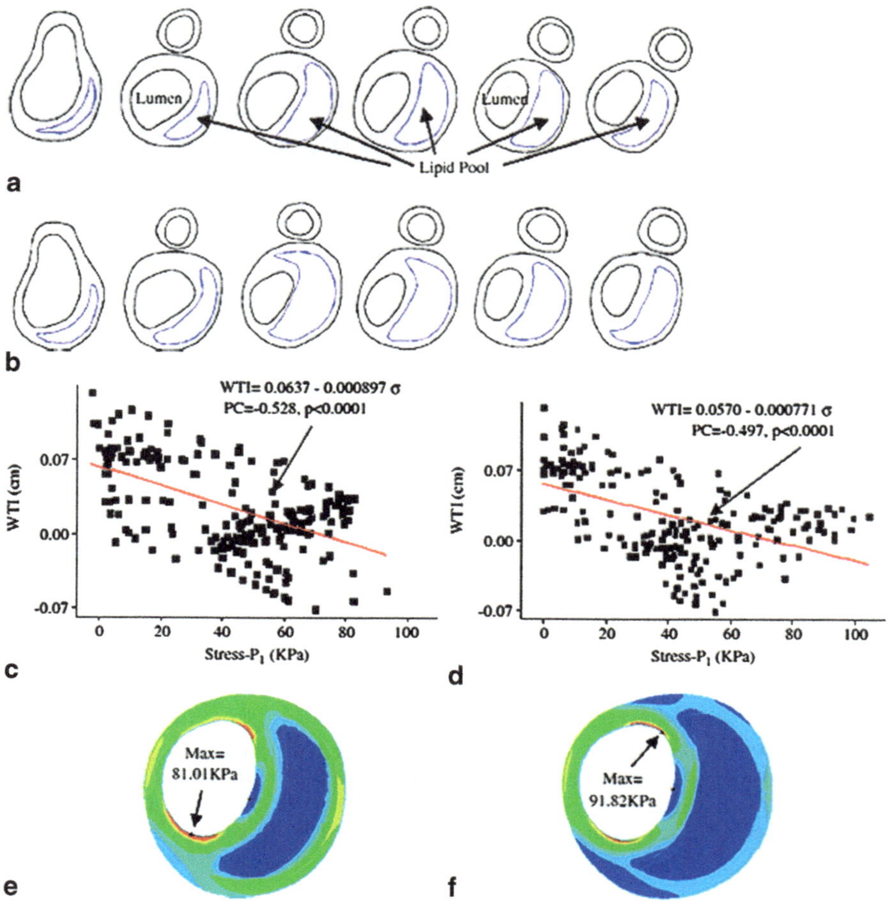

Fig. 8.44 Results from 3D FSI model show negative correlation between wall thickness increase (*WTI*) and plaque wall stress Stress-P1 (*PWS*), supporting 2D correlation results. (**a**) Real patient scan at Time 1; (**b**) Patient scan at Time 2 (*10 month later*); (**c**) WTI vs. Stress-P1 (*3D*); (**d**) WTI vs. Stress-P1 (*2D*) (**e**) Stress-P1 from 3D FSI Model; (**f**) Stress-P1 from 2D Model, the same slice. (Image from Tang et al. 2008)

intimal thickening and atherosclerosis initiation. Despite this, intermediate and advanced plaques continue to grow under high shear stress conditions. The next example is by Yang et al. (2010) which investigated the correlation of flow shear stress with advanced plaque progression. The 3D models with FSI were created similarly to the previous example where serial MRI data of carotid atherosclerotic plaques were acquired 2–4times over an interval of 18 months with an example of the models shown in Fig. 8.45. The material properties and boundary conditions make use of the same profiles given earlier in Fig. 8.42.

FSI models were constructed for 32 cases of the common carotid artery and internal carotid artery while the external carotid artery (ECA) was omitted because it

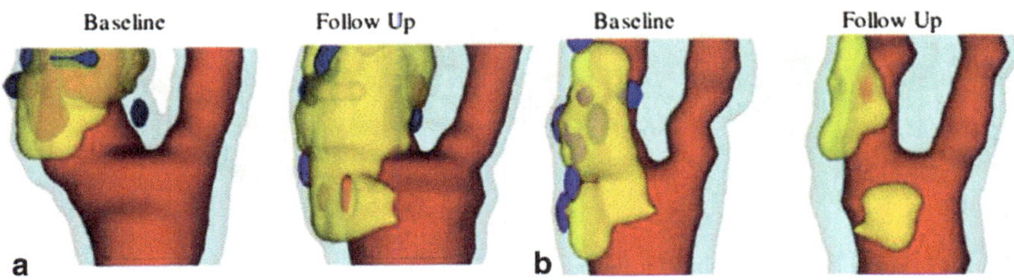

Fig. 8.45 3D plaque samples re-constructed from in vivo MR images showing progression and regression (**a**) One sample showing plaque growth. (**b**)One sample showing plaque reduction. Scan time interval: 18-months. *Red*:lumen; *yellow*:lipid; *dark blue*:calcification; *light blue*:outerwall. (Image from Yang et al. 2010)

is less prone to atherosclerosis. Flow shear stress and plaque wall stress fields were obtained for correlation analysis. These were taken at baseline time and follow-up scan time (18-months later). Their results found that wall thickness increase correlates positively with flow shear stress, the overall correlation was rather weak. There were 26 cases that showed the negative correlation with plaque wall stress, and it was suggested that both fluid and structural forces should be considered as possible mechanisms governing plaque progression. There was increased plaque progression in the posterior quadrants of the ICA, corresponding to disturbed flow region, and lipid rich necrotic core regions were associated with higher flow shear stress values.

In an attempt to reduce the computational modelling burden of creating the models, the lipid core can be treated as a singular entity referred to as a no-component lipid pool. Figure 8.46 compares plaque wall stress and flow shear stress distributions from the no-component model and the model with components us-

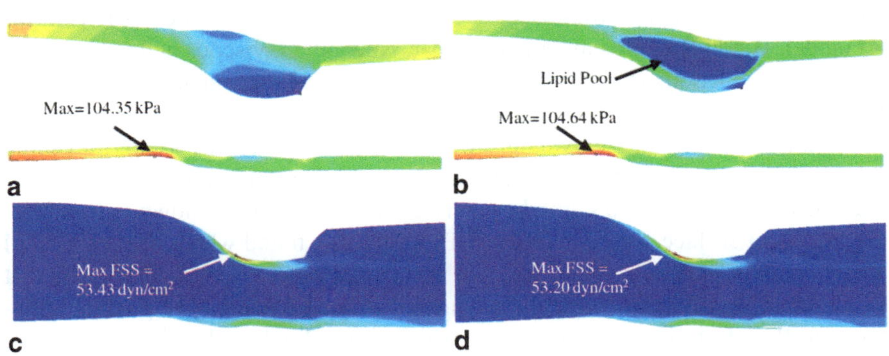

Fig. 8.46 Comparison of plaque wall stress (*PWS*) and flow shear stress distributions on a bifurcation cut surface and a longitudinal cut surface from no-component model and model with components showing good agreement for flow shear stress and reasonable agreement for PWS (**a**) PWS on a Longitudinal Cut, No-Component Model, (**b**) PWS on a Longitudinal Cut, Model with Component, (**c**) flow shear stress on a Longitudinal Cut, No-Component Model and (**d**) PWS on a Longitudinal Cut, Model with Component. (Image adapted from Yang et al. 2010)

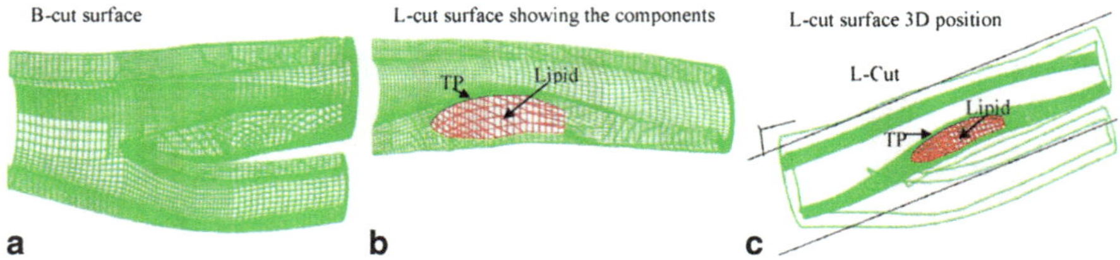

Fig. 8.47 Finite element meshes for the computational model showing different cut-surfaces. (**a**) Mesh for the solid domain showing the position of a bifurcation-cut (*B-cut*) surface; (**b**) A longitudinal-cut (*L-cut*) surface showing the lipid core and thin plaque cap; (**c**) Location of the L-cut surface. (Image adapted from Yang et al. 2010)

ing sliced planes showing bifurcation and another sliced plane showing the lipid core.

Flow shear stress distributions from both models were similar on the two planes, since the no-component modelling simplification does not change the fluid domain and its effect on the overall plaque deformation. However there is a difference in the plaque wall stress distribution on the longitudinal which contained a lipid pool. Yang et al. (2010) suggests that the errors introduced by the no-component approach depend heavily on the number of components, component size and the total "thin-cap" lumen area.

A further example studying patient specific 3D FSI models of carotid plaques is by Huang et al. (2009). Their work presented a method to quantify human carotid artery axial and inner circumferential shrinkages, based in vivo MRI data. However for this example we review the plaque stress, strain, flow velocity, and shear stress behaviour results. Scans of ten patients were used to reconstruct the carotid bifurcation using the CASCADE software while the computational mesh used the ADINA software (Fig. 8.47). The artery wall and plaque components were assumed hyperelastic, isotropic, incompressible and homogeneous. The vessel wall and plaque component material properties used the 3D nonlinear modified Mooney-Rivlin (M-R) model. Blood flow was assumed laminar, Newtonian, viscous and incompressible and the equations were discretised to the Arbitrary-Lagrangian-Eulerian (ALE) formulation.

Results showing the maximum principal stress (Stress-P1) and maximum principal strain (Strain-P1) distributions, flow velocity, pressure and maximum-shear-stress (MSS) on two sagittal cut surfaces for one case are shown in Fig. 8.48. On the B-cut surface, maximum Stress-P1 value was found where the vessel wall was thin and a minimum value occurred at the lipid pool. Figure 8.48b shows that maximum Strain-P1 was located at plaque cap. Figure 8.48c–f give plots of structure and flow features on L-cut surface which shows lipid pool and cap thickness. The distribution of Stress-P1 shows a maximum value located at the lipid cap position. Flow velocity is higher at the stenosis narrowing of internal carotid artery (ICA). A maximum value of Maximum-Shear Stress occurred at the plaque throat where narrowing occurs.

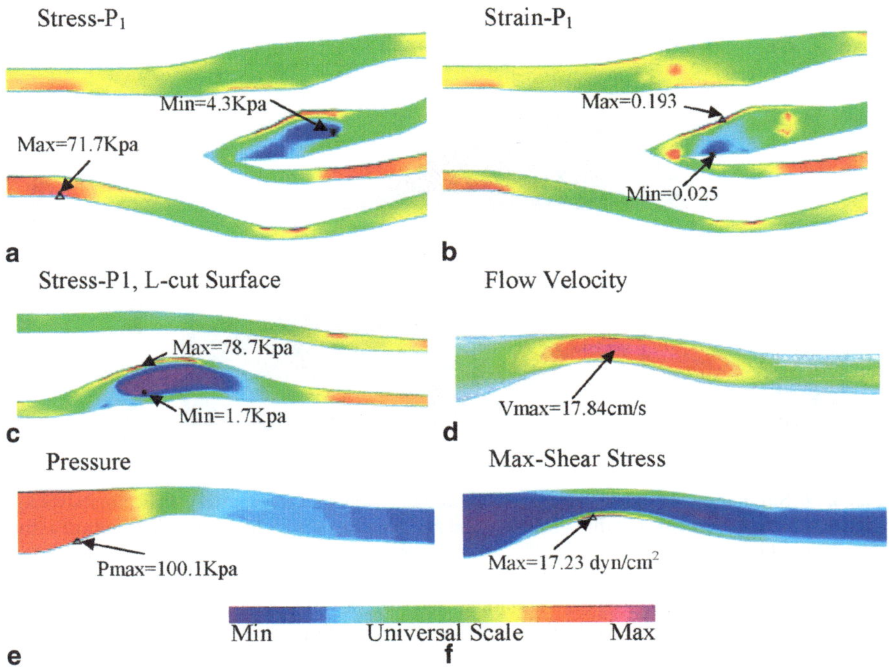

Fig. 8.48 Stress/strain distributions and flow characteristics from a 3D FSI model (*10% axial stretch, 7.8 inner circumferential shrinkage*), *Pin*=100 mmHg. (**a**) Plot of maximum principal stress (*Stress-P1*) distribution on B-cut surface; (**b**) Plot of maximum principal strain (*Strain-P1*) distribution on B-cut surface; (**c**) Stress-P1 on L-cut surface; (**d**) Flow velocity reaching its maximum in the stenotic region; (**e**) Pressure plot on L-cut surface; (**f**) Flow maximum shear stress plot on L-cut surface showing a maximum at the stenosis throat. (Image from Yang et al. 2010)

8.5.6 Mechanical Stresses in 2D Carotid Plaque

In this next group of examples, we review recent work in the literature that focusses on the mechanical stresses in simplified and patient specific 2D geometries models. Li et al. (2006) used a fluid–structure-interaction model to simulate the blood flow through a vessel with an atheromatous plaque. This work evaluated the effect of luminal stenosis and fibrous cap thickness on plaque vulnerability. An idealised 2D model with a lipid core was used.

The modelling approach conforms to the previous examples presented where the flow was assumed laminar, Newtonian, viscous, and incompressible and applied within an Arbitrary Lagrangian-Eluerian formulation using the COMSOL software package (see Appendix for list of software). The inflow boundary condition in this case was a fully developed laminar velocity profile on the left boundary, but the amplitude of the flow varied with time. The outflow, on the right boundary was a pressure condition with $P=0$. The lipid core of the plaque was assumed incompressible and nonlinear using a 2-term Ogden strain energy formulation to represent the

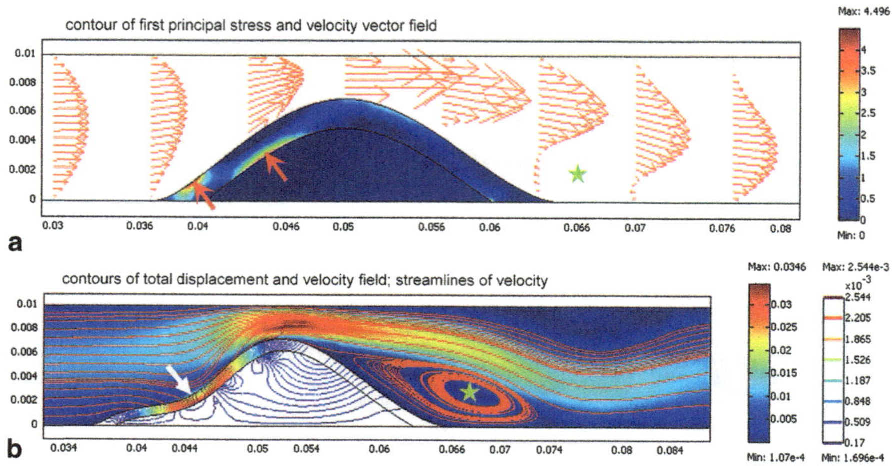

Fig. 8.49 Simulation results of pulsatile flow through a stenotic carotid artery. (**a**) velocity vectors showing the flow velocities within the artery and surface plot of principal stress distribution within the plaque. (*Luminal stenosis was 70 % and the fibrous cap thickness was 1 mm*). (**b**) Surface plot showing flow velocities within the artery and flow streamlines. The contour plot within the plaque shows the strain distribution. (Luminal stenosis was 80 %, and the fibrous cap thickness was 0.5 mm. *White arrow* shows the maximal plaque deformation, and *star* indicates the large flow recirculation zone). (Image from Li et al. 2006)

hyperelasticity of the human tissue. Luminal stenosis (ϕ) was varied from 10 to 95 %, and the fibrous cap thickness (d) was varied from 0.1 to 2 mm. The flow field and stress/strain distribution within the plaque were calculated for every degree of luminal stenosis and fibrous cap thickness used.

Velocity vectors and stress distribution in a vessel with a 70 % degree of luminal stenosis are shown in Fig. 8.49a. The fibrous cap thickness is 1.0 mm. Reverse flow occurs distal to the plaque, which is a fluid dynamics characteristic produced by separated flows (see Chap. 4). The principal stress contours are shown in the fibrous cap and the lipid pool, and high stress concentrations are found within the fibrous cap. This occurs at the shoulder regions of the plaque where the oncoming fluid tends to build up pressure. In Fig. 8.49b the luminal stenosis is 80 %, and the fibrous cap thickness is 0.5 mm.

The plaque is deformed with a maximum deformation of 2.44 mm because of the large luminal stenosis (80 %) and thin fibrous cap (0.5 mm). Severe stenosis and 100 % eccentricity lead to high flow velocity, high pressure at the throat of the stenosis, and a large recirculation region.

Li et al. (2006) showed that when the plaque changed from concentric to eccentric, the maximal plaque stress was slightly increased. This effect was smaller when compared with the effect of thinning the fibrous cap. It is suggested that plaque eccentricity does not play a major role in plaque vulnerability.

Kock et al. (2008) studied longitudinal 2D FSI models from MRI scans to predict plaque rupture risk and examinations of correlations between local stress variation and morphology. Blood-flow was simulated in a Navier–Stokes model, and treated

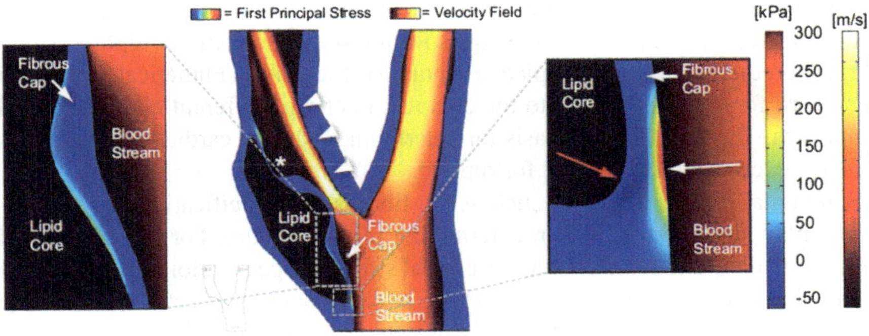

Fig. 8.50 First principal stresses are presented in the surrounding tissues along with blood velocities. The fibrous cap is seen to exhibit severe stresses, maximal in the inflow region (*white arrow*). A pressure zone with negative first principal stresses (*red arrow*) was visible adjacent to the area exposed to maximal stresses. The severe luminal narrowing resulted in a marked velocity jet (*arrowheads*) and area of recirculating blood (*asterisk*). In contrast to findings at the inflow region, stresses at the position depicted in the left inset were maximal at the interfacial region between lipid core and fibrous cap. (Image from Kock et al. 2008)

as an incompressible, homogeneous, Newtonian, however Kock et al. (2008) used a turbulent model to predict the fluid flow citing that carotid bifurcations with large degrees of stenosis are more accurately depicted using *k*-ω models than both laminar flow and *k*–ε models. Tissues were simulated as isotropic homogenous entities with non-linear stress/strain dependency of human tissues based on a Neo-Hookean hyper-elastic model. Their results showed first-principal stress distribution (max. 254.1 kPa, mean 181.4 kPa) and flow velocities for a patient with a systolic blood pressure of 160 mmHg, severe carotid stenosis, and a large lipid core located immediately below the carotid bifurcation beneath the internal carotid artery (Fig. 8.49).

The severe stenosis reduces the cross-sectional area, which produces a flow acceleration characterized by a marked velocity jet (max. velocity 0.784 m/s). Following the minimum cross-sectional area, the flow separates and large areas of recirculating blood flow are produced. The soft lipid core deformed significantly generating severe stresses in the overlaying fibrous cap, most prominent in the upstream and downstream shoulder regions of the wall-adjacent fibrous cap region. These regions experienced the highest principal stresses of 175.4 (upstream shoulder) and 254.1 kPa (downstream shoulder). Kock et al. (2008) suggests that maximal first-principal stress level, approaching established criteria for plaque rupture, are located in the upstream and downstream shoulder regions of the fibrous cap (Fig. 8.50).

8.5.7 Closure

Medical imaging modalities are able to characterize the atherosclerotic plaque in terms of their morphological and mechanical properties. Non-invasive imaging techniques not only identify flow-limiting vascular stenosis, but also detect calcified and non-calcified plaque, measure atherosclerotic plaque burden and its response to treatment, and differentiate stable plaques from those which tend to rupture. However, the prediction of high-risk plaque rupture still requires a numerical simulation framework for verification due to the complex matrix of different elastic material composites. This can form the basis for determining adverse cardiovascular events that have exceeded the threshold for rupture.

Subintimal plaque structures such as the fibrous cap, calcification gap and lipid core play an important role in determining plaque rupture. For a non-calcified plaque with constant luminal area, the critical stress and peak deformation increase as the fibrous cap becomes thinner. These two mechanical effects reduce in the presence of calcification agglomerates. For a thin fibrous cap and a large calcification gap, the stress levels will be significant and results in high vulnerability of the plaque despite the fact that they may show angiographically insignificant. Therefore, the subintimal structure should be used as the basis for determining plaque vulnerability instead of information on stenotic severity that is based on medical image visualisation.

Calcification clusters plays a major role in plaque rupture demonstrated by structural analysis on a continuous calcification agglomerate structure. The agglomerate model is a linear combination of microcalcification, fibrous plaque and lipid at specific percentages and assumed a uniform property based on this homogenous mixture, which may be adjusted depending on patient-specific density of calcium in plaque. This material has varying degrees of elasticity, and attached to the distensible artery. As such, a FSI framework is required to simulate the blood-vessel-plaque model.

8.6 Flow in a Realistic Aortic Aneurysm

Understanding of the underlying processes that lead to the growth and structural weakening of an abdominal aortic aneurysm (AAA) is of critical importance in both diagnosis of the lesion progression and design of the patient-specific intervention. AAAs have been associated with local and systemic alterations of the aorta, influenced by age as well as genetic factors (Baxter et al. 1994; Goodall et al. 2001). Marked reduction of the elastin content in AAA tissues has been reported in several studies (Campa et al. 1987; He and Roach 1994; Menashi et al. 1987). It has been suggested that elastin degradation is attributed to the elevated activation of proteolytic matrix metalloproteinases that can be induced by various factors such as the abnormal distribution of wall shear stress (WSS) (Hoshina et al. 2003; Nakahashi et al. 2002; Sho et al. 2004), inflammatory responses (Choke et al. 2005; Middleton et al. 2007), and intraluminal thrombus formation (Fontaine et al. 2002; Vorp et al. 2001).

To improve the understanding of blood flow patterns and stress distribution within the aneurysmal wall of AAA, one subject-specific model was constructed from CT data and physiologic reasonable boundary conditions were adopted representing situations under dynamic exercise. For simplicity, only the main aortic vessel was considered in the simulation, while its daughter branches were neglected.

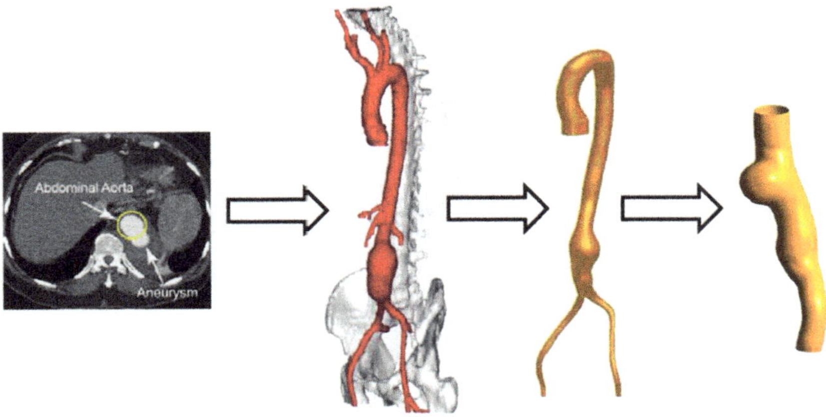

Fig. 8.51 Reconstruction of an abdominal aortic aneurysm based on multi-slice CT angiography

8.6.1 Geometric Models and Material Properties

An anatomical realistic model was reconstructed from multi-slice CT angiography of an abdominal aorta with a moderate aneurysm. The reconstruction technique has been described in the previous chapter. In this case study, only the artery vessel with aneurysm (Fig. 8.51) was selected for numerical simulation.

The aneurysm wall was assumed as hyperelastic, homogenous, incompressible and isotropic. Similar to the carotid artery vessel, a widely accepted strain energy density function, the Mooney-Rivlin model was used to express its elasticity (see Eqn 8.1). The model parameters were set to C_{10}=0.070 MPa, C_{20}=3.2 MPa, C_{21}=0.0716 MPa, D_1=0.1 MPa, and C_{ij}=0 MPa for the others. The arterial vessel wall thickness is 0.5 mm thickness. Using multi-blocking with O-grid strategy, the computational fluid domain was meshed into structured elements, and uniform thickness was assigned on the fluid domain surface through extruding the arterial inner wall outwards. All mesh results are shown in Fig. 8.52. Again, mesh refinement was imposed at the near wall regions of the fluid domain to increase the spatial resolution.

Representative flow waveforms shown in Fig. 8.53 were used as boundary conditions, at the inlet a pressure inlet was assigned, while at the outlet, a velocity outlet boundary condition was applied. Both inlet and outlet were constrained with fixed support (Fig. 8.54), and no movements and deformation were allowed at these locations.

8.6.2 Haemodynamics Inside the Abdominal Aortic Aneurysm

Velocities at the cross-sectional plane of the aneurysm were plotted for the CFD and FSI modelling approaches (Fig. 8.55). The CFD model using a rigid wall assumption

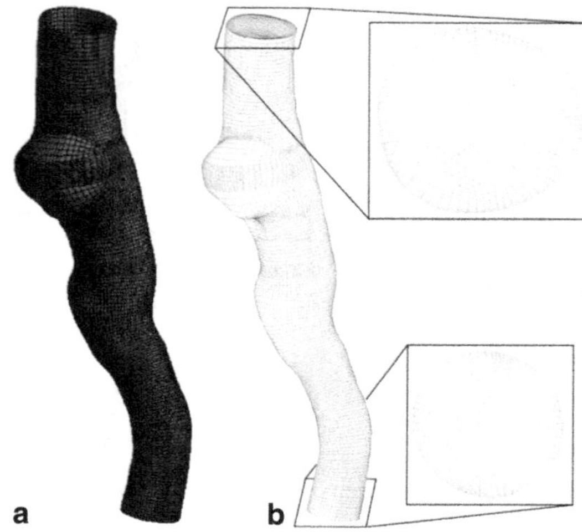

Fig. 8.52 Computational mesh results of the truncated abdominal aneurysm model: (**a**) structure domain, (**b**) fluid domain

predicts a larger low velocity region than the FSI model. This means that when the arterial wall is rigid, less blood flow enters the aneurysm. However, when the arterial wall is elastic, the aneurysm cross-sectional plane expands considerably from

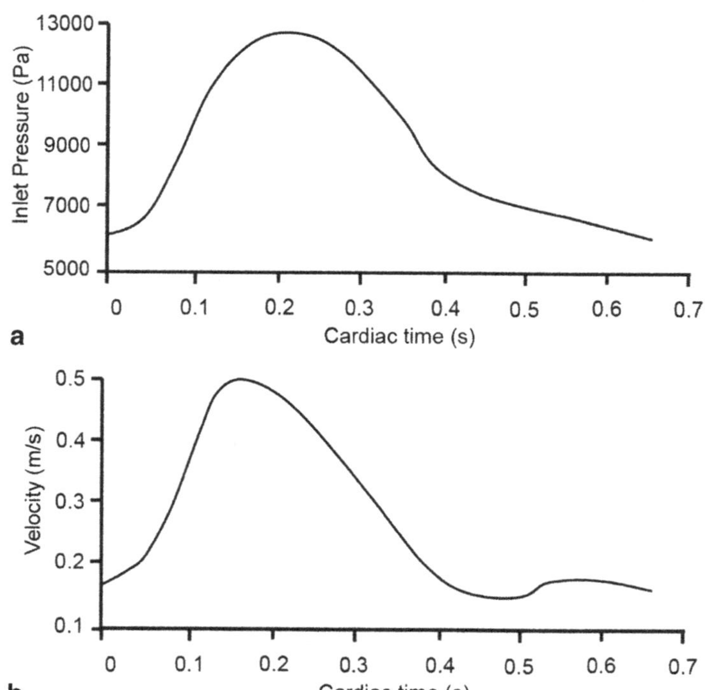

Fig. 8.53 Physiologically reasonable flow waveforms representing a patient under exercised situation: (**a**) inlet pressure waveform, (**b**) outlet velocity waveform

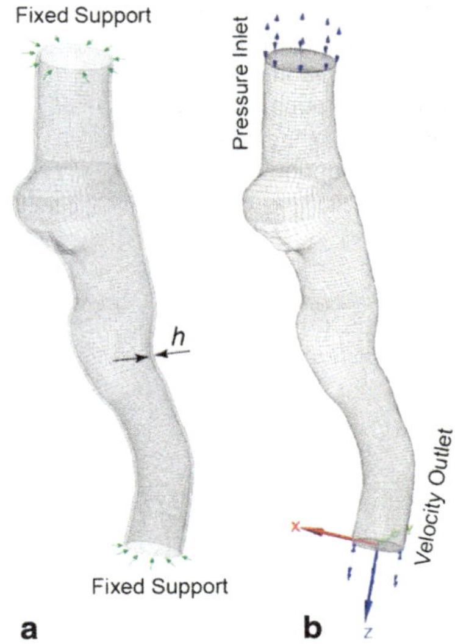

Fig. 8.54 Boundary conditions setup: (**a**) fixed supports were placed at terminals of the structure domain, (**b**) pressure inlet and velocity outlets were prescribed at the inlet and outlet of the fluid domain

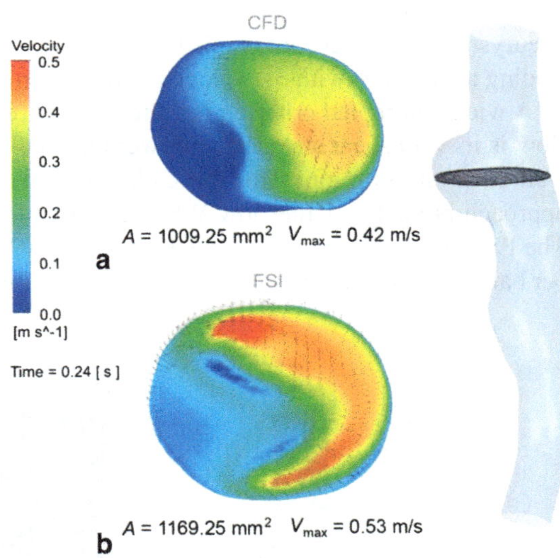

Fig. 8.55 Plane velocity comaprison: (**a**) CFD simulation using rigid wall assumption, (**b**) FSI simulation using compliant wall assumption

1009.25 mm^2 to 1169.25 mm^2, and more blood flow with high momentum enters the aneurysm. This is different from the velocity findings from the carotid artery case study presented earlier in Sect. 8.3, where the CFD model under predicts the peak velocity by 21 %. The main reason is the strengthened secondary flow in this region, leading to an increased acceleration in the flow field.

Flow vortices characterized by its swirling strength (Fig. 8.56) were generated to examine the secondary flow patterns between these two approaches. The vortex swirl inside the aneurysm is much greater for the FSI simulation compared with the CFD simulation. Almost twice the influence region of the vortex was predicted by

Fig. 8.56 Vortex swirling strength comparison: (**a**) CFD simulation using rigid wall assumption, (**b**) FSI simulation using compliant wall assumption

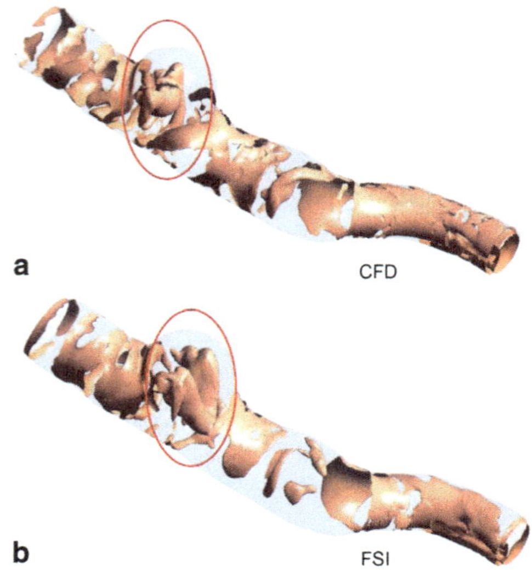

the FSI simulation, which means the interaction between the main stream and intra-aneurysmal flow were intensified compared with the rigid wall simulation. This finding also agrees with the velocity comparison as depicted by Fig. 8.55.

A widely accepted criterion for arterial wall remodelling from exposure to fluid flow is its wall shear stress being lower than 0.4 Pa. Figure 8.57 compares the low wall shear stress regions for the CFD and FSI simulations. Anteriorly, the CFD approach predicts a larger low WSS region covering almost the whole aneurysm. The FSI simulation predicts a much smaller low WSS region only covering the upper half of the aneurysm. Posteriorly, the CFD simulation does not produce any low

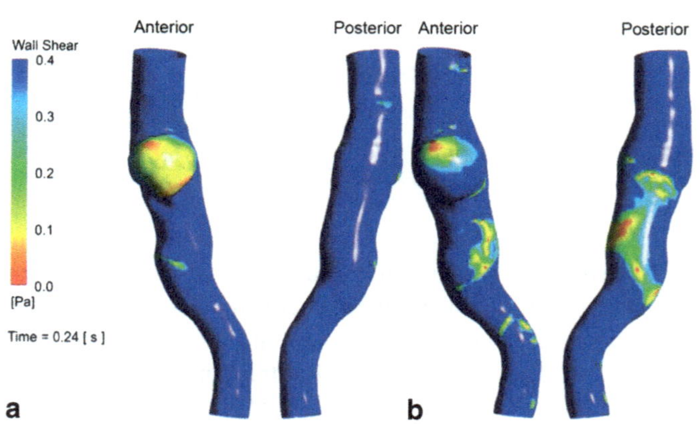

Fig. 8.57 Wall shear stress distribution (**a**) CFD simulation using rigid wall assumption, (**b**) FSI simulation using compliant wall assumption

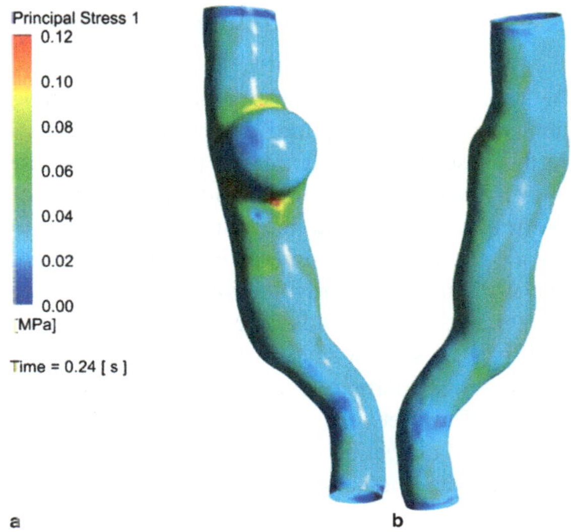

Fig. 8.58 First principal stress distribution: (**a**) *front view*, (**b**) *back view*

wall shear stress regions, while the FSI simulation gives a large area of low wall shear stress right downstream of the aneurysm.

One advantage of using FSI is the ability to capture the mechanical stress distribution within the arterial wall. This complements the fluid flow analysis to gain a better understanding of the vascular disease. In this case study, the first principal stress was used for mechanical stress analysis. Figure 8.58 presents the stress distribution of the AAA model at the peak systole event. The upper and lower shoulder of the aneurysm experiences extremely high principal stress, which is greater than 0.1 MPa, and peaks at the lower shoulder of the aneurysm at a value of 0.15 MPa, while all other regions are exposed to a first principal stress under 0.04 MPa. This indicates the lower shoulder of the aneurysm is exposed to a high mechanical loading due to the pulsatile blood flow, and faces a high risk of rupture.

8.6.3 Closure

Aneurysmal disease and its progression is a complex multifactorial process, and the FSI method is able to offer a representation of the aneurysmal wall deformation behaviour and blow flow haemodynamics inside the aneurysm. By implementing proper flow boundary conditions and arterial wall elasticity properties, high risk rupture locations can be accurately predicted. This modelling approach can contribute towards a non-invasive diagnosis of aneurysm rupture examination. Such diagnostic capabilities can provide a refined diagnostic and decision assistance for treating physicians when performing vascular disease diagnosis.

8.7 Coronary and Abdominal Arterial Bypass Grafts

In the previous examples we demonstrated a method to predict high risk and vulnerability to aneurysm rupture. A method in treating aneurysms is to use bypass grafts. In this case study we summarise two similar studies. The first is by Kabinejadian and Ghista (2011) for coronary arterial bypass grafts (CABG). This study analyses the design and demonstrate the superiority of the CABG over the conventional end-to-side anastomosis, which can be demonstrated based on their blood flow patterns and wall shear stress distributions. The wall compliance using a two-way fluid–structure interaction and non-Newtonian rheology were applied for physiological realism. The second study is by Qiao et al. (2015) for a DeBakey III type aortic dissection. An aortic dissection is when a tear in the inner wall of the aorta causes blood to flow between the layers of the wall of the aorta, forcing the layers apart. The type of dissection is classified based on the DeBakey system (Fig. 8.59) which considers the original tear location and extent of the dissection (localized to either the ascending aorta or descending aorta, or involves both the ascending and descending aorta).

Two different strategies for bypassing the dissection were undertaken. The first is to create the bypass between the ascending aorta and abdominal aorta, and the other is a bypass between the left subclavian artery and abdominal aorta. Their results showed that blood flow velocity, pressure and vessel wall displacement of false lumen all reduced after bypassing.

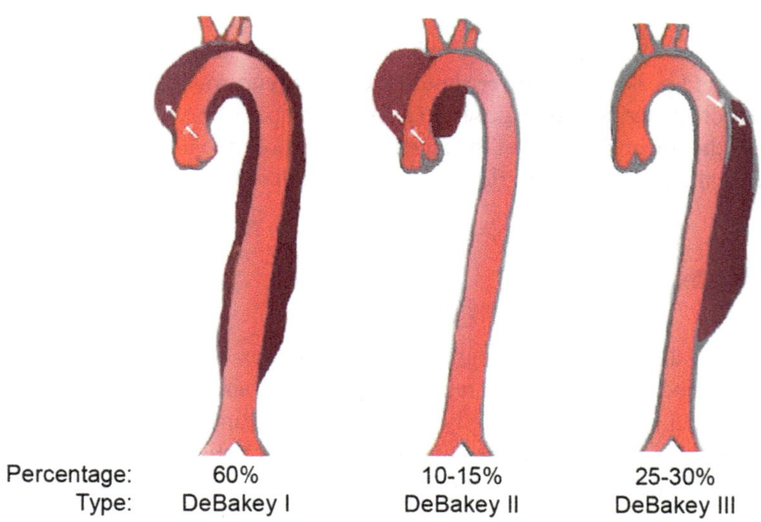

Fig. 8.59 The DeBaky classification of aortic dissection. Type I—Originates in ascending aorta, propagates at least to the aortic arch and often distally. Type II—Originates in ascending aorta and is confined to the ascending aorta. Type III—Originates in descending aorta, rarely extends proximally but will extend distally. (Adapted from Wikicommons)

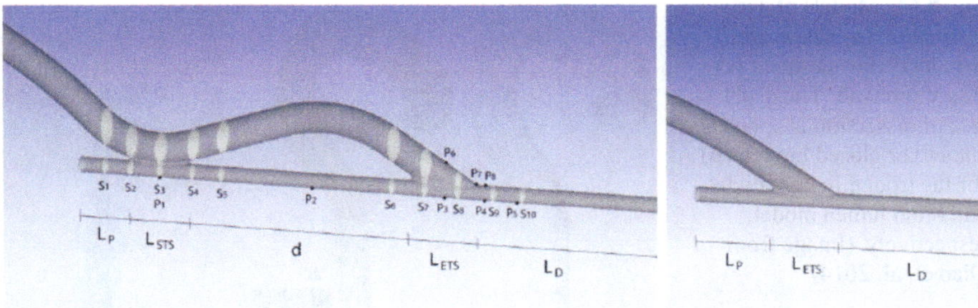

a b

Fig. 8.60 Configuration of the coupled-sequential anastomosis (*SQA*) and conventional end-to-side (*ETS*) models. Based on the coupled side-to-side:end-t-side sequential anastomoses model. Sections, S_1 to S_{10} and points, P_1 to P_7 indicate monitor the velocity profiles and WSS variations. The conventional ETS anastomosis model is based on only a single connection to the main artery. (Image from Kabinejadian and Ghista 2011)

8.7.1 Geometric Configurations and Computational Details

For the coronary artery, the coupled sequential anastomosis model and the conventional end-to-side model are shown in Fig. 8.60 based on the following dimensions: graft diameter, D_G=4 mm, coronary artery diameter, D_A=2 mm, L_{STS}=9 mm, d=20 mm, L_{ETS}=10 mm for the sequential anastomosis model and L_D=48 mm, L_P=6.5 mm and 10 mm for the end-to-side model. Both models exhibit an end-to-side anastomotic angle of 30°.

Patient-specific models of DeBakey III aortic dissection were reconstructed using commercial tool MIMICS (Materialise Europe, Leuven, Belgium). MRI serial slices were integrated by medical image-processing techniques such as threshold segmentation and region growing methods. The thickness of arterial wall was built by offsetting the inner vessel wall 2×10^{-3} m outwards according to the statistical value of human aortic wall (Fig. 8.61). Models with blind false lumen and through false lumen were established separately. The model with blind false lumen was constructed artificially by cutting the false lumen for the sake of comparison between different models (Fig. 8.61b).

Two bypass graft strategies were applied to these dissection models: graft bypassing between ascending aorta and abdominal aorta; and graft bypassing between left subclavian artery and abdominal aorta. The four models of bypassed aortic dissection with different grafting procedures are shown in Fig. 8.62. The aortic dissection models with the through and blind false lumens were also reserved to perform numerical simulation for comparison with the bypassed models. Thus, totally there are six models for the simulation.

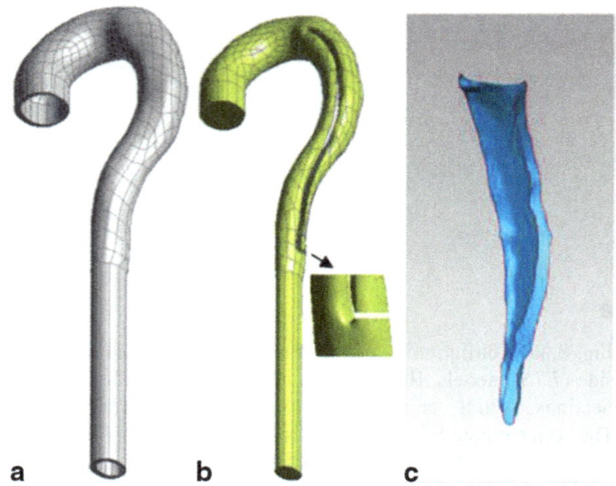

Fig. 8.61 Models of aortic dissection: (**a**) vessel wall; (**b**) blood flow domain; (**c**) dissection wall. The distal exit of dissection is open (*not shown*) or closed/broken (**b**) for the trough lumen model and blind lumen model, respectively. (Image from Qiao et. al. 2014)

8.7.2 Flow Patterns and Wall Deformation in Coronary Bypass Graft

Figure 8.63 illustrates the mesh displacement contour plot of the sequential anastomosis model at the peak pressure (t/T=0.43). The same trend of wall deformation occurs throughout the cardiac cycle. The maximum mesh displacement of approximately 0.5 mm exists at the suture line edge of the side-to-side anastomosis. At the end-to-side anastomosis, maximum displacement occurs on its suture line edge nearer to the heel in comparison to the toe. These findings may be used to analyse the effectiveness of suturing at the anastomosis.

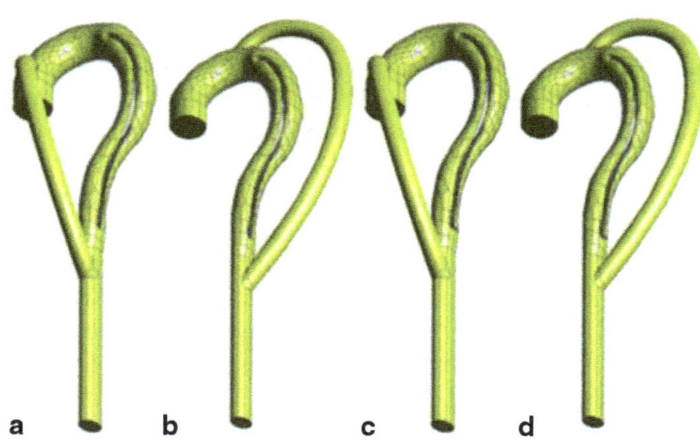

Fig. 8.62 Models of blood flow domain after bypassing. (**a**) Through lumen model bypassing between ascending aorta and abdominal aorta; (**b**) through lumen model bypassing between left subclavian artery and abdominal aorta; (**c**) blind lumen model bypassing between ascending aorta and abdominal aorta; (**d**) blind lumen model bypassing between left subclavian artery and abdominal aorta. (Image from Qiao et. al. 2014)

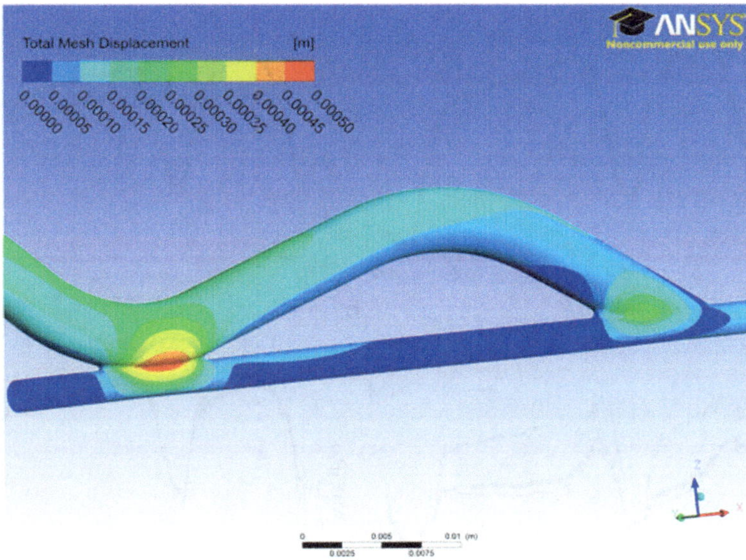

Fig. 8.63 Total mesh displacement contour plot of sequential anastomosis model. The results is captured at the peak pressure where $t/T = 0.43$. The maximum displacement exists at the side walls of the side-to-side anastomosis and a relatively higher displacement occurs at the ETS connection to the main artery. Note that the mesh deformation scale is in m. (Image from Kabinejadian and Ghista 2011)

Transient WSS for the non-Newtonian fluid and one cardiac cycle were investigated at monitor points, P_1 to P_8, highlighted from Fig. 8.60. Comparison of this parameter for the rigid and compliant models were made and shown in Fig. 8.64. In particular, WSS variations generated at point P_7 (that is located at the toe of the end-to-side anastomosis) and point P_6 (that is located on the graft outer wall proximal to the end-to-side anastomosis) demonstrate, respectively, the minimum and maximum disparity between the rigid and compliant models.

Compared to distal points, proximal point P_1 generates a WSS variation that resembles more like the calculated pressure waveform. The WSS magnitudes based on rigid and distensible models did not show a great discrepancy from their comparison, however, the WSS is generally higher for the rigid-wall model. This is due to a higher velocity in the absence of dilation in vessel diameter. However, for point P_8 that is located distal to the toe of the end-to-side anastomosis, its WSS variation is lower for the rigid model compared to the compliant model as a result of the higher flow separation resulting in a significant drop in WSS magnitude. As such, we deduce that FSI modelling of distensible flow channels can influence the haemodynamics results and analysis significantly.

Based on the rigid and distensible wall comparison, time-averaged wall shear stress (TAWSS) decreased at monitored locations for the distensible vessel, ranging from 3 % (on the suture line) to 32 % (on the heel) for the conventional end-to-side model, while from 4 % (at the toe of the side-to-side anastomosis) to 27 % (on the suture line of the side-to-side anastomosis) for the sequential anastomosis model.

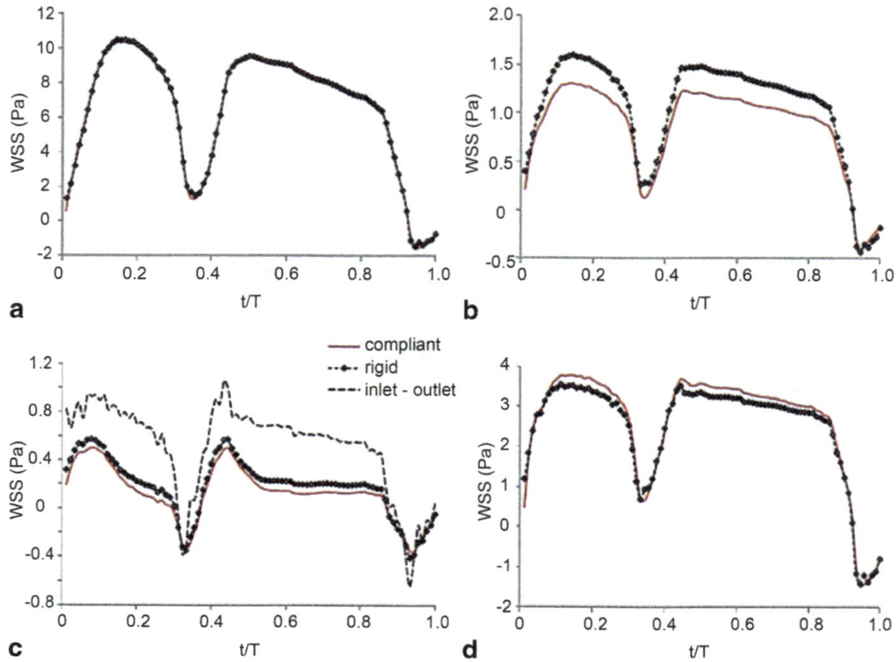

Fig. 8.64 WSS variations for one cardiac cycle at monitor points (P_1, P_6, P_7 *and* P_8). The WSS parameter that pertains to the rigid and compliant models is shown by the *red straight line* and the *dotted dashed line* respectively. The calculated pressure waveform is shown by *dashed line* in sub-figure **c**. (Image from Kabinejadian and Ghista 2011)

Comparison of Newtonian and non-Newtonian fluid models showed that the TAWSS decreased at the toe, suture line, and on the bed of the anastomosis. There was an increase at the heel for the end-to-side anastomosis that for the conventional and sequential anastomosis models. At the side-to-side anastomosis, an increase in TAWSS was found at all critical monitor points. This can be explained by the shear thinning of the non-Newtonian fluid, which presents a higher shear rate at most of the points of the end-to-side anastomosis than at the side-to-side anastomosis.

The non-Newtonian velocity profile is flattened compared to the parabolic velocity profile of a Newtonian fluid shown in Fig. 8.65. This implies that the TAWSS is reduced at high shear rates, shown in Table 8.7, which is attributed to the blood shear thinning behaviour. The velocity change becomes higher near the wall region.

8.7.3 *Flow Patterns and Wall Deformation in Abdominal Arterial Bypass Grafts*

The haemodynamics before and after each bypass strategy and different types of false lumen were compared. Velocity vectors at time t = 0.04 s during the acceleration segment of systolic phase is shown in Fig. 8.66. The mean velocity in

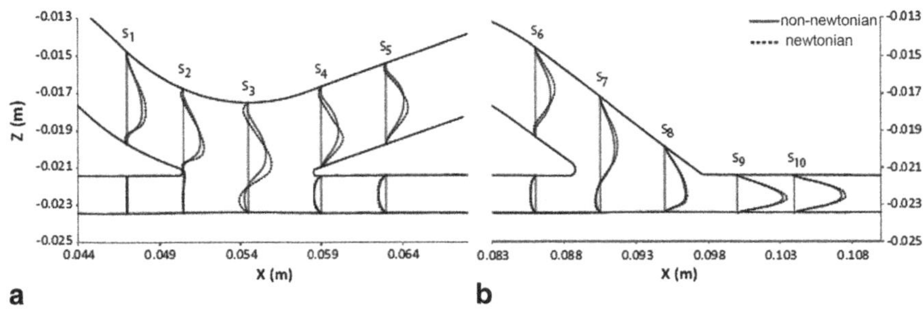

Fig. 8.65 Comparison of axial velocity profiles for different blood flow behaviour in the symmetry plane of the sequential anastomosis model. Based on the end of systole for the non-Newtonian (*continuous line*) and Newtonian (*dashed line*) fluid models, comparison was performed for the (**a**) side-to-side anastomosis and (**b**) end-to-side anastomosis of the sequential anastomosis model. (Image from Qiao et. al. 2014)

the false lumen after bypass is generally less than that before bypass. The greater the velocity, the more obvious the reduction. The change is not obvious when the velocity is low, but this does not affect the decreasing trend. Velocity decreases more significantly in the model with a blind lumen bypassed between the ascending aorta and the abdominal aorta. For the lumen model, the mass flow ratios of the bypass graft are 18.7% for bypass between the ascending aorta and abdominal aorta; and 8.4 % for bypass between the left subclavian artery and abdominal aorta. Similarly for the blind lumen model, the corresponding mass flow ratios are 52.8 and 51.3 % respectively.

Figure 8.67 shows the distribution of blood pressure at time of 0.04 s where the values are compared with the reference pressure at the abdominal aorta outlet. The mean blood pressure on the vessel wall after bypass is generally less than that without bypass. Peak values of inlet pressure decrease for the lumen models by 1.09×10^3 Pa for the ascending to abdominal aorta bypass, and 6.84×10^2 Pa for the left subclavian artery to abdominal aorta bypass. Similarly peak values of inlet pressure decrease for the blind lumen models by 3.13×10^3 Pa for the ascending to abdominal aorta bypass, and 2.47×10^3 Pa for the left subclavian artery to abdominal aorta bypass.

8.7.4 Closure

For the coronary artery bypass, flow phenomenon through the coronary arterial bypass graft models is characterized by blood flow that deforms structures to an extent influencing flow patterns through modified geometries. The importance of FSI in analysing anastomosis models can be emphasized as follows. Distensible vessels and non-Newtonian rheology influence the haemodynamics of anastomsis models and improve physiological realism in the simulation. The use of haemodynamic parameters such as flow patterns, wall deformation, wall shear stress, and time-averaged wall shear stress can serve as key haemodynamic indicators.

Table 8.7 Table of comparison of segmental averages of haemodynamic parameters. This table compares the rigid and distensible wall models, and the Newtonian and non-Newtonian flow models using haemodynamic parameters. (Adapted from Kabinejadian and Ghista 2011)

Comparison of the segmental averages of <HP>s.

Anastomosis model	Location	Rigid[a]						Distensible[a]					
		Newtonian[b]			Non-Newtonian[b]			Newtonian[b]			Non-Newtonian[b]		
		<OSI>	<TAWSS> (Pa)	<TAWSSG>	<OSI>	<TAWSS> (Pa)	<TAWSSG>	<OSI>	<TAWSS> (Pa)	<TAWSSG>	<OSI>	<TAWSS> (Pa)	<TAWSSG>
Conventional ETS model	1- Toe	0.01	6.69	34.02	0.01	5.65	28.47	0.01	6.14	31.00	0.02	5.23	26.34
	2- Heel	0.09	0.56	8.29	0.07	0.71	9.57	0.11	0.38	6.10	0.09	0.52	7.97
	3- Bed	0.09	1.67	5.46	0.08	1.55	4.64	0.11	1.47	5.01	0.09	1.40	4.29
	4- Suture line	0.04	2.12	15.09	0.04	2.06	14.94	0.05	2.05	13.30	0.04	1.99	13.66
Coupled sequential anastomoses model													
ETS component	1- Toe	0.01	6.33	36.95	0.02	5.36	31.23	0.01	5.85	29.71	0.02	5.01	25.24
	2- Heel	0.08	0.49	4.39	0.07	0.69	6.26	0.08	0.40	3.52	0.07	0.61	5.50
	3- Bed	0.05	1.86	4.93	0.05	1.76	4.05	0.06	1.76	5.02	0.06	1.59	3.88
	4- Suture line	0.03	2.51	15.12	0.03	2.39	15.00	0.03	2.10	11.44	0.04	2.05	11.61
STS component	5- Toe-STS	0.04	0.70	6.03	0.05	0.80	6.00	0.04	0.67	7.32	0.05	0.77	7.44
	6- Heel-STS	0.08	0.41	4.42	0.08	0.54	5.38	0.09	0.32	3.44	0.07	0.45	4.89
	7- Bed-STS	0.22	0.17	0.40	0.18	0.25	0.50	0.24	0.15	0.45	0.18	0.22	0.55
	8- Sutureline-STS	0.05	0.65	5.94	0.05	0.80	6.54	0.05	0.47	5.37	0.05	0.62	6.39

[a] Wall model.
[b] Fluid model.

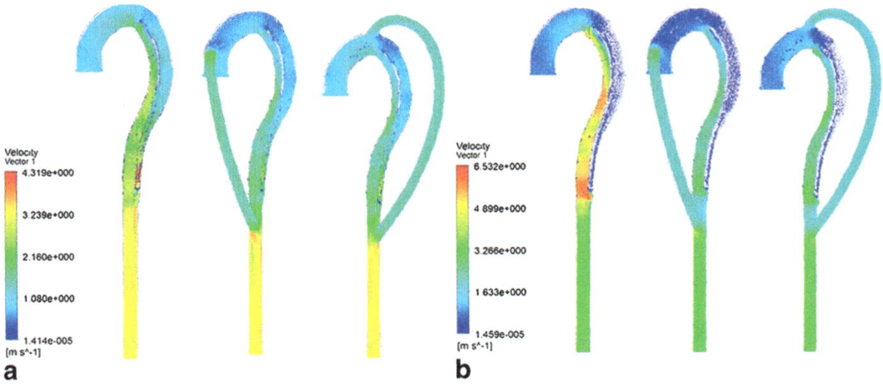

Fig. 8.66 Velocity vectors contour of blood flow at a time of 0.04 s. (**a**) Through lumen model; (**b**) blind lumen model. In each figure, *left*: before bypassing; *middle*: bypassing between ascending aorta and abdominal aorta; *right*: bypassing between left subclavian artery and abdominal aorta. (Image from Qiao et. al. 2014)

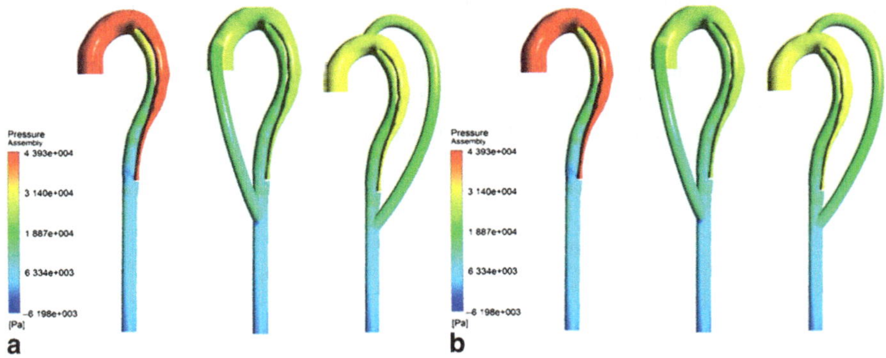

Fig. 8.67 Distribution of blood pressure at a time of 0.04 s. (**a**) Through lumen; (**b**) blind lumen. In each figure, *left*: before bypassing; *middle*: bypassing between ascending aorta and abdominal aorta; *right*: bypassing between left subclavian artery and abdominal aorta. (Image from Qiao et. al. 2014)

Bypass graft, as an alternative approach in treating aortic dissection, has many advantages compared with general thoracic aortic replacement, such as smaller trauma, simpler operation, shorter operation time and less surgical risk. It is also an alternative treatment for patients unable to withstand major surgery. The simulation results demonstrated that bypass grafting can effectively divert blood flow from the aortic dissection. The pressure and velocity in the false lumen, the mass flow rate in the dissected aorta, and the inlet pressure near the ascending aorta are greatly reduced after bypass grafting. These changes can reduce the blood pressure of the upper limbs and reduce the impact and stress of aortic dissection. It is deduced that bypass grafting is beneficial to reduce and prevent the expansion and rupture of aortic dissection, and also promotes the closure of the dissection layer and improve blood perfusion of the lower limbs. This treatment is an effective procedure to treat DeBakey III aortic dissection.

8.8 Mitral Valve Dynamics

8.8.1 Introduction

The mitral valve is composed of the anterior and posterior leaflets, the chordae tendineae, and the papillary muscles; separating the left ventricle and left atrium in the heart. It behaves as a valve, controlling the flow of blood. During diastole it allows blood flow from the left atrium to the left ventricle. During ventricular systole, the mitral valve closes and prevents backflow of the left atrium. The mitral valve action is caused by blood flow associated with pressure differences over the leaflets and leaflet muscle fibre activation resulting from a complex interaction of wall tension and flow pattern in the atrium and ventricle compartments (Skallerud et al. 2011; Fig. 8.68).

Mitral valve function requires the coordinated action of its interrelated components, and alterations in its structure caused by remodeling, stenosis, or weakening leading to regurgitation, may lead to eccentric ventricular hypertrophy, pulmonary

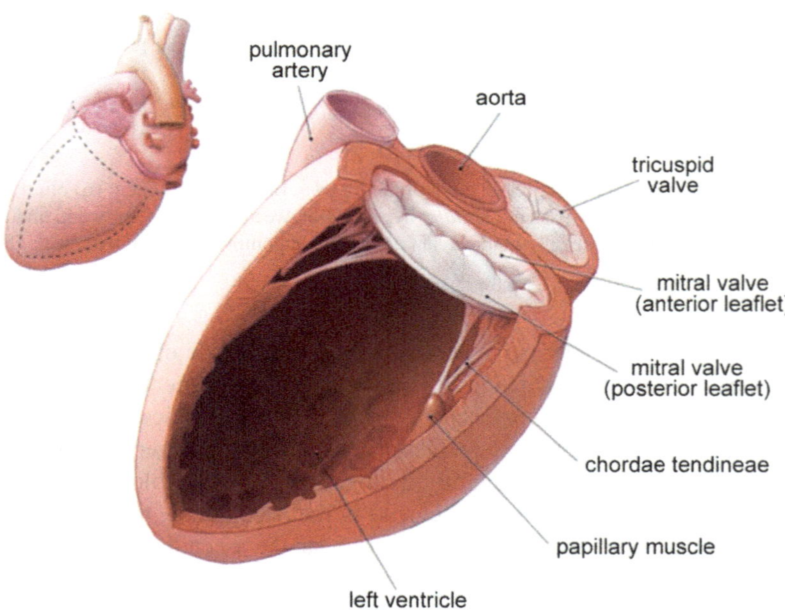

Fig. 8.68 Mitral valve is composed of the anterior and posterior leaflets, the chordae tendineae, and the papillary muscles. (adapted from Wikicommons)

oedema and instances of heart attack. Computational models have been used to gain a better understanding of mitral valve biomechanics which can lead to improved medical and surgical planning to restore normal mitral valve function. In this section we provide an example of FSI modelling for investigating the asymmetric mitral valve dynamics during diastolic filling by Dahl et al. (2010).

8.8.2 Asymmetric Mitral Valve Dynamics During Diastolic Filling

During diastole, the left atrium directs flow from the pulmonary veins towards the mitral valve. Simultaneously the left ventricle relaxes, increasing the pressure difference between the two chambers. This assists in the valve opening and allows blood to from the left atrium into the left ventricle. As the ventricle fills it becomes stiffer, and an atrial contraction at end-diastole is needed to force blood into the distended ventricle.

Dahl et al. (2010) implemented an implicit ALE procedure for the rigid body motion of the mitral leaflets. The leaflets were simulated with an asymmetric geometry and their motion computed with an FSI coupling scheme that accounts for the mutual interaction of the leaflets. To achieve physiological realism in the simulations, ultrasound recordings with speckle tracking were used to render the subject-specific left ventricular wall movement of a healthy young person. The specific movement was imposed as a boundary condition in a 2D transient simulation of diastolic filling.

Ultrasound recordings of the left-ventricle wall geometry moving were obtained using the speckle tracking algorithm in EchoPAC PC (version 6.0.0, GE Vingmed Ultrasound, Norway) and a sample image is shown in Fig. 8.69. The green vertical line marks the start of the simulation, i.e. early diastole. The short red line depicts

Fig. 8.69 Ultrasound recording of the left-ventricle wall. ECG shown in the *lower right corner*. (Image from Dahl et al. 2010)

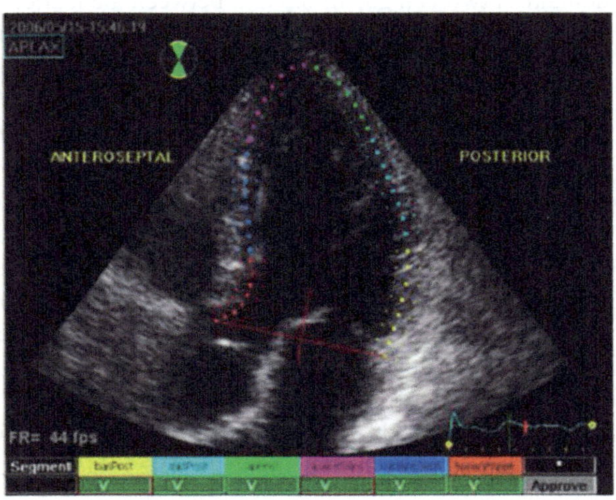

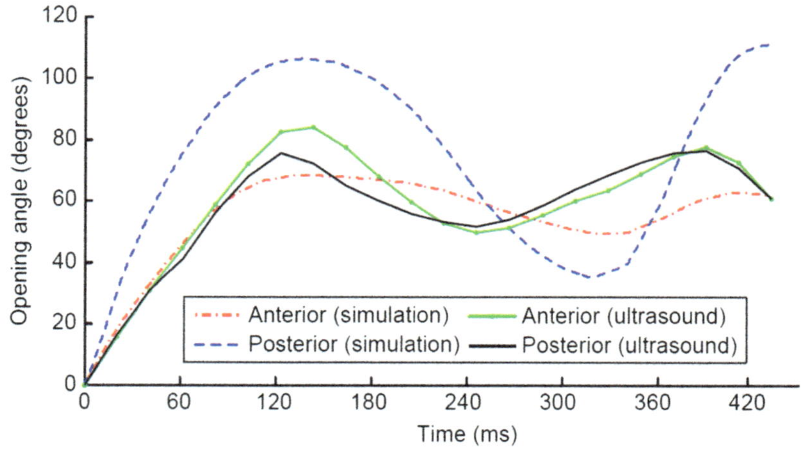

Fig. 8.70 Opening angle versus time of simulated and measured mitral leaflets. (Image from Dahl et al. 2010)

the instant when the displayed image was taken. The inflow region was modelled as a typical, but fixed, atrium with two inlets in the upper part, each reflecting an orifice of a pulmonary vein. The leaflets were modelled as two separate rigid bodies, rotating around their annulus attachment points. The anterior and posterior valve lengths were taken as $l_a = 19.9$ mm and $l_p = 6.6$ mm, respectively. The choice of lengths was based on ultrasound recordings. The thickness was set to $t = 1.0$ mm uniformly for both leaflets.

A deforming mesh under the Arbitrary-Lagrangian-Eulerian formulation was used where grid velocities are included in the momentum and continuity equations. The domain was discretised with triangular cells with increased density around the leaflets to allow for their large motions. Due to limitations in the dynamic mesh module, a gap is required between the moving parts to maintain a continuous fluid domain. A two-cell gap of 0.5 mm between the leaflets and the wall, and a two-cell gap of 0.6 mm between the two leaflets in their closed position were included in the model. The blood was modelled as an incompressible, laminar, Newtonian fluid, with properties representative for healthy human blood, a density of 1056 kg/m^3 and a dynamic viscosity of 3.5×10^{-3} kg/ms. The diastole time of 0.43 s was discretised with 2150 time steps (e.g. time step of 0.2 ms).

Figure 8.70 shows the opening angle of the simulated valve plotted against the opening of a natural mitral valve measured in the same ultrasound recordings during left ventricle wall movement. The three stages of diastole can be identified in the graph. These are: early diastole, that depicts, the rapid or early filling phase; mid-diastole or the diastasis, which shows the partial closing; and end-diastole, which represents the second opening due to atrial contraction. During early diastolic filling, the simulated anterior leaflet's opening angle coincides relatively well with the measurements. In mid- and end-diastole, there are some deviations between the simulation and the measurements. The posterior leaflet, on the other hand, has

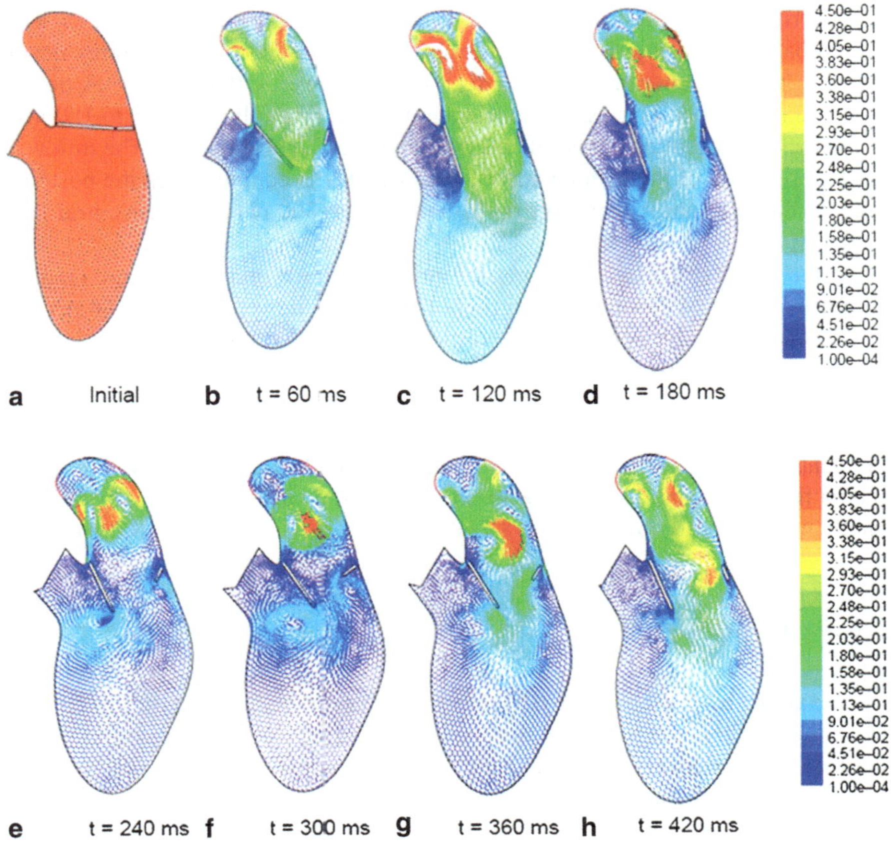

Fig. 8.71 Initial grid and velocity vectors at seven different times during diastole, the velocity scale to the *right* is given in (*m/s*). (Image from Dahl et al. 2010)

higher angular velocities throughout the simulation. However, despite some variations, the simulated leaflets follow the same main pattern as in the measurements. In the ultrasound recordings, the opening velocities of the posterior and anterior leaflets do not deviate much from each other. In the simulation, the two leaflets have significantly different velocities.

Figure 8.71 shows the initial grid and seven sequential velocity-vector plots at time intervals of 60 ms. When the ventricular pressure falls below the atrial pressure, the rapidly changing atrium-ventricle pressure gradient leads to an opening of the valve. In the first filling phase, (t=60 ms, and 120 ms), the transmitral flow is almost uniform. No vortices are observed in the ventricle, whereas a vortex has started to develop at each pulmonary valve orifice in the atrium.

As the blood flows into the ventricle, the atrio-ventricular pressure gradient decreases and reaches a minimum value in mid-diastole/diastasis (t=180 ms to 300 ms). This causes a deceleration of the transmitral flow and the leaflets now drift back towards the annulus. This is called partial closure. Two large vortices develop in the atrium during diastasis. A vortex also starts developing at the tip of each leaflet, probably enhancing the partial closing seen in mid-diastole. The vortex developing behind the anterior leaflet is larger and more intense than the vortex behind the posterior leaflet. During the deceleration of the first filling wave (t=180 ms), the velocity distribution across the mitral annulus is slightly skewed. The inflow at the anterior side of the annulus has ceased, while it is continuous at the posterior side. In the two subsequent plots (t=240 ms and t=300 ms), the transmitral flow is almost zero.

At the end of diastasis, t=330 ms, the atrium contracts and the pressure gradient increases causing a second acceleration of flow into the ventricle (t=360 ms and t=420 ms). The valve opens and the vortices at the leaflet tips become less intense. The vortices developed in the atrium during diastasis disappear through the mitral valve opening, resulting in a prominent asymmetric transmitral flow with a high velocity at the posterior side. At the very end of diastole, the atrio-ventricular pressure gradient decreases again. The flow is hindered as the leaflets drift back, but due to inertia the blood continues to flow through the mitral valve (t=420 ms), predominantly near the posterior side (Fig. 8.71).

8.8.3 Closure

This example by Dahl et al. (2010) showed the prescribed motion of the left ventricle during filling. It is evident that variations in shape and volume occur during diastole. The results also indicate that important features of the flow field may not be predicted by the use of symmetric leaflets or in the absence of an adequate model for the left atrium, particularly during diastasis and atrial contraction.

8.9 Summary

FSI is being used as a demonstration of practicality of achieving a higher physiological realism in haemodynamics analysis in applications such as atherosclerosis in carotid bifurcation, calcified plaque rupture, aortic aneurysm, as well as coronary artery bypass graft. The contributions from analysing these applications can be summarized below.

- The presence of severe stenosis produces smaller flow reversal at advanced stages of the disease. This leads to diminishing low wall shear stress in the stenotic region. The stenosis apex experiences compressive stress which can intensify as it grows which explains why plaque can still grow at advanced stages when the wall shear stress is higher. The relationship between pulse pressure, maximum displacement, and maximum principal stress suggest that elevated heart rates can lead to a higher risk of stroke due to longer exposure to high vessel wall displacement and stresses.

- The comparative studies between the CFD and FSI approach for different geometries showed significant differences caused vessel deformation. This led to different flow patterns such as a decrease in blood flow velocity and different wall shear stress values.
- The study of the left coronary artery showed that patients with a bifurcation angle $\theta > 80°$ was significantly larger than that measured in patients with bifurcation angle $\theta < 80°$ due to the presence of atherosclerotic plaques, and wider bifurcation angles are closely related to the development of atherosclerosis, thus leading to coronary artery disease.
- Calcification clusters plays a major role in plaque rupture demonstrated by structural analysis on a continuous calcification agglomerate structure. The analysis of calcified plaque showed that subintimal plaque structures such as the fibrous cap, calcification gap and lipid core play an important role in determining plaque rupture. This data should be used in place of information on stenotic severity that is based on medical image visualisation.
- For coronary artery bypass, flow was characterized by the deforming structures that influenced the flow pattern. This meant that the use of FSI modelling is an integral in haemodynamic study of coronary arterial bypass graft. For aortic dissections, bypass graft simulations showed that it can divert blood flow from the aortic dissection. The bypass graft reduces the expansion and rupture of aortic dissection, to improve the closure of the dissection layer to improves blood perfusion of the lower limbs.
- Finally FSI simulations were able to model the complex mitral valve in the heart. The simulations prescribed motion of the left ventricle during diastolic filling. The results showed the influence of a pressure difference to drive the valve motion and induce blood filling in the left ventricle.

Chapter 9
Advanced Topics and Future Trends

9.1 Introduction

This chapter considers the latest research capabilities in haemodynamics. Significant progress in computational modelling has been made in the last few decades. These advances along with increased computational power will further enable modelling realistic physiological scenarios of haemodynamic flows. The material presented in this book thus far serves as an introduction to some of the current trends and modeling capabilities. In this chapter the latest developments of some research fields of haemodynamics including microflow analysis of blood, medical maging for flow analysis, advanced medical devices, and heart valves are explored.

9.2 Blood Rheology

9.2.1 Multiphase Flow

Single phase flow refers to a one phase flow where the fluid does not change state during temperature changes whereas a multiphase flow displays a variety of pattern behaviours such as droplets, bubbles, slugs as well as films. Single-phase modelling may be applicable in large arteries such as aorta and some larger coronary arteries in the heart. However it is not appropriate for explaining blood flow behaviour when blood vessel diameters are small and within the same order as the red blood cells ($d \lesssim 10$ μm).

Blood at physiologic conditions is a dense suspension of cells and platelets dominated by red blood cells. In most cases, they are the blood component principally responsible for its rheology.

The flow regime is termed sub-Stokesian (Re ≤ 0.01) where the local environment remains predominately viscous. However at these flows, the red blood cells are significantly distorted due to their flexibility leading to a change in its effective viscosity and frictional resistance.

J. Tu et al., *Computational Hemodynamics—Theory, Modelling and Applications*,
Biological and Medical Physics, Biomedical Engineering,
DOI 10.1007/978-94-017-9594-4_9

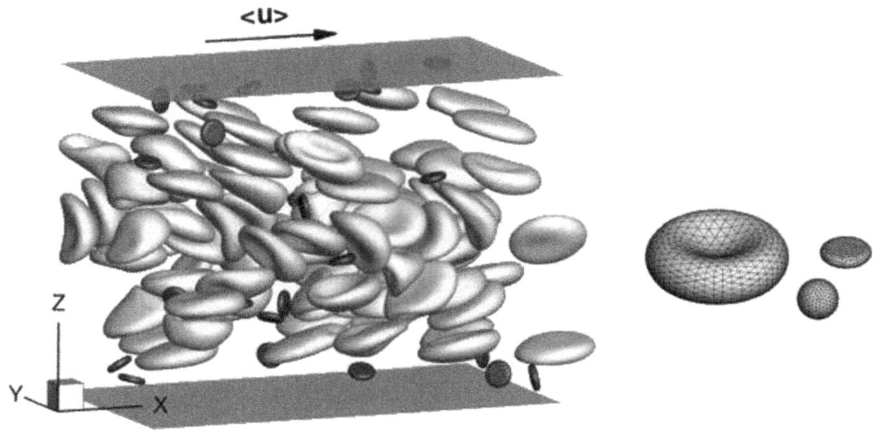

Fig. 9.1 Modelling red blood cells and platelets in a microchannel of 34 μm in height. (Image from Zhao et al 2012)

For a multiphase flow, blood is represented by a two-phase macroscopic model, that is, a suspension of red cells (particles, also called erythrocytes) in plasma (fluid) which is a Newtonian viscous incompressible fluid. Discrete elements can be used to construct and model the elastic cardiovascular structures effectively and will be the next state-of-the-art technique in this aspect. The disturbance of walls in a flow channel is structurally influenced by flow velocity or by hydrodynamic drag force. The use of nodal elements can allow us to compute the force two ways at all nodes representing the vessel.

The term multiphase flow suggests that the red blood cells can be treated as a deformable discrete phase moving through a flowing continuous phase in the form of plasma. The Eulerian and Lagrangian perspectives introduced earlier in Chap. 6 describe how the Arbitrary-Lagrangian-Eulerian mesh could be used for FSI problems. In this same context the discrete red blood cells are treated in a Lagrangian approach, tracked individually, while the continuous phase is in the Eulerian approach. Zhao et al. (2012) used this approach to produce a suspension of red blood cells and platelets flowing between two flat plates of a microchannel of height 34 μm (Fig. 9.1).

Their results showed the red blood cell suspensions under bulk shear and how this shear induced particle migration. The particle velocity fluctuations were closely related to the alignment of the red blood cells in suspension. It was observed that greater shear was found in the near wall where the red blood cells became more elliptic than those in the centre of the microchannel. These behaviours demonstrate that a continuous single phase description fails to capture the micro flow dynamics.

Srivastava and Srivastava (2009) also studied particle-fluid blood flow such as the cell-plasma mixture. The accurate modelling of the real nature of blood in the flow through the cardiovascular structure as well as the flow development of patterns allows better understanding of the cardiac physiology. An accurate model can confirm experimental observations made by earlier studies (Srivastava and Srivastava 1983). For instance, the effective viscosity, the frictional resistance and other flow characteristics are influenced by hematocrit (Srivastava and Srivastava 2009). Multiphase haemodynamics strengthen understanding of how the effective viscosity and the frictional resistance is influenced by the concentration of the hematocrit.

The review by Freund (2014) discusses some of the modelling challenges to account for the large deformations in red blood cells. When this is coupled with flow turbulence, the range of scales from submicron to micron flow structures needs to be considered. A suggested practice to alleviate this problem is to perform mesh-refinement studies to ensure quality in the results. The other challenge Freund (2014) discusses is the near incompressibility of the membrane, which introduces numerical stiffness, and thus a significant time-step restriction for explicit time integration algorithms. Implicit methods, which have no restriction on the time-step, for blood cells have been peformed by Dimitrakopolos (2007). Another approach is to use immersed boundary methods to employ membrane forces that distribute on the local mesh. Here the immersed interface enforces membrane jump conditions built into discrete operators.

A third phase can be introduced to isolate the leukocytes with the red blood cells. Jung and Hassanein (2008) used a three-phase computational fluid dynamics approach including plasma, RBCs, and leukocytes to numerically simulate the local haemodynamics in such a flow regime. The model tracked wall shear stress, phase distributions, and flow patterns for each phase in a concentrated suspension shear flow of blood. Higher leukocyte concentration was correlated with relatively low wall shear stress near a stenosis having a high wall shear stress. Such flow behaviour demonstrates the use of the three-phase haemodynamic analysis in to vulnerable plaque formation in arteries within vivo complex flow conditions.

9.2.2 *Direct Numerical Simulations of Blood Cells*

By strict definition, a direct numerical simulation of blood flow means to account for every single blood cell in the domain of interest. In a true physiological flow the number of red blood cells is in the millions in a small vessel. One microliter of blood (1×10^{-6} L) contains about 4 million red blood cells. The surrounding plasma, which is a viscous fluid, interacts with every red blood cell while there are also interactions between the cells. This means that accounting for each individual deformable cell and its interaction with the surrounding fluid requires an extensive computational resources.

The first fundamental stage of direct numerical simulation of blood cells is to account for the individual interactions between blood cells, and platelets in a vessel. The work by Ii et al. (2012) provides the application of a full Eulerian fluid-membrane coupling method with a smoothed volume-of-fluid approach for spherical membrane deformation problems, and is applied to a pressure-driven flow with red blood cells. The advantage of their work compared with traditional immersed boundary methods and with the Lagrangian method is to avoid instability occurring from mesh distortion with large deformation. Another advantage is that mass is strictly conserved, and the full Eulerian approach can be easily parallelized for high level supercomputing facilities.

The multiphysics, and multiscales related to blood flow requires numerical methods for fluid-structure and fluid-membrane interactions to be accounted for. Matsumoto et al. (2014) presented a multiscale framework for modelling a deformable vesicle problem dealing with red blood cells and platelets. A supercomputer of 10 Peta flops speed under the Next-Generation Supercomputer Project in Japan was used. The researchers developed a full Eulerian Fluid-Structure Interaction solver without mesh generation, to allow simulations to be conducted directly from medical images.

The conventional FSI method has been to define the fluid domain in the Eulerian approach, while the structure domain is described in the Lagrangian approach, which makes up the Arbitrary-Lagrangian-Eulerian, or the Immersed Boundary method, based on how the kinematic and dynamic interactions are coupled on the moving interface. However, for a system involving complex geometry of a large number of bodies, it requires extreme computational resources to generate the mesh and to reconstruct the mesh topology for the moving bodies. Matsumoto et al. (2014) formulated an efficient numerical scheme to account for geometrical flexibility and avoids a breakdown in a large deformation owing to the absence of the mesh distortion. The equations of the system consist of a Newtonian fluid and hyperelastic structure/membrane and are applied to three-dimensional blood flows including blood cells in a capillary vessel and in a microchannel, and the relevance of the red blood cells thrombus formation.

This method allows a fully coupled approach between the red blood cells, platelets and vessel. Figure 9.2 shows the multiphase blood flow at three time instants, where the red blood cells rotate and mix. Its deformed shape takes the form of a parachute, and migrates towards the vessel wall. Matsumoto et al. (2014) work showed that the cell-free layer (in which the number density of the red blood cell is very low) is confirmed to form near the wall and to be thicker over time.

Rahimian et al. (2010) presented a fast, petaflop-scalable algorithm for direct simulation of blood, modelling a mixture of a Stokesian fluid (plasma) and red blood cells. Their work simulated up to 260 million deformable red blood cells while preserving the physics related to nonlinear solid mechanics to capture the

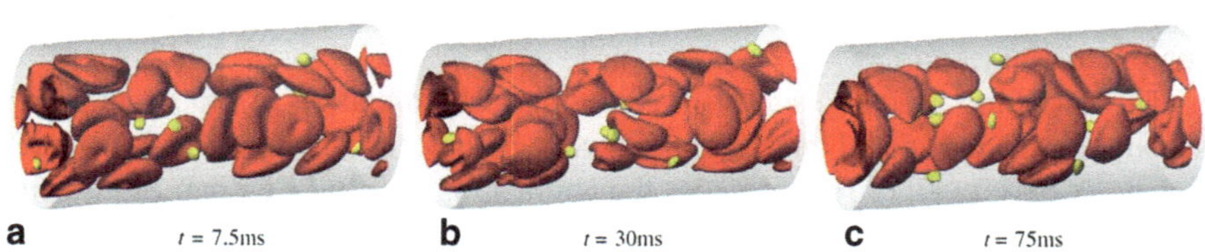

Fig. 9.2 Results from the work by Matsumoto et al. (2014) showing snapshots of the red blood cells (in *red*) and platelets (in *yellow*)

deformations of each individual red blood cell, and the hydrodynamic interactions between the red blood cells and the surrounding plasma. Software named MoBo (for "Moving Boundaries") was developed and applied on a supercomputer cluster, namely the Teragrid's Lincoln cluster that had 200,000 computer processor cores on the Oak Ridge National Laboratory's Jaguar PF system.

9.2.3 Blood Rheology in Large Arteries Using Lattice Boltzman

Simulating 3D deformable particle suspensions using the Lattice Boltzmann method with discrete external boundary force is an alternative approach being used by a number of researchers (Melchionna et al. 2013; Reasor et al. 2012; van Wyk et al. 2013; Wu and Aidun 2010). When studying disease development in arteries, it is important to understand local variations in blood rheology. Blood flow in large arteries is often assumed to behave as a homogeneous fluid, an assumption that is not entirely correct. The local viscosity changes, concentration of red blood cells, and the rate of shear strongly influences the wall shear stress and its gradients. The red blood cell flow behaviour is also influenced by the flow environment geometry. Experimentally, rheological properties across a tube cross-section are difficult to measure if non-invasive techniques are used. The Lattice Boltzmann method can be used to model blood as a particle suspension of red blood cells. The multiphase mixture is made up of plasma (55 % volume fraction), red blood cells (45 % volume fraction), white blood cells (< 1 % volume fraction), and platelets (< 1 % volume fraction). The plasma is accurately modelled as a Newtonian fluid, with a temperature dependent viscosity. The whole blood rheology depends on the interaction between red blood cells and the plasma (e.g. dependency on hematocrit, shear rate, red blood cell deformability, and vessel geometry). Viscosity models are usually based on steady state rheometer measurements. The particles are represented in Lagrangian coordinates and the fluid on a regular Eulerian grid. Particle-particle interactions are only resolved down to the grid spacing, where the idea is to resolve the far-field interactions and use a subgrid model at short ranges. Results from van Wyke et al. (2013) are given in Fig. 9.3.

9.3 Medical Imaging for Flow Validation and Analysis

9.3.1 Imaging for Flow Validation

The recent developments in the fields of CT and MRI mapping systems, velocity-encoded measurements, computational modelling and simulation, experimental testing instruments, as well as manufacturing processes are giving new dimensions to our understanding of the pathology and the impact of cardiovascular disease. Experimental and clinical verification complements the existing computational re-

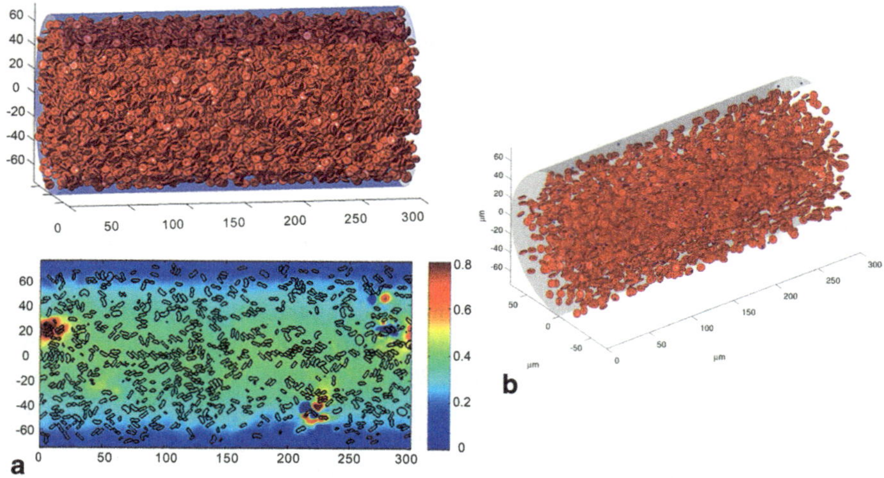

Fig. 9.3 Lattice Boltmann method for direct numerical simulations of red blood cells in large vessels. (Image adapted from van Wyk et al. 2013). **a** Hematocrit 15 % red blood cell distribution and velocity cutplane. **b** Hematocrit 5 % red blood cell distribution with platelets

sults based on the system performance and vice versa. For flow validation, there are well-established velocimetry systems to generate accurate time-resolved vector fields up to three spatial dimensions. Such flow tracking can be classified as optical-, magnetic resonance-and ultrasonic-image velocimetry.

Phase-sensitive flow MRI (velocity mapping) (Stahlberg et al. 1995) or velocity-encoded cine magnetic resonance imaging (VEC-MR) (Hartiala et al. 1993) are well-established methods for quantifying flow in the cardiovascular system. These methods form a class of nuclear-based imaging known as magnetic resonance imaging velocimetry (MRIV). Optical-based imaging using particle image velocimetry (PIV) provides a validation tool for verifying MRIV (Markl et al. 2003b; Elkins et al. 2004). Technically, phase contrast MRI can also be compared with ultrasonic imaging in various studies (Jung et al. 2004; Seitz et al. 2006).

9.3.2 Imaging for Flow Analysis

Diagnosis of heart conditions can help to save lives. For this purpose, a concatenation of medical imaging systems and software for extracting medical information and deciphering can be specifically directed at developing and investigating a novel framework for post-processing medical images.

High-resolution imaging techniques have been used to demonstrate the location and appearance of atherosclerosis in blood vessels. These imaging technologies include intravascular ultrasound, multi-detector CT and MRI. We can use patient-specific geometries for the three-dimensional computational models in CFD simulations. Obtaining real-time CFD solutions to biomedical flows in human vascular systems can improve the speed and effectiveness of assessing the cardiac condition. Diagnostic accuracy and therapeutic effect may be significantly improved if patient-specific engineering information could be provided in a timely manner in clinical practices.

The first step would typically involve high-resolution imaging techniques to determine the location of cardiovascular problems, such as artherosclerosis in blood vessels. Despite the significant advances achieved, obtaining real-time CFD solutions to biomedical flow problems are still not well developed due to difficulties in modelling the physics, in image processing and geometric modelling.

There have been various studies of medical imaging used to support CFD simulation of cardiovascular structures. For instance, a surgical remedy named surgical ventricular restoration aims to restore abnormal ventricular shape and size for subjects that are diagnosed with left ventricular dysfunction. The type of medical imaging and CFD platform used to assess the success of surgical intervention in this instance can quantify the left ventricle flow behaviour pre-and post-surgical ventricular restoration (eg. usually a combination of flow patterns, pressure differences in the chamber, vorticity analysis, as well as left ventricle ejection fraction). Studies have shown that surgical ventricular restoration can reduce left ventricle volumes and augmented left ventricle ejection fraction (Khalafvand et al. 2014).

Wong et al. (2012b) reviews flow analysis based on computer-aided evaluation of magnetic resonance intensity images, in comparison with other commonly used medical imaging modalities. CT and MRI provide excellent anatomical information on myocardial structures, but fail to show the cardiac flow and detect heart defects in vivo condition. The computerized technique for fluid motion estimation by pixel intensity tracking based on magnetic resonance signals represents a promising technique for functional assessment of cardiovascular disease, as it can provide functional information of the heart in addition to analysis of its anatomy. Cardiovascular flow characteristics can be measured in both normal controls and patients with cardiac abnormalities such as atrial septal defect, thus enabling identification of the underlying causes of these flow phenomena.

9.3.3 MRI Flow Imaging

Cardiac MRI possesses superior resolution, good blood and tissue contrast and offers a wide topological field of view. Unlike X-ray based techniques, it is a nonionizing modality. Cardiac MRI allows multiple contiguous slices at various orientations scanned at various phases of one cardiac cycle. Typically, the scan can either be short axis (SA) or long axis (LA), and based on the two-chamber (2C), three-chamber (3C) or four-chamber (4C) configurations, illustrated in Fig. 9.4.

Estimating motion is achieved by harnessing image signals of varying spatio-temporal intensity. This theory is based on observations which we can deduce from visual motion registration. In the context of electromagnetic signals which is analogous to magnetic resonance signals, and based on our knowledge of identifying

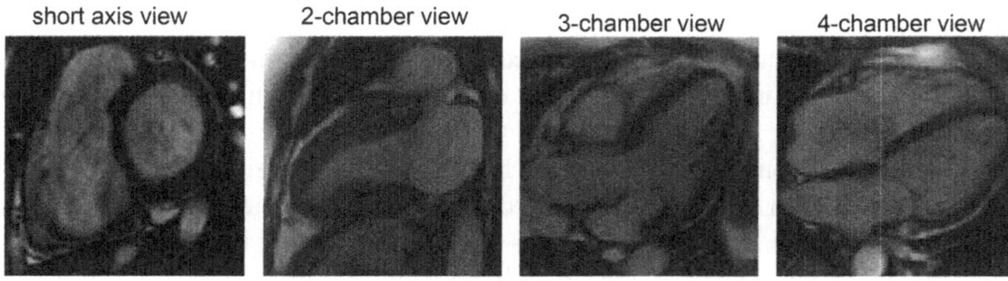

Fig. 9.4 Cardiac magnetic resonance image views. Display of cardiac magnetic resonance images is based on the short axis and also on the long axis (with 2-chamber, 3-chamber and 4-chamber views). Each configuration is selected to show the relevant chambers and cardiovascular structures with clarity. Different orientations display the chamber of interest at various positions and bisections

motion by referencing contrasting moving light sources, we state Observations1 and 2 as follows:

- **Observation 1:** The movement of light lines can be perceived as the movement of shadow lines on the illuminated background. We perceive light and darkness as opposing signals and this is used to determine motion.
- **Observation 2:** When observing blood movement in MRI scans of the heart, it is a mixture of observing moving dark shadows of asynchronous proton spins among the bright imaged blood, and moving bright shadows of synchronous spins among the contrasting imaged blood.

The computerized motion tracking technique makes use of the moving accumulation of asynchronous spin phase shifts. The incoherent spins cause the blood to appear dark in steady-state free precession (SSFP) MRI. Apart from passing blood through heterogeneous magnetic fields, turbulence in blood flow entering into the scanned slice may consist of different spin shifts from the imaged region of interest. It is the in-plane tracking of such asynchronous spins that allows the velocity field of blood to be developed. Since the dark regions represented by the irregular spins correspond to the general flow path of the blood, the perception of motion developed from their intensity shifts may be correlated to the actual cardiac flow speed and direction (Fig. 9.5). If the blood containing a collection of spins (X) that has coherent phases leaves the slice of interest perpendicularly, and a new region of blood with incoherent spins (Y) with multiple degrees of phase shift enters the same slice at the same point, the pixel region at that point turns from white to black. Therefore, the motion estimation algorithm identifies the blood in that region as stationary in-plane. There is no specific motion in-plane but a velocity component through-plane.

The aim is to track flow patterns that relate to the in-plane orientation. This is achieved by estimating the pixel displacement presented by the movements of a collective group X within a region filled with Y or vice versa. If we assume a collective group of spins Y moving in the in-plane direction then this appears in the image as

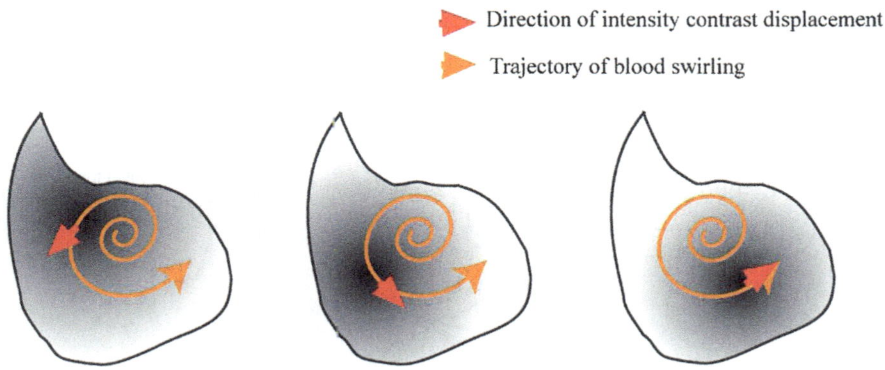

Fig. 9.5 Motion estimation of in-plane MR-signals. Based on schematic display of a right atrial flow, the ensembles of asynchronous proton spins that show up as contrasting signal intensity on the cine-magnetic resonance images are represented by *grey* patches of varying intensity. Based on motion estimation, motion pertaining to the blood flow images of arbitrary $(n-1)$, n and $(n+1)$ phases in a cardiac cycle of N phases can be predicted

a dark stream of moving pixels across the region. This time, the algorithm is able to track the movement in the plane using the pre-and post-sequence of cine-MR images. The flow 'tracking' fails only when we have a group of X and no Y group spins that are imaged. The movement of white pixels against a white background cannot be picked up. However, this will never occur in a cardiac chamber as the flow within definitely comprises of a variety of different spins phases. Traditionally, the visual deduction of blood movement using steady-state free precession MRI has been carried out by doctors to examine patients.

9.4 Ventricular Assist Devices

Harnessing CFD can spur forth computational haemodynamics research and assist surgical interventional planning or development of cardiac assist devices. This can support design of medical prototypes under investigation prior to its finalization and development. Simulation of the flow dynamics in the device can help us to understand the suitability of any particular patient prior to surgery.

During heart failure, the heart is unable to pump blood at a sufficient rate to meet the demands of the body. The ventricular assist device can provide mechanical circulatory support to restore this function. The ventricular assist device can be categorized into the right and left devices to support its corresponding ventricular functions. Where both chambers need support a bi-ventricular assist device is used. The pulmonary arterial resistance that contributes to a load on the heart chamber is

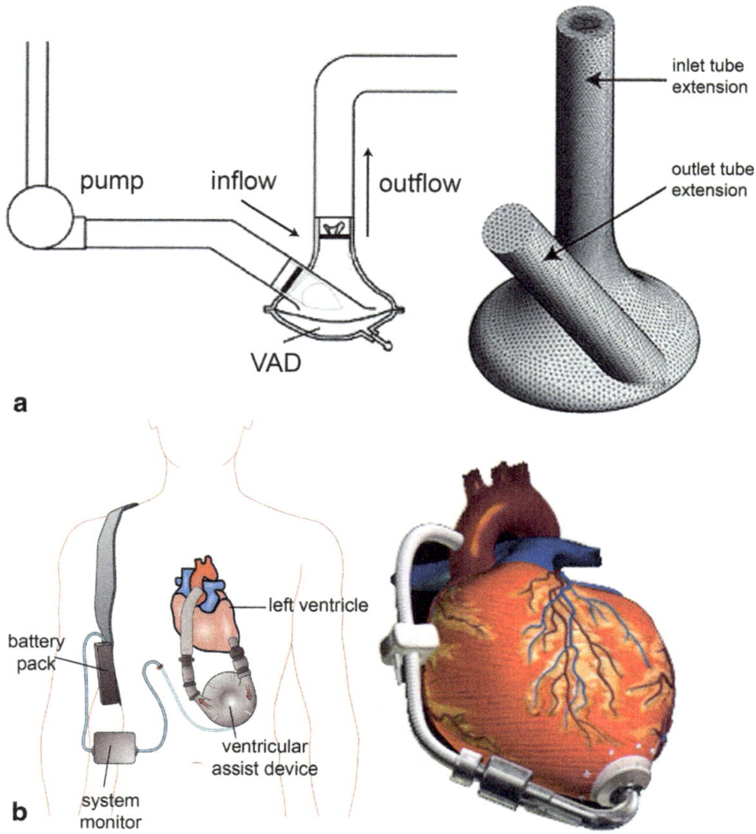

Fig. 9.6 **a** spiral vortex ventricular device commonly used during heart surgery. **b** left ventricular assist device pumping blood from the left ventricle to the aorta, connected to an externally worn control unit and battery pack. (Adapted from Wikicommons)

the determining factor for implantation of the device to relieve ventricular hydraulic burden.

Examples of ventricular assist devices studied through computational modelling are given in Fig. 9.6. The spiral vortex ventricular assist device is a pneumatically-driven pulsatile device used during heart surgery to reproduce the blood pumping action while surgery takes place on the heart itself. It consists of a conical-shaped housing, an outflow tract located toward the apex, and an inflow conduit directed tangentially to the diaphragm housing junction with an angle of 35 degrees downward. During the diastolic phase, air is extracted from the device via the air connector, causing the diaphragm to deflect upwards. The low pressure generated induces blood to flow into the conical housing via a valve. During systole, air is injected into the device, and the diaphragm deflects in the opposite direction into the housing, resulting in a higher pressure that causes an outflow via a valve at the top of the housing.

The electromechanical ventricular assist device is used by patients directly to pump blood through the body. These devices are designed to assist the ventricle to overcome some of the pulmonary arterial resistance is high during heart disease. Computational simulations are used to determine if the design produces high shear stress, that leads to significant platelet activation, microparticle formation, and/or to reduce the risk of thrombosis.

9.5 Simulation-Based Virtual Surgery

Ageing populations, coupled with increased cardiovascular diseases and advancements in new innovative technologies, are driving growth of cardiac-based clinical management systems. CFD simulation and analysis is clinically important in providing the basis and causes of flow resistance assessment causing pathological lesions. This provides a basis for surgical design and construction of coronary arterial bypass grafting for maximum patency.

A simulation-based virtual surgery system is a unique combination of cardiac diagnosis, medical imaging, computer visualisation, computational simulation, cardiac flow analysis and mechanical prototyping, which are undertaken prior to their integration. Development of the CFD and FEA platforms provides the potential to study diseased vascular systems, offering appropriate treatment solutions. The current visualisation modelling platform applies sequences of medical images from CT or MRI and can rapidly reconstruct the geometry of the diseased vessel. The integration of these approaches can provide effective assessment of diseases such as stenosis and aneurysm.

A simulation-based virtual reality platform with a web-based interface to incorporate multiple imaging, computational and visualisation modules with an efficient coupling is illustrated in Fig. 9.7. Each successful treatment of the diseased vessel can be saved into a database of patient records.

An example using this platform is a, stent design based on the flow and patient specific information is evaluated using the virtual reality platform. The stents are customized to the patient modelled and evaulated for its haemodynamic consequences (Fig. 9.8). An optimal stent design is determined based on best fit at the appropriate location along the diseased section of the blood vessel. Furthermore the design needs to account for surgical implantation. Effectiveness in preventing re-stenosis and its design based on the patient-specific condition can be updated in a database for the reference and future case studies.

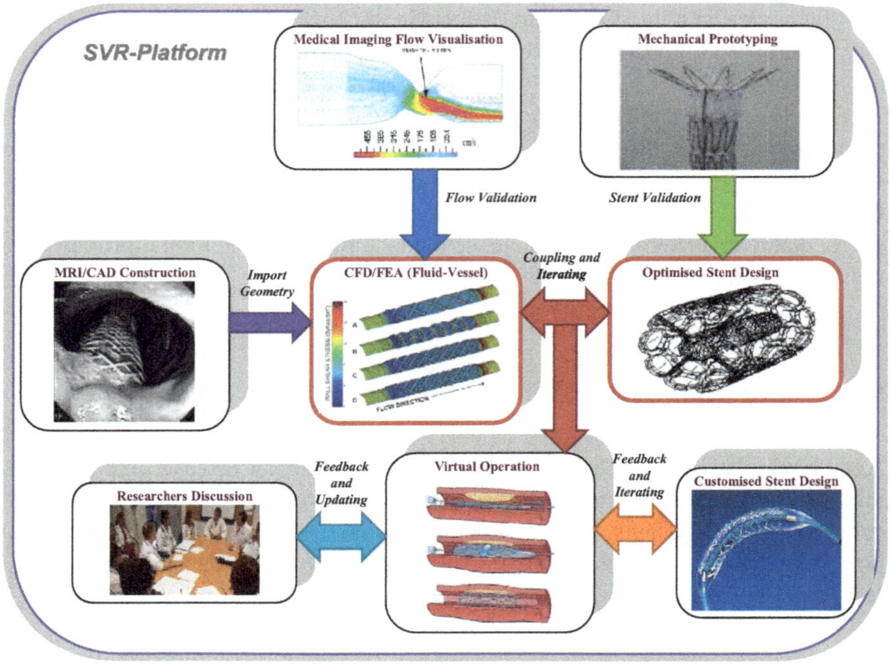

Fig. 9.7 A simulation-based virtual reality platform integrating multidisciplinary techniques and systems. In this exemplified system, with regards to treatment of atherosclerosis, effectiveness in preventing re-stenosis and its design based on the patient-specific condition of a blocked artery can be updated in a database for the reference and future case studies

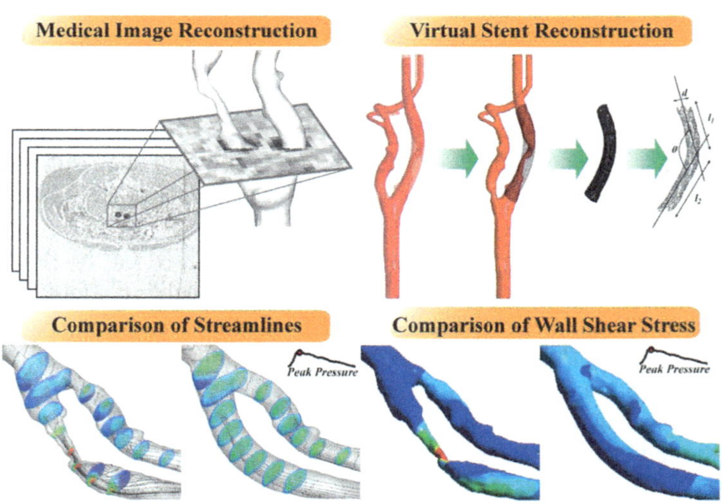

Fig. 9.8 Virtual stent design and analysis procedure. Based on these already available research capabilities, the framework will provide a simulation-based virtual stenting environment for clinical management, simulation of flow conditions due to cardiovascular diseases, planning of stent designs and virtual treatment of the human vascular system

9.6 Advanced Heart Valve Modelling

Advanced modelling techniques have allowed researchers to investigate the fluid and structural dynamics of the heart and heart valve. This section we summarise the latest advanced research outcomes from this field.

Yun et al. (2014) investigated the flow dynamics for pulsatile flow through bi-leaflet mechanical heart valves. The Lattice-Boltzmann method was used to simulate a pulsatile, high-Reynolds-number flow with the inclusion of reverse leakage flow through prosthetic devices. The flowrate and leaflet motion were derived from experimental data. Comparison of the numerical simulations with experimental digital particle image velocimetry (DPIV) showed good validation quantitatively as well as qualitatively.

Le and Sotiropoulos (2013) developed a large scale kinematic model for animating the left ventricle wall and used this model to drive the fluid–structure interaction between blood flow and a mechanical heart valve prosthesis (Fig. 9.9). The kinematic model simulated the left ventricle wall motion in response to an excitation wavefront propagating along the heart wall. The large-scale left ventricle wall motion exhibits complex contractile mechanisms that include contraction (twist) and expansion (untwist). The fluid–structure interaction between the leaflets of the mechanical heart valve and the blood flow driven by the dynamic left ventricle wall motion and mitral inflow is simulated using the curvilinear immersed boundary method and implemented in conjunction with a domain decomposition approach.

Similar work include Borazani (2013) who coupled a sharp-interface immersed boundary incompressible Navier–Stokes solver for the fluid domain with a non-

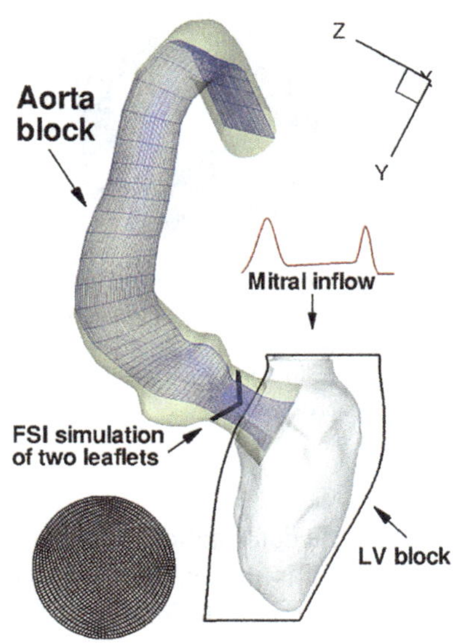

Fig. 9.9 Computational grid of the left ventricle block and the aorta block. The left ventricular block is a structured grid of size $161 \times 281 \times 161$. The aorta block is a body fitted mesh of size $161 \times 161 \times 401$. At the mitral position, uniform pulsatile flow is specified as boundary condition and the mitral valve is assumed to be fully open during diastole. (Image from Le and Sotiropoulos 2013)

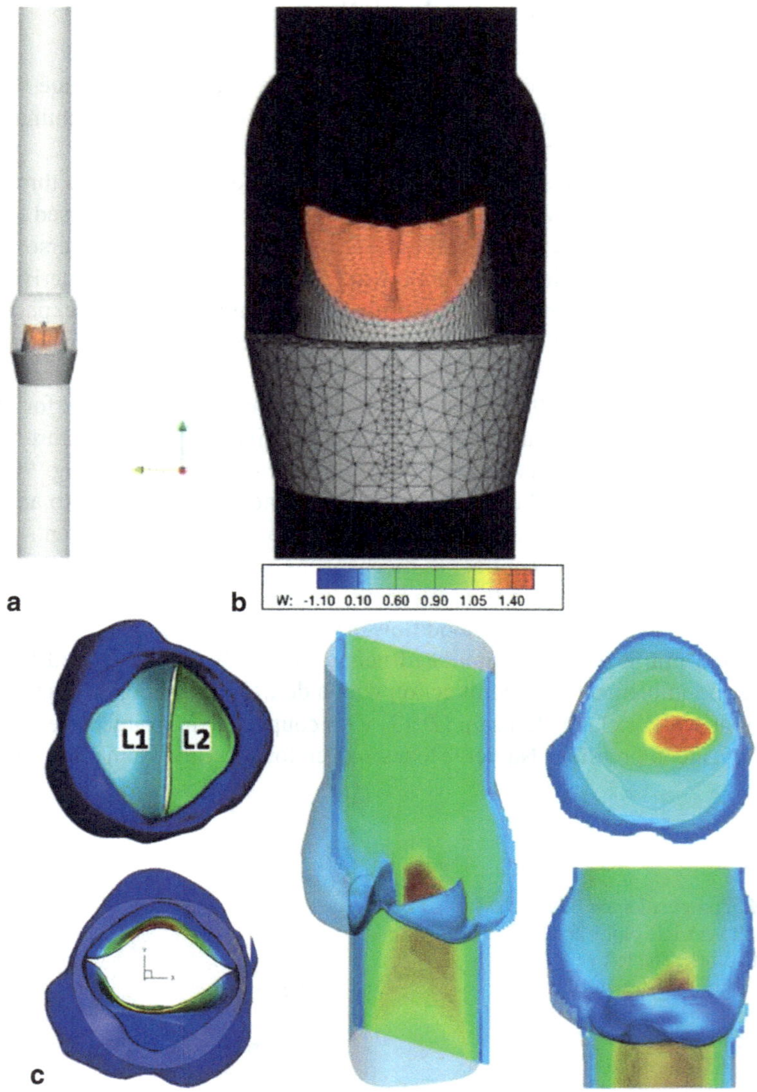

Fig. 9.10 Computationl model result from Borazjani (2013) showing the domain for the BHV simulations. **a** the computational domain geometry and the bio-prosthetic heart valve implanted in the sinus region. **b** Side view showing the background curvilinear grid in the sinus region. **c** Results from Chandran and Vigmostad (2013) for a full FSI simulation of a patient-specific bicuspid aortic valve

linear large deformation finite element method for the structural soft tissue domain. This model is shown in Fig. 9.10a, b. The results showed that the flow's threefold symmetry breaks during the early systole, questioning the threefold symmetry assumption of other simulations. The flow fields created by the tissue and mechanical valves, showed breakdown of vortices into small-scale vortical structures before peak systole in mechanical heart valves, which is not observed in the bio-prosthetic heart valves.

Chandran and Vigmostad (2013) discuss several studies suggesting mechanical stresses induced on the valve leaflets and the abnormal flow development in the ascending aorta may be an important factor in the diseases of the valve and the aortic root. They present details of the computational method in creating a patient-specific geometry of the normal tri-cuspid and bicuspid valves from real-time 3D ultrasound images and the dynamic analyses performed to determine the potential effects of mechanical stresses on the valve leaflet and aortic root pathology. Their preliminary results are shown in Fig. 9.10c.

Such high level analysis provides insight and fine details of the flow patterns which are not easily visualised with in vivo observations. The simulations also reveal complex kinematics of the valve leaflets, thus, underscoring the need for patient-specific simulations of heart valve prosthesis and other cardiac devices, which can form part of the simulation-based virtual platform discussed in Sect. 9.5.

9.7 Summary

Haemodynamics plays an important role in the development and progression of cardiovascular disease. Many researchers are at the frontiers of different topics, and this chapter provides just a sample of the high level research currently being undertaken. In blood rheology studies, the computational resources being used are phenomenal reaching teraflops in computing speeds, and accounting for millions of red blood cells simultaneously. Advanced medical imaging modalities are at a level where they can complement fluid flow analysis providing insight into, and validation of the computational model algorithms. This unique method is highly beneficial for blood flow that is otherwise difficult to replicate in-vivo (unlike for other fluids) due to its physiological nature. Ventricular devices and artificial heart valves are also designed based on haemodynamics simulations to ensure the designs produce sufficient shear stress for platelet activation, and reduce the risk of thrombosis. The discussions in this section provide a basis for understanding the advanced topics that will continue to push the boundaries of computational modelling as new algorithms and computational power become available.

Appendix

List of Software

DICOM Viewers

The listed DICOM viewers have similar functionality

- Acculite www.accuimage.com
- MicroDicom www.microdicom.com
- DICOM Works www.dicomworks.com
- Sante DICOM Viewer www.santesoft.com
- Mango *ric.uthscsa.edu/mango*
- OsiriX (MAC) www.osirix-viewer.com
- AMIDE amide.sourceforge.net
- Irfanview www.irfanview.com
- XNView www.xnview.com

Open Source Medical Imaging and Segmentation

CVIPTools	A UNIX/Win32-based package and contains a collection of C and C++ computer imaging tools that includes edge/line detection, segmentation, and many other functions www.ee.siue.edu/CVIPtools
Fiji/ImageJ	A Java-based image processing package that uses additional plugins for a variety of functionalities including segmentation algorithms pacific.mpi-cbg.de/wiki/index.php/Fiji
GemIdent	An interactive program that is designed for colour segmentation in images with few colours, and the objects of interest look alike with small variation www.gemident.com

J. Tu et al., *Computational Hemodynamics—Theory, Modelling and Applications*,
Biological and Medical Physics, Biomedical Engineering,
DOI 10.1007/978-94-017-9594-4

ITK-SNAP	An interactive software application that allows users to navigate three-dimensional medical images, manually delineate anatomical regions of interest, and perform automatic image segmentation www.itksnap.org
Megawave 2	Made up of C library modules, that contains original algorithms written by researchers and is run using Unix/Linux (megawave.cmla.ens-cachan.fr)
MITK & 3Dmed	Made up of C++ library for integrated medical image processing, segmentation, and registration algorithms www.mitk.net/download.htm
Slicer	Has a GUI that allows manual and automatic segmentation, registration, and three-dimensional visualization. It is a modular platform which means that it allows addition of new modules www.slicer.org
VXL	A collection of C++ libraries designed for computer vision research and implementation vxl.sourceforge.net

Commercial Medical Imaging and Segmentation

3D Doctor	An image processing and measurement software for MRI, CT, PET, microscopy, scientific, and industrial imaging applications www.ablesw.com/3d-doctor
Amira	Includes custom modules through C++. There is also a research version called ZIBAmira (http://amira.zib.de) which provide licenses for joint research collaboration (www.visageimaging.com/amira.html)
Analyse	A software package for multi-dimensional display and segmentation www.analyzedirect.com)
Mimics	An interactive tool for the visualization, 3D rendering, and segmentation of CT/MRI images. It also has a built in mesh program for CFD and structural analysis models www.materialise.com
SliceOmatic	Targeted at imaging of soft-tissue and for this, the use of MRI images is an advantage, however it can be used with CT. (www.tomovision.com)
Vida Diagnostics	A lung analysis tool for chronic obstructive pulmonary disease, emphysema and asthma www.vidadiagnostics.com
CASCADE	Computer- Aided System for Cardiovascular Disease Evaluation (CASCADE) developed by the Vascular Imaging Laboratory (VIL) at the University of Washington http://www.washington.edu/news/2006/05/04/uw-licenses-innovative-cardiovascular-software-to-vpdiagnostics

Open Source Computer Aided Design Software

FreeCAD	3D computer assisted design program sourceforge.net/projects/free-cad
Open CASCADE	Allows for 3D surface and solid modeling, visualization, data exchange and rapid application development www.opencascade.org
BRL CAD	Has interactive editing for 3D solid modeling. Can also use for image processing and analysis brlcad.org
OpenSCAD	Used for creating solid 3D CAD objects www.openscad.org

Commercial Computer Aided Design Software

The listed CAD software have similar functionality in their ability to create 3D solid models ready for importing into a CFD meshing program.

- Geomagic www.geomagic.com
- CATIA www.3ds.com
- Autodesk www.autodesk.com
- Solidworks www.solidworks.com
- PRO/Engineer www.ptc.com
- IronCAD www.ironcad.com

CFD-FSI Packages

OpenFOAM	Open source CFD software package built using C++ and compiled under UNIX. The code is open and therefore allows full customisation and extensions to its standard capability www.openfoam.com
ADINA	ADINAR & D, Inc., Watertown, MA, USA). A commercial finite element software used to solve FSI models
ANSYS	Includes ICEM meshing, CFX and Fluent CFD solvers, and CFD-Post for post processing. Also includes multi-physics for structural and FSI www.ansys.com
CD-Adapco	Includes STAR-CD and STAR-CCM for simulations involving flow of fluids and solids, heat transfer and stress analysis www.cd-adapco.co
	The simulation was performed using a finite element solver (FEMLAB 3.1; COMSOL, Inc

Numeca	Provides for fluid dynamics simulations for industrial applications www.numeca.be
Phoenics	Handles CFD simulations for fluid flow, heat or mass transfer, chemical reaction and combustion in engineering equipment and the environment www.cham.co.uk

Third Party Post Processing Software

ParaView	An open-source, multi-platform data analysis and visualization application www.paraview.org
GNU Plot	An open source portable command-line graphing utility www.gnuplot.info
OpenDX	Uses IBM's visualisation data explorer interface for data input and output www.opendx.org
Ensight	Visualisation for most CFD data file formats www.ensight.com
Tecplot	Visualisation for most CFD data file formats www.tecplot.com
Plot3D	Interactive graphics program for visualizing CFD results www.openchannelfoundation.org

Bibliography

AIHW Board (2008) Australia's health 2008. Australian Institute of Health and Welfare
Antiga L, Ene-Iordache B, Caverni L, Cornalba GP, Remuzzi A (2002) Geometric reconstruction for computational mesh generation of arterial bifurcations from CT angiography. Comput Med Imaging Graph 26:227–235
Arcilla AS, Häuser J, Eiseman PR, Thompson JF (1991) Numerical grid generation in computational fluid dynamics and related fields. North-Holland, Amsterdam
Baltes C, Kozerke S, Hansen MS, Pruessmann KP, Tsao J, Boesiger P (2005) Accelerating cine phase-contrast flow measurements using k-t BLAST and k-t SENSE. Magn Reson Med 54:1430–1438
Banerjee RK, Back LH, Back MR, Cho YI (2000) Physiological flow simulation in residual human stenoses after coronary angioplasty. J Biomech Eng 122:310–322
Baskurt OK, Hardeman M, Rampling MW, Meiselman HJ (2007) Handbook of hemorheology and hemodynamics, vol 69 of Biomedical and Health Research, IOS Press, Amsterdam, Netherlands. ISBN: 978-1-58603-771-0, p 468
Batchelor GK (1967) An Introduction to fluid dynamics. Cambridge University Press, Cambridge
Baxter BT, Davis VA, Minion DJ, Wang YP, Lynch TG, McManus BM (1994) Abdominal aortic-aneurysms are associated with altered matrix proteins of the nonaneurysmal aortic segments. J Vasc Surg 19:797–803
Beattie D, Xu C, Vito R, Glagov S, Whang MC (1998) Mechanical analysis of heterogeneous, atherosclerotic human aorta. J Biomech Eng 120:602–607
Beaussier H, Masson I, Collin C, Bozec E, Laloux B, Calvet D, Zidi M, Boutouyrie P, Laurent S (2008) Carotid plaque, arterial stiffness gradient, and remodeling in hypertension. Hypertension 52:729
Berger SA, Jou LD (2000) Flows in stenotic vessels. Annu Rev Fluid Mech 32:347–382
Bernaschi M, Bisson M, Fatica M, Melchionna S, Succi S (2013) Petaflop hydrokinetic simulations of complex flows on massive GPU clusters. Comput Phys Commun 184:329–341
Bertolotti C, Deplano V, Fuseri J, Dupouy P (2001) Numerical and experimental models of post-operative realistic flows in stenosed coronary bypasses. J Biomech 34:1049–1064
Beythien C, Gutensohn K, Bau J, Hamm CW, Kühnl P, Meinertz T, Terres W (1999) Influence of stent length and heparin coating on platelet activation: a flow cytometric analysis in a pulsed floating model. Thromb Res 94:79–86
Bezdek JC, Hall LO, Clarke LP (1993) Review of MR image segmentation techniques using pattern recognition. Med Phys 20:1033–1048
Bhatti MS, Shah RK (1987) Turbulent and transition flow convective heat transfer in ducts. In: Kakac S, Shah RK, Aung W (eds) Handbook of single-phase convective heat transfer. Wiley Interscience, New York
Bieniarz MC, Delgado R (2007) The financial burden of destination left ventricular assist device therapy: who and when? Curr Cardiol Rep 9:194–199
Birchall D, Zaman A, Hacker J, Davies G, Mendelow D (2006) Analysis of haemodynamic disturbance in the atherosclerotic carotid artery using computational fluid dynamics. Eur Radiol 16:1074–1083

J. Tu et al., *Computational Hemodynamics—Theory, Modelling and Applications*,
Biological and Medical Physics, Biomedical Engineering,
DOI 10.1007/978-94-017-9594-4

Bird RB, Armstrong RC, Hassager O (1987) Dynamics of polymer liquids. Wiley, New York

Bluestein D, Alemu Y, Avrahami I, Gharib M, Dumont K, Ricotta JJ, Einav S (2008) Influence of microcalcifications on vulnerable plaque mechanics using FSI modeling. J Biomech 41:1111–1118

Borazjani I (2013) Fluid-structure interaction, immersed boundary-finite element method simulations of bio-prosthetic heart valves. Comput Methods Appl Mech Eng 257:103–116

Botnar R, Rappitsch G, Scheidegger MB, Liepsch D, Perktold K, Boesiger P (2000) Hemodynamics in the carotid artery bifurcation: a comparison between numerical simulations and in vitro MRI measurements. J Biomech 33:137–144

Boutsianis E, Dave H, Frauenfelder T, Poulikakos D, Wildermuth S, Turina M, Ventikos Y, Zund G (2004) Computational simulation of intracoronary flow based on real coronary geometry. Eur J Cardiothorac Surg 26:248–256

Brady AR, Thompson SG, Fowkes FG, Greenhalgh RM, Powell JT (2004) Abdominal aortic aneurysm expansion: risk factors and time intervals for surveillance. Circulation 110:16–21

Burke AP, Farb A, Malcom GT, Liang Y, Smialek J, Virmani R (1997) Coronary risk factors and plaque morphology in men with coronary disease who died suddenly. N Engl J Med 336:1276–1282

Burke AP, Kolodgie F, Farb A, Weber D, Virmani R (2002) Morphological predictors of arterial remodeling in coronary atherosclerosis. Circulation 105:297–303

Byun HS, Rhee K (2003) Intraaneurysmal flow changes affected by clip location and occlusion magnitude in a lateral aneurysm model. Med Eng Phys 25:581–589

Calcagno C, Ramachandran S, Izquierdo-Garcia D, Mani V, Millon A, Rosenbaum D, Tawakol A, Woodward M, Bucerius J, Moshier E, Godbold J, Kallend D, Farkouh M, Fuster V, Rudd JF, Fayad Z (2013) The complementary roles of dynamic contrast-enhanced MRI and 18F-fluorodeoxyglucose PET/CT for imaging of carotid atherosclerosis. Eur J Nucl Med Mol Imaging 40:1884–1893

Campa JS, Greenhalgh RM, Powell JT (1987) Elastin degradation in abdominal aorticaneurysms. Atherosclerosis 65:13–21

Campbell IC, Ries J, Dhawan SS, Quyyumi AA, Taylor WR, Oshinski JN (2012) Effect of inlet velocity profiles on patient-specific computational fluid dynamics simulations of the carotid bifurcation. J Biomech Eng 134:051001–051001

Canny J (1986) A computational approach to edge detection. IEEE Trans Pattern Anal Mach Intell 8:679–698

Cebral JR, Löhner R, Soto O, Choyke PL, Yim PJ (2001) Patient-specific simulation of carotid artery stenting using computational fluid dynamics. Lecture notes in computer science, medical image computing and computer-assisted intervention—MICCAI 2001 2208:153–160

Chaichana T, Sun Z, Jewkes J (2011) Computation of hemodynamics in the left coronary artery with variable angulations. J Biomech 44:1869–1878

Chaichana T, Sun Z, Jewkes J (2012) Computational fluid dynamics analysis of the effect of plaques in the left coronary artery. Comput Math Methods Med 2012:504367

Chandran KB, Vigmostad SC (2013) Patient-specific bicuspid valve dynamics: overview of methods and challenges. J Biomech 46:208–216

Chandran KB, Mun JH, Choi KK, Chen JS, Hamilton A, Nagaraj A, McPherson DD (2003) A method for in vivo analysis for regional arterial wall material property alterations with atherosclerosis: preliminary results. Med Eng Phys 25:289–298

Chandran KB, Wahle A, Vigmostad SC, Olszewski ME, Rossen JD, Sonka M (2006) Coronary arteries: imaging, reconstruction, and fluid dynamic analysis. Crit Rev Biomed Eng 34:23–103

Chen J, Lu XY (2006) Numerical investigation of the non-Newtonian pulsatile blood flow in a bifurcation model with a non-planar branch. J Biomech 39:818–832

Cheng GC, Loree HM, Kamm RD, Fishbein MC, Lee RT (1993a) Distribution of circumferential stress in ruptured and stable atherosclerotic lesions. A structural analysis with histopathological correlation. Circulation 87:1179–1187

Cheng Y, Oertel H, Schenkel T (2005) Fluid-structure coupled CFD simulation of the left ventricular flow during filling phase. Ann Biomed Eng 33:567–576

Cheung SCP, Wong KKL, Yang W, Yeoh GH, Tu JY, Beare R, Phan T (2010a) Experimental and numerical study on the hemodynamics of stenosed carotid bifurcation. Australas Phys Eng Sci Med 33:319–328

Chien S (1982) Hemorheology in clinical medicine. Clin Hemorheol 2:137–142

Cho YI, Kensey KR (1991) Effects of the non-Newtonian viscosity of blood on flows in a diseased arterial vessel. Part 1: steady flows. Biorheology 28:241–262

Choke E, Cockerill G, Wilson WRW, Sayed S, Dawson J, Loftus I (2005) A review of biological factors implicated in abdominal aortic aneurysm rupture. J Vasc Endovasc Surg 30:227–244

Chua LP, Song GL, Yu SCM, Zhou TM (2005) Computational fluid dynamics of gap flow in a biocentrifugal blood pump. Artif Organs 29:620–628

Clarke LP, Velthuizen RP, Phuphanich S, Schellenberg JD, Arrington JA, Silbiger M (1993) MRI: stability of three supervised segmentation techniques. Mag Res Imaging 11:95–106

Clarke LP, Velthuizen RP, Camacho MA, Heine JJ, Vaidyanathan M, Hall LO, Thatcher RW, Silbiger ML (1995) MRI segmentation: methods and applications. Mag Res Imaging 13:343–368

Colebrook CF (1939) Turbulent flow in pipes, with particular reference to the transition region between smooth and rough pipe laws. J Inst Civ Eng 11:133–156

Colombo A, Zahedmanesh H, Toner DM, Cahill PA, Lally C (2010) A method to develop mock arteries suitable for cell seeding and in-vitro cell culture experiments. J Mech Behav Biomed Mater 3:470–477

Condon BR, Patterson J, Wyper D, Jenkins A, Hadley DM (1987) Image non-uniformity in magnetic resonance imaging: its magnitude and methods for its correction. Br J Radiol 60:83–87

Craig RR, Kurdila AJ (2006) Fundamentals of structural dynamics. John Wiley & Sons. Hoboken, New Jersey, USA. ISBN: 9781118174449, p 744

Cunningham KS, Gotlieb AL (2005) The role of shear stress in the pathogenesis of atherosclerosis. Lab Invest 85:9–23

Dahl SK, Vierendeels J, Degroote J, Annerel S, Hellevik LR, Skallerud B (2010) FSI simulation of asymmetric mitral valve dynamics during diastolic filling. Comput Methods Biomech Biomed Engin 15:121–130

Dahl SK, Fagerholt E, Kiss G, Prot V, Amundsen B, Hellevik LR, Skallerud B (2011) 3D moving boundary conditions for heart CFD simulations—from echocardiographic recordings to discretised surface. MekIT'11: Sixth National Conference on Computational Mechanics:33–46

Dahl SK, Thomassen E, Hellevik L, Skallerud B (2012) Impact of pulmonary venous locations on the intra-atrial flow and the mitral valve plane velocity profile. Cardiovasc Eng Technol 3:269–281

Davies MJ (1996) Stability and instability: two faces of coronary atherosclerosis. Circulation 94:2013–2020

Davies MJ, Richardson PD, Woolf N, Katz DR, Mann J (1993) Risk of thrombosis in human atherosclerotic plaques: role of extracellular lipid, macrophage, and smooth muscle cell content. Br Heart J 69:377–381

de Berg M, Ogtfried C, van Kreveld M, Overmars M (2008) Computational geometry: algorithms and applications. Springer-Verlag, Berlin, Heidleberg, New York. ISBN: 978-3540779735, p 386

de Rochefort L, Vial L, Fodil R, Maitre X, Louis B, Isabey D, Caillibotte G, Thiriet M, Bittoun J, Durand E, Sbirlea-Apiou G (2007) In vitro validation of computational fluid dynamic simulation in human proximal airways with hyperpolarised 3He magnetic resonance phase-contrast velocimetry. J Appl Physiol 102:2012–2023

Demirkaya O, Asyali MH, Sahoo PK (2009) Image processing with Matlab—Application in medicine and biology. CRC Press, Broken Sound Parkway

Dimitrakopoulos P (2007) Interfacial dynamics in stokes flow via a three-dimensional fully-implicit interfacial spectral boundary element algorithm. J Comput Phys 225:408–426

Ding Z, Wang K, Li J, Cong X (2001) Flow field and oscillatory shear stress in a tuning-fork-shaped model of the average human carotid bifurcation. J Biomech 34:1555–1562

Donea J, Huerta A, Ponthot J-P, Rodríguez-Ferran A (2004) Arbitrary Lagrangian-Eulerian methods. Encyclopedia of Computational Mechanics. 1:14

Dong J, Wong KK, Tu J (2013a) Hemodynamics analysis of patient-specific carotid bifurcation: a CFD model of downstream peripheral vascular impedance. Int J Numer Method Biomed Eng 29:476–491

Dong JL, Inthavong K, Tu JY (2013b) Image-based computational hemodynamics evaluation of atherosclerotic carotid bifurcation models. Comput Biol Med 43:1353–1362

Duraiswamy N, Schoephoerster RT, Moore JE (2009) Comparison of near-wall hemodynamic parameters in stented artery models. ASME J Biomech Eng 131:061006

Ecabert O, Peters J, Schramm H, Lorenz C, von Berg J, Walker MJ, Vembar M, Olszewski ME, Subramanyan K, Lavi G, Weese J (2008) Automatic model-based segmentation of the heart in CT images. IEEE Trans Med Imaging 27:1189–1201

Elkins CJ, Markl M, Iyengar A, Wicker R, Eaton J (2004) Full field velocity and temperature measurements using magnetic resonance imaging in turbulent complex internal flows. Int J Heat Fluid Flow 25:702–710

Fan Y, Jiang W, Zou Y, Li J, Chen J, Deng X (2009) Numerical simulation of pulsatile non-Newtonian flow in the carotid artery bifurcation. Acta Mech Sin 25:249–255

Farmakis TM, Soulis JV, Giannoglou GD, Zioupos GJ, Louridas GE (2004) Wall shear stress gradient topography in the normal left coronary arterial tree: possible implications for atherogenesis. Curr Med Res Opin 20:587–596

Fayad ZA, Fuster V (2001) Clinical imaging of the high-risk or vulnerable atherosclerotic plaque. Circ Res 89:305–316

Fontaine V, Jacob MP, Houard X, Rossignol P, Plissonnier D, Angles-Cano E (2002) Involvement of the mural thrombus as a site of protease release and activation in human aortic aneurysms. Am J Pathol 161:1701–1710

Freund JB (2014) Numerical simulation of flowing blood cells. Annu Rev Fluid Mech 46:67–95

Friedman MH, Brinkman AM, Qin JJ, Seed WA (1993) Relation between coronary artery geometry and the distribution of early sudanophilic lesions. Atherosclerosis 98:193–199

Fry DL (1969) Certain histological and chemical responses of the vascular interface to acutely induced mechanical stress in the aorta of the dog. Circ Res 24:93–108

Fuessl RT, Kranenberg E, Kiausch U, Baer FM, Sechtem U, Höpp HW (2001) Vascular remodeling in atherosclerotic coronary arteries is affected by plaque composition. Coron Artery Dis 12:91–97

Fung YC (1993) Biomechanics: mechanical properties of living tissues, 2nd edn. Springer, New York. ISBN: 9780387979472, p 568

Gerstenberger A, Wall WA (2008) An eXtended finite element method/Lagrange multiplier based approach for fluid-structure interaction. Comput Methods Appl Mech Eng 197:1699–1714

Gertz SD, Roberts WC (1990) Hemodynamic shear force in rupture of coronary arterial atherosclerotic plaques. Am J Cardiol 66:1368–1372

Gertz SD, Uretzky G, Wajnberg RS, Navot N, Gotsman MS (1981) Endothelial cell damage and thrombus formation after partial arterial constriction: relevance to the role of coronary artery spasm in the pathogenesis of myocardial infarction. Circulation 63:476–486

Ghalichi F, Deng XY (2003) Turbulence detection in a stenosed artery bifurcation by numerical simulation of pulsatile blood flow using the low-Reynolds number turbulence model. Biorheology 40:637–654

Giannoglou GD, Soulis JV, Farmakis TM, Giannakoulas GA, Parchardidis GE, Louridas GE (2005) Wall pressure gradient in normal left coronary artery tree. Med Eng Phys 27:455–464

Gibbs P, Buckley DL, Blackband SJ, Horsman A (1996) Tumour volume determination from MR images by morphological segmentation. Phys Med Biol 41:2437–2446

Gijsen FJH, van de Vosse FN, Janssen JD (1999a) The Influence of the non-Newtonian properties of blood on the flow in large arteries: steady flow in a carotid bifurcation model. J Biomech 32:601–608

Gijsen FJH, van de Vosse FN, Janssen JD (1999b) The Influence of the non-Newtonian properties of blood on the flow in large arteries: unsteady flow in a 90° curved tube. J Biomech 32:705–713

Gijsen FJ, Wentzel JJ, Thury A, Lamers B, Schuurbiers JC, Serruys PW, van der Steen AF (2007) A new imaging technique to study 3-D plaque and shear stress distribution in human coronary artery bifurcations in vivo. J Biomech 40:2349–2357

Girasis C, Serruys PW, Onuma Y, Colombo A, Holmes DRJ, Feldman TE, Bass EJ, Leadley K, Dawkins KD, Morice MC (2010) 3-Dimensional bifurcation angle analysis in patients with left main disease: a substudy of the SYNTAX trial (SYNergy Between Percutaneous Coronary Intervention with TAXus and Cardiac Surgery). JACC: Cardiovasc Interv 3:41–48

Glasbey CA (1993) An analysis of histogram-based thresholding algorithms. Graph Models Image Process 55:532–537

Goldstein T, Bresson X, Osher S (2010) Geometric applications of the split bregman method: segmentation and surface reconstruction. J Sci Comput 45:272–293

Gonzalez HA, Moraga NO (2005) On predicting unsteady non-Newtonian blood flow. Appl Math Comput 170:909–923

Gonzalez RC, Woods RE (2002) Digital image processing. Prentice-Hall Inc, New Jersey

González RC, Woods RE (2011) Digital image processing, 3rd edn. Pearson-Prentice Hall, USA ISBN: 9780133002324, p 976

Goodall S, Crowther M, Hemingway DM, Bell PR, Thompson MM (2001) Ubiquitous elevation of matrix metalloproteinase-2 expression in the vasculature of patients with abdominal aneurysms. Circulation 104:304–309

Gordon PC, Friedrich SP, Piana RN, Kugelmass AD, Leidig GA, Gibson CM, Cohen DJ, Carrozza JP, Kuntz RE, Baim DS (1994) Is 40 to 70 % diameter narrowing at the site of previous stenting or directional coronary atherectomy clinically significant? Am J Cardiol 74:26–32

Görgülü Ş. (2007) Stent Implants do not Totally Reset the Heart. www.acibademinternational.com/Stent_implants.asp. Accessed 12 March 2014

Greenemeier L (2008) Next-Gen heart stents may feature toothlike coating, Scientific American. http://www.scientificamerican.com/article/next-gen-heart-stents/. Accessed 14 March 2014

Gundert TJ, Shadden SC, Williams AR, Koo BK, Feinstein JA, Ladisa JF (2011) A rapid and computationally inexpensive method to virtually implant current and next-generation stents into subject-specific computational fluid dynamics models. Ann Biomed Eng 39:1423–1437

Hall CW, Liotta D, Henly WS, Crawford ES, Debakery ME (1964) Development of artificial intrathoracic circulatory pumps. Am J Surg 108:685–692

Hall LO, Bensaid AM, Clarke LP, Velthuizen RP, Silbiger MS, Bezdek JC (1992) A comparison of neural network and fuzzy clustering techniques in segmenting magnetic resonance images of the brain. IEEE T Neural Netw 3:672–682

Hansegard J, Urheim S, Lunde K, Malm S, Rabben S (2009) Semi-automated quantification of left ventricular volumes and ejection fraction by real-time three-dimensional echocardiography. Cardiovasc Ultrasound 7:18

Harada K, Amano T, Uetani T, Funahashi H, Arai K, Okada K, Hirashiki A, Hayashi M, Oshima S, Ishii H, Izawa H, Matsubara T, Murohara T (2010) Accuracy of 64-slice multidetector computed tomography for classification and quantitation of coronary plaque: Comparison with integrated backscatter intravascular ultrasound. Int J Cardiol 149:95–101

Haralick RM, Shapiro LG (1985) Image segmentation techniques. Comput Vis Graph Im Proc 29:100–132

Hartiala JJ, Mostbeck GH, Foster E (1993) Velocity-encoded cine MRI in the evaluation of left ventricular diastolic function: measurement of mitral valve and pulmonary vein flow velocities and flow volume across the mitral valve. Am Heart J 125:1054–1066

He CM, Roach MR (1994) The composition and mechanical-properties of abdominal aortic-aneurysms. J Vasc Surg 20:6–13

He XJ, Ku DN (1996) Pulsatile flow in the human left coronary artery bifurcation: average conditions. J Biomech Eng-TransASME 118:74–82

He Y, Duraiswamy N, Frank AO, Moore JE (2005) Blood flow in stented arteries: a parametric comparison of strut design patterns in three dimensions. ASME J Biomech Eng 127:637–647

Helft G, Worthley SG, Fuster V, Zaman AG, Fayad ZA, Osende JI, Fallon JT, Badimon JJ (2002) Serial noninvasive magnetic resonance imaging documents progression and regression of individual plaques. Circulation 105:993–998

Hernandez AF, Shea AM, Milano CA (2008) Long-term outcomes and costs of ventricular assist devices among medicare beneficiaries. JAMA 300:2398–2406

Hodgson JM, Reddy KG, Suneja R, Nair RN, Lesnefsky EJ, Sheehan HM (1993) Intracoronary ultrasound imaging: correlation of plaque morphology with angiography, clinical syndrome and procedural results in patients undergoing coronary angioplasty. J Am Coll Cardiol 21:35–44

Hoi Y, Wasserman BA, Lakatta EG, Steinman DA (2010) Effect of common carotid artery inlet length on normal carotid bifurcation hemodynamics. J Biomech Eng 132:121008–121014

Holzapfel GA, Stadler M, Schulze-Bauer CA (2002) A layer-specific three-dimensional model for the simulation of balloon angioplasty using magnetic resonance imaging and mechanical testing. Ann Biomed Eng 30:753–767

Hopkins LM, Kelly JT, Wexler AS, Prasad AK (2000) Particle image velocimetry measurements in complex geometries. Exp Fluids 29:91–95

Hoshina K, Sho E, Sho M, Nakahashi TK, Dalman RL (2003) Wall shear stress and strain modulate experimental aneurysm cellularity. J Vasc Surg 37:1067–1074

Huang X, Yang C, Yuan C, Liu F, Canton G, Zheng J, Woodard PK, Sicard GA, Tang D (2009) Patient-specific artery shrinkage and 3D zero-stress state in multi-component 3D FSI models for carotid atherosclerotic plaques based on in vivo MRI data. Mol Cell Biomech 6:121–134

Hubbard BJ, Chen HC (1994) A chimera scheme for incompressible viscous flows with applications to submarine hydrodynamics. AIAA Paper 94-2210

Hubbard BJ, Chen HC (1995) Calculation of unsteady flows around bodies with relative motion using a chimera RANS method, in Proc. 10th ASCE Engineering Mechanics Conference, vol II, University of Colorado at Boulder, Boulder, CO, 7832–7785

Hutton D (2005) Fundamentals of finite element analysis. McGraw-Hill, Education, India, ISBN: 978-0072922363, p 494

Ii S, Gong X, Sugiyama K, Wu J, Huang H, Takagi S (2012) A full Eulerian fluid-membrane coupling method with a smoothed volume-of-fluid approach. Commun Comput Phys 12:544–576

Jessup M (2001) Mechanical cardiac-support devices—dreams and devilish details. N Engl J Med 345:1490–1493

Johnston PR, Kilpatrick D (1997) The effect of branch angle on human coronary artery blood flow. In: McDonald AD, McAleer M (eds) International congress on modelling and simulation (MODSIM97). Hobart, Tasmania, pp1029–1034

Joshi AK, Leask RL, Myers JG, Ojha M, Butany J, Ethier CR (2004) Intimal thickness is not associated with wall shear stress patterns in the human right coronary artery. Arterioscler Thromb Vasc Biol 24:2408–2413

Juhan V, Nazarian B, Malkani K, Bulot R, Bartoli JM, Sequeira J (1997) Geometrical modelling of abdominal aortic aneurysms. In: Troccaz J, Grimson E and R. Mösges R (eds) CVRMed-MRCAS'97. Springer Berlin Heidelberg, pp 243–252

Jung J, Hassanein A (2008) Three-phase CFD analytical modeling of blood flow. Med Eng Phys 30:91–103

Jung B, Schneider B, Markl M, Saurbier B, Geibel A, Hennig J (2004) Measurement of left ventricular velocities: phase contrast MRI velocity mapping versus tissue-Doppler-ultrasound in healthy volunteers. J Cardiovasc Magn Reson 6:777–783

Kabinejadian F, Ghista DN (2011) Compliant model of a coupled sequential coronary arterial bypass graft: effects of vessel wall elasticity and non-Newtonian rheology on blood flow regime and hemodynamic parameters distribution. Med Eng Phys 34(7):860–872

Kallergi M, Woods K, Clarke LP, Qian W, Clark RA (1992) Image segmentation in digital mammography: comparison of local thresholding and region growing algorithms. Comput Med Imaging Graph 16:323–331

Kalvin AD, Taylor RH (1996) Superfaces: polygonal mesh simplification with bounded error. Comput Graph Appl IEEE 16:64–77

Kays WM, Crawford ME (1993) Convective heat and mass transfer, 3rd edn. McGraw-Hill, New York

Kefayati S, Holdsworth DW, Poepping TL (2014) Turbulence intensity measurements using particle image velocimetry in diseased carotid artery models: effect of stenosis severity, plaque eccentricity, and ulceration. J Biomech 47:253–263

Kenner T (1989) The measurement of blood density and its meaning. Basic Res Cardiol 84:111–124

Khalafvand SS, Zhong L, Ng EYK (2014) Three-dimensional CFD/MRI modeling reveals that ventricular surgical restoration improves ventricular function by modifying intraventricular blood flow. Int J Numer Method Biomed Eng 30(10):1044–1056

Kherani AR, Maybaum S, Oz MC (2004) Ventricular assist devices as a bridge to transplant or recovery. Cardiology 101:93–103

Kim YH, Xu X, Lee JS (2010) The effect of stent porosity and strut shape on saccular aneurysm and its numerical analysis with lattice Boltzmann method. Ann Biomed Eng 38:2274–2292

Kiousis DE, Rubinigg SF, Auer M, Holzapfel GA (2009) A methodology to analyze changes in lipid core and calcification onto fibrous cap vulnerability: the human atherosclerotic carotid bifurcation as an illustratory example. J Biomech Eng 131:121002

Kitagawa T, Yamamoto H, Horiguchi J, Ohhashi N, Tadehara F, Shokawa T, Dohi Y, Kunita E, Utsunomiya H, Kohno N, Kihara Y (2009) Characterization of noncalcified coronary plaques and identification of culprit lesions in patients with acute coronary syndrome by 64-slice computed tomography. J Am Coll Cardiol Img 2:153–160

Klabunde RE (2005) Cardiovascular physiology concepts. Lippincott Williams & Wilkins, Phildelphia USA, ISBN: 978-1451113846, P 256

Kleinstreuer C, Nazemi M, Archie JP (1991) Hemodynamics analysis of a stenosed carotid bifurcation and its plaque-mitigating design. J Biomech Eng 113:330–335

Kock SA, Nygaard JV, Eldrup N, Fründ E-T, Klærke A, Paaske WP, Falk E, Yong Kim W (2008) Mechanical stresses in carotid plaques using MRI-based fluid-structure interaction models. J Biomech 41:1651–1658

Koshiba N, Ando J, Chen X, Hisada T (2007) Multiphysics simulation of blood flow and LDL transport in a porohyperelastic arterial wall model. J Biomech Eng 129:374–385

Ku DN (1997) Blood flow in arteries. Annu Rev Fluid Mech 29:399–434

Ku DN, Giddens DP (1987) Laser doppler anemometer measurements of pulsatile in a model carotid bifurcation. J Biomech 20:407–421

Ku DN, Giddens DP, Zarins CK, Glagov S (1985) Pulastile flow and atherosclerosis in the human carotid bifurcation: positive correlation between plaque location and low and oscillating shear stress. Atherosclerosis 5:293–302

Kuntz RE, Baim DS (1993) Defining coronary restenosis: newer clinical and angiographic paradigms. Circulation 88:1310–1323

LaDisa JF, Olson LE, Guler I, Hettrick DA, Audi SH, Kersten JR, Warltier DC, Pagel PS (2004) Stent design properties and deployment ratio influence indexes of wall shear stress: a three-dimensional computational fluid dynamics investigation within a normal artery. J Appl Physiol 97:424–430

LaDisa JF, Olson LE, Hettrick DA, Warltier DC, Kersten JR, Pagel PS (2005a) Axial stent strut angle influences wall shear stress after stent implantation: analysis using 3d computational fluid dynamics models of stent foreshortening. Biomed Eng Online 4:59

LaDisa JF, Olson LE, Molthen RC, Hettrick DA, Pratt PF, Hardel MD, Kersten JR, Warltier DC, Pagel PS (2005b) Alterations in wall shear stress predict sites of neointimal hyperplasia after stent implantation in rabbit iliac arteries. Am J Physiol Heart Circ Physiol 288:H2465–H2475

Lagerstrand KM, Vikhoff-Baaz B, Starck G, Forssell-Aronsson E (2006) Quantitative phase-contrast flow MRI measurements in the presence of a second vessel closely positioned to the examined vessel. J Magn Reson Imaging 23:156–162

Lai WM, Rubin DH, Rubin D, Krempl E (2009) Introduction to continuum mechanics. Butterworth-Heinemann, Boston, ISBN: 978-0750685603, P 608

Le TB, Sotiropoulos F (2013) Fluid-structure interaction of an aortic heart valve prosthesis driven by an animated anatomic left ventricle. J Comput Phys 244:41–62

Le SU, Chung SY, Park RH (1990) A comparative performance study of several global thresholding techniques for segmentation. Graph Models Image Process 52:171–190

Lee SW, Steinman DA (2007) On the relative importance of rheology for image-based CFD models of the carotid bifurcation. J Biomech Eng 129:273–278

Lee S, Lee S, Fischer P, Bassiouny H, Loth F (2007) Intravascular turbulence in a diseased carotid bifurcation: numerical study. World Congress on medical physics and biomedical engineering 2006. IFMBE Proc 14:3411–3414

Lee SE, Lee S-W, Fischer PF, Bassiouny HS, Loth F (2008) Direct numerical simulation of transitional flow in a stenosed carotid bifurcation. J Biomech 41:2551–2561

Lee SW, Antiga L, Steinman DA (2009) Correlations among indicators of disturbed flow at the normal carotid bifurcation. J Biomech Eng 131:0610131–0610137

Lei M, Giddens DP, Jones SA, Loth F, Bassiouny H (2001) Pulsatile flow in an end-to-side vascular graft model: comparison of computations with experimental data. J Biomech Eng 123:80–87

Lendon CL, Davies MJ, Born GVR, Richardson PD (1991) Atherosclerotic plaque caps are locally weakened when macrophages density is increased. Atherosclerosis 87:87–90

Leon MB, Kandzari DE, Eisenstein EL, Anstrom KJ, Mauri L, Cutlip DE, Nikolsky E, O'Shaughnessy C, Overlie PA, Kirtane AJ, McLaurin BT, Solomon SL, Douglas JS, Popma JJ, Investigators EI (2009) Late Safety, efficacy, and cost-effectiveness of a zotarolimus-eluting stent compared with a paclitaxel-eluting stent in patients with de novo coronary lesions: 2-year follow-up from the endeavor iv trial (randomized, controlled trial of the medtronic endeavor drug [abt-578] eluting coronary stent system versus the taxus paclitaxel-eluting coronary stent system in de novo native coronary artery lesions). JACC Cardiovasc Inter 2:1208–1218

Li ZY, Howarth SPS, Tang T, Gillard JH (2006) How critical is fibrous cap thickness to carotid plaque stability? A flow-plaque interaction model. Stroke 37:1195

Liang Z (1993) Tissue classification and segmentation of MR images. Med Biol Mag 12:81–85

Liou TM, Liou SN (2004) Pulsatile flows in a lateral aneurysm anchored on a stented and curved parent vessel. Exp Mech 44:253–260

Liseikin VD (1999) Grid generation methods. Springer, Berlin

Liu SQ, Goldman J (2001) Role of blood shear stress in the regulation of vascular smooth muscle cell migration. IEEE Trans Biomed Eng 48:474–483

Lo SH (1985) A new mesh generation scheme for arbitrary planar domains. Int J Numer Method Eng 21:1403–1426

Long Q, Xu XY, Ariff B, Thom SA, Hughes AD, Stanton AV (2000) Reconstruction of blood flow patterns in a human carotid bifurcation: a combined CFD and MRI study. J Magn Reson Imaging 11:299–311

Long Q, Xu XY, Kohler U, Robertson MB, Marshall I, Hoskins P (2002) Quantitative comparison of CFD predicted and MRI measured velocity fields in a carotid bifurcation phantom. Biorheology 39:467–474

Loree HM, Kamm RD, Atkinson CM, Lee RT (1991) Turbulent pressure fluctuations on surface of model vascular stenoses. Am J Physiol Heart Circ Physiol 261:H644–H650

Loree HM, Kamm RD, Stringfellow RG, Lee RT (1992) Effects of fibrous cap thickness on peak circumferential stress in model atherosclerotic vessels. Circ Res 71:850–858

Loree HM, Grodzinsky AJ, Park SY, Gibson LJ, Lee RT (1994) Static circumferential tangential modulus of human atherosclerotic tissue. J Biomech Eng 27:195–204

Lowe GD, Lee AJ, Rumley A, Price JF, Fowkes FG (1997) Blood viscosity and risk of cardiovascular events: the Edinburgh artery study. Br J Haematol 96:168–173

Malek AM, Alper SL, Izumo S (1999) Hemodynamic shear stress and its role in atherosclerosis. J Am Med Assoc 282:2035–2042

Malve M, Garcia A, Ohayon J, Martinez MA (2012) Unsteady blood flow and mass transfer of a human left coronary artery bifurcation: FSI vs. CFD. Int Commun Heat Mass Trans 39:745–751

Manniesing R, Viergever MA, van der Lugt A, Niessen WJ (2008) Cerebral arteries: fully automated segmentation from CT angiography-a feasibility study. Radiology 247:841–846

Manousakas IN, Undrill PE, Cameron GG, Redpath TW (1998) Split-and-merge segmentation of magnetic resonance medical images: performance evaluation and extension to three dimensions. Comput Biomed Res 31:393–412

Marcum DL, Weatherill NP (1995) Unstructured grid generation using iterative point insertion and local reconnection. AIAA 33:1619–1625

Markl M, Chan FP, Alley MT, Wedding KL, Draney MT, Elkins CJ, Parker DW, Wicker RR, Taylor CA, Herfkens RJ, Pelc NJ (2003) Time-resolved three-dimensional phase-contrast MRI. J Magn Reson Imaging 17:499–506

Markl M, Harloff A, Bley TA, Zaitsev M, Jung B, Weigang E, Langer M, Hennig J, Frydrychowicz A (2007) Time-resolved 3D MR velocity mapping at 3T: improved navigator-gated assessment of vascular anatomy and blood flow. J Magn Reson Imaging 25:824–831

Marr D, Hildreth E (1980) Theory of edge detection. Proc R Soc London 207:187–217

Marshall I, Papathanasopoulou P, Wartolowska K (2004a) Carotid flow rates and flow division at the bifurcation in healthy volunteers. Physiol Meas 25:691–697

Marshall I, Zhao SZ, Papathanasopoulou P, Hoskins P, Xu XY (2004b) MRI and CFD studies of pulsatile flow in healthy and stenosed carotid bifurcation models. J Biomech 37:679–687

Matsumoto Y, Ii S, Shiozaki S, Sugiyama K, Takagi S (2014) Toward the multi-scale simulation for a human body using the next-generation supercomputer. Procedia IUTAM 10:193–200

Mavriplis DJ (1997) Unstructured grid techniques. Annu Rev Fluid Mech 29:473–514

Meijboom WB, Meijs MF, Schuijf JD, Cramer MJ, Mollet NR, van Mieghem CA, Nieman K, van Werkhoven JM, Pundziute G, Weustink AC, de Vos AM, Pugliese F, Rensing B, Jukema JW, Bax JJ, Prokop M, Doevendans PA, Hunink MG, Krestin GP, de Feyter PJ (2008) Diagnostic accuracy of 64-slice computed tomography coronary angiography: a prospective multicenter, multivendor study. J Am Coll Cardiol 52:2135–2144

Melchionna S, Pontrelli G, Bernaschi M, Bisson M, Halliday I, Spencer T, Succi S (2013) The lattice Boltzmann method as a general framework for blood flow modelling and simulations. In: Collins MW, Koenig CS (eds) Micro and nano flow systems for bioanalysis. Springer, New York, pp 153–170

Menashi S, Campa JS, Greenhalgh RM, Powell JT (1987) Collagen in abdominal aortic-aneurysm-typing content, and degradation. J Vasc Surg 6:578–582

Merrifield R, Long Q, Xu XY, Kilner PJ, Firmin DN, Yang G-Z (2004) Combined CFD/MRI analysis of left ventricular flow, in Proceedings of the medical image and augmented reality 2004, pp 229–236

Meyer F, Beucher S (1990) Morphological segmentation. J Visual Commun Image Represent 1:21–46

Michler C, Hulshoff SJ, van Brummelen EH, de Borst R (2004) A monolithic approach to fluid-structure interaction. Comput Fluids 33:839–848

Middleton RK, Lloyd GM, Bown MJ, Cooper NJ, London NJ, Sayers RD (2007) The pro-inflammatory and chemotactic cytokine microenvironment of the abdominal aortic aneurysm wall: a protein array study. J Vasc Surg 45:574–580

Morbiducci U, Gallo D, Massai D, Consolo F, Ponzini R, Antiga L, Bignardi C, Deriu MA, Redaelli A (2010) Outflow conditions for image-based hemodynamic models of the carotid bifurcation: implications for indicators of abnormal flow. J Biomech Eng 132:0910051–09100511

Moreno PR, Falk E, Palacios IF, Newell JB, Fuster V, Fallon JT (1994) Macrophage infiltration in acute coronary syndromes. Implications for plaque rupture. Circulation 90:775–778

Murphy JB, Boyle FJ (2010) A full-range, multi-variable, cfd based methodology to identify abnormal near-wall hemodynamics in a stented coronary artery. Biorheology 47:117–132

Nakahashi TK, Hoshina K, Tsao PS, Sho E, Sho M, Karwowski JK (2002) Flow loading induces macrophage antioxidative gene expression in experimental aneurysms. Arterioscler Thromb Vasc Biol 22:2017–2022

Nerem RM, Seed WA (1983) Coronary artery geometry and its fluid mechanical implications. In: Schettler G, Nerem R, Schmid-Schönbein H, Mörl H, Diehm C (eds) Fluid dynamics as a localizing factor for atherosclerosis. Springer Berlin Heidelberg, pp 51–59

Nichols W, O'Rourke M, Vlachopoulos C (2011) McDonald's blood flow in arteries, 6th edn. Theoretical, experimental and clinical principles. CRC Press Taylor & Francis Florida, ISBN:978034098501, P 768

O'Rourke RA, Brundage BH, Froelicher VF, Greenland P, Grundy SM, Hachamovitch R, Pohost GM, Shaw LJ, Weintraub WS, Winters WL (2000) American college of cardiology/American heart association expert consensus document on electron-beam computed tomography for the diagnosis and prognosis of coronary artery disease. Circulation 102:126–140

Ojha M (1993) Spatial and temporal variations of wall shear stress within an end-to-side arterial anastomosis model. J Biomech 26:1377–1388

Paz M, Leigh WE (2004) Structural Dynamics: theory and computation, updated with SAP2000. Kluwer Academic Publishers, Springer Science and Business New York, ISBN: 978-1-4613-5098-9, P 812

Peacock J, Hankins S, Jones T, Lutz R (1995) Flow instabilities induced by coronary artery stents: assessment with an in vitro pulse duplicator. J Biomech Eng 28:17–26

Perktold K, Peter RO, Resch M, Langs G (1991) Pulsatile non-Newtonian blood flow in three-dimensional carotid bifurcation models: a numerical study of flow phenomena under different bifurcation angles. J Biomech Eng 13:507–515

Peters TM, Collins DL, Evans AC (1993) Image segmentation and 3-D applications. In: Sprawls P, Bronskill MJ (eds) The Physics of MRI: 1992 AAPM Summer School Proceedings. American Association of Physicists in Medicine by the American Institute of Physics, Woodbury, NY, pp 607–646

Pham DL, Xu C, Prince JL (2000) A survey of current methods in medical image segmentation. Annu Rev Biomed Eng 2:315–338

Philips Medical Systems Clinical Education (1984) Basic principles of MR imaging. http://theonlinelearningcenter.com/Catalog/Product.aspx?mid=2249. Accessed 20 March 2014

Plank MJ, Comerford A, Wall DJN, David T (2007) Modelling the early stages of atherosclerosis. In: Mathematical modeling of biological systems, vol. I, Modeling and Simulation in Science, Engineering and Technology 2007, pp 263–274

Popel AS, Johnson PC (2005) Microcirculation and Hemorheology. Annu Rev Fluid Mech 37:43–69

Powell AJ, Maier SE, Chung T, Geva T (2000) Phase-velocity cine magnetic resonance imaging measurement of pulsatile blood flow in children and young adults: in vitro and in vivo validation. Pediatr Cardiol 21:104–110

Pries AR, Neuhaus D, Gaehtgens P (1992) Blood viscosity in tube flow: dependence on diameter and hematocrit. Am J Physiol Heart Circ Physiol 263:1770–1778

Pries AR, Secomb TW, Gaehtgens P (1995) Design principles of vascular beds. Circ Res 77:1017–1023

Qiao A, Yin W, Chu B (2014) Numerical simulation of fluid-structure interaction in bypassed DeBakey III aortic dissection. Comput Methods Biomech Biomed Engin 18:1173–1180

Raback P, Ruokolainen J, Lyly M, Järvinen E (2001) Fluid-structure interaction boundary conditions by artificial compressibility. ECCOMAS Computational Fluid Dynamics Conference 2001, Swansea, Wales, UK, 4–7 September 2001

Rahimian A, Lashuk I, Veerapaneni S, Chandramowlishwaran A, Malhotra D, Moon L, Sampath R, Shringarpure A, Vetter J, Vuduc R, Zorin D, Biros G (2010) Petascale direct numerical simulation of blood flow on 200k cores and heterogeneous architectures, in Proceedings of the 2010 ACM/IEEE International Conference for High Performance Computing, Networking, Storage and Analysis, IEEE Computer Society, pp 1–11

Reasor DA, Clausen JR, Aidun CK (2012) Coupling the lattice-Boltzmann and spectrin-link methods for the direct numerical simulation of cellular blood flow. Int J Numer Methods Fluids 68:767–781

Reddy JN (2013) An introduction to continuum mechanics. Cambridge University Press, New York, ISBN: 9781107025431, P 470

Regan M, Sheppard S (1996) Interactive multimedia courseware and the hands-on learning experience: an assessment study. J Eng Educ 85:123–131

Richardson PD, Davies MJ, Born GV (1989) Influence of plaque configuration and stress distribution on fissuring of coronary atherosclerotic plaques. Lancet 2:941–944
Richter GM, Palmaz JC, Noeldge G, Tio F (1999) Relationship between blood flow, thrombosis and neointima in stents. J Vasc Intervent Radiol 10:598–604
Rogers C, Edelman ER (1995) Endovascular stent design dictates experimental restenosis and thrombosis. Circulation 91:2995–3001
Rose JC, Hufnagel CA, Freis ED, Harvey WP, Partenope EA (1954) The hemodynamic alterations produced by a plastic valvular prosthesis for severe aortic insufficiency in man. J Clin Invest 33:891–900
Rose E, Moskowitz A, Packer M et al (1999) The REMATCH trial: rationale, design, and end points. Ann Thorac Surg 67:723–730
Ross R (1999) Atherosclerosis—an inflammatory disease. N Engl J Med 340:115–127
Saber NR, Gosman AD, Wood NB, Kilner PJ, Charrier CL, Firmin DN (2001) Computational flow modeling of the left ventricle based on in vivo MRI data. Ann Biomed Eng 29:275–283
Sahoo PK, Soltani S, Wong AKC, Chen Y (1988) A survey of thresholding techniques. Comput Graph Image Process 41:233–260
Salzar RS, Thubrikar MJ, Eppink RT (1995) Pressure-induced mechanical stress in the carotid artery bifurcation: a possible correlation to atherosclerosis. J Biomech 28:1333–1340
Sanz J, Fayad ZA (2008) Imaging of atherosclerotic cardiovascular disease. Nature 451:953–957
Schirmer CM, Malek AM (2011) Patient based computational fluid dynamic characterization of carotid bifurcation stenosis before and after endovascular revascularization. J Neurointerv Surg 4:1–7
Schmidt B, Ernst S, Ouyang F, Chun KR, Broemel T, Bansch D, Kuck KH, Antz M (2006) External and endoluminal analysis of left atrial anatomy and the pulmonary veins in three-dimensional reconstructions of magnetic resonance angiography: the full insight from inside. J Cardiovasc Electrophysiol 17:957–964
Schuijf JD, Beck T, Burgstahler C, Jukema JW, Dirksen MS, de Roos A, van der Wall EE, Schroeder S, Wijns W, Bax JJ (2007) Differences in plaque composition and distribution in stable coronary artery disease versus acute coronary syndromes; non-invasive evaluation with multislice computed tomography. Acute Card Care 9:48–53
Seitz J, Strotzer M, Schlaier J, Nitz WR, Völk M, Feuerbach S (2006) Comparison between MR phase contrast imaging and transcranial doppler ultrasound with regard to blood flow velocity in intracranial arteries: work in progress. J Neuroimaging 11:121–128
Sezgin M, Sankur B (2004) Survey over image thresholding techniques and quantitative performance evaluation. J Electron Imaging 13:146–165
Shephard MS, Georges MK (1991) Automatic three-dimensional mesh generation by the finite octree technique. Int J Numer Methods Eng 32:709–749
Shewchuk JR (2002) Delaunay refinement algorithms for triangular mesh generation. Comput Geom 22:21–74
Sho E, Sho M, Hoshina K, Kimura H, Nakahashi TK, Dalman RL (2004) Hemodynamic forces regulate mural macrophage infiltration in experimental aortic aneurysms. Exp Mol Pathol 76:108–116
Siauw WL, Ng EY, Mazumdar J (2000) Unsteady stenosis flow prediction: a comparative study of non-newtonian models with operator splitting scheme. Med Eng Phys 22:265–277
Silvera SS, Aidi HE, Rudd JHF, Mani V, Yang L, Farkouh M, Fuster V, Fayad ZA (2009) Multimodality imaging of atherosclerotic plaque activity and composition using FDG-PET/CT and MRI in carotid and femoral arteries. Atherosclerosis 207:139–143
Simmons A, Tofts PS, Barker GJ, Arridge SR (1994) Sources of intensity nonuniformity in spin echo images at 1.5T. Mag Res Med 32:121–128
Sivewright GJ, Elliott PJ (1994) Interactive region and volume growing for segmenting volumes in MR and CT images. Med Inform (Lond) 19:71–80
Skallerud B, Prot V, Nordrum IS (2011) Modeling active muscle contraction in mitral valve leaflets during systole: a first approach. Biomech Model Mechanobiol 10:11–26

Sled JG, Pike GB (1998) Standing-wave and RF penetration artifacts caused by elliptic geometry: an electrodynamic analysis of MRI. IEEE T Med Imaging 17:653–662
Sloop GD (1996) A unifying theory of atherogenesis. Med Hypotheses 47:321–325
Spiegel M, Redel T, Zhang YJ, Struffert T, Hornegger J, Grossman RG, Doerfler A, Karmonik C (2010) Tetrahedral vs. polyhedral mesh size evaluation on flow velocity and wall shear stress for cerebral hemodynamic simulation. Comput Methods Biomech Biomed Engin 14:9–22
Sprague EA, Luo J, Palmaz JC (2000) Endothelial cell migration onto metal stent surfaces under static and flow conditions. J Long Term Eff Med Implants 10:97–110
Srivastava LM, Srivastava VP (1983) On two-phase model of pulsatile blood flow with entrance effects. Biorheology 20:761–777
Srivastava VP, Srivastava R (2009) Particulate suspension blood flow through a narrow catheterized artery. Comput Math Appl 58:227–238
Stahlberg F, Sondergaard L, Thomsen C (1995) MR flow quantification with cardiovascular applications: a short overview. Acta Paediatr Suppl 410:49–56
Stary HC (2003) Atlas of atherosclerosis—progression and regression. Parthenon, New York
Stasa FL (1985) Applied finite element analysis for engineers. Holt, Rinehart, and Winston , USA/Canada ISBN: 9780030627378, P 657
Steinman DA, Rutt BK (1998) On the nature and reduction of plaque mimicking flow artifacts in black blood MRI of the carotid bifurcation. Magn Reson Med 39:635–641
Steinman DA, Poepping TL, Tambasco M, Rankin RN, Holdsworth DW (2000) Flow patterns at the stenosed carotid bifurcation: effect of concentric versus eccentric stenosis. Ann Biomed Eng 28:415–423
Steinman DA, Thomas JB, Ladak HM, Milner JS, Rutt BK, Spence JD (2002) Reconstruction of carotid bifurcation hemodynamics and wall thickness using computational fluid dynamics and MRI. Magn Reson Med 47:149–159
Stone GW, Rizvi A, Sudhir K, Newman W, Applegate RJ, Cannon LA, Maddux JT, Cutlip DE, Simonton CA, Sood P, Kereiakes DJ, Investigators SI (2011) Randomized comparison of everolimus- and paclitaxel-eluting stents. 2-year follow-up from the SPIRIT (clinical evaluation of the XIENCE V everolimus eluting coronary stent system) IV trial. J Am Coll Cardiol 58:19–25
Sukavaneshvar S, Rosa GM, Solen KA (2000a) Enhancement of stent-induced thromboembolism by residual stenoses: contribution of hemodynamics. Ann Biomed Eng 28:182–193
Sun Z, Cao Y (2011) Multislice CT angiography assessment of left coronary artery: correlation between bifurcation angle and dimensions and development of coronary artery disease. Eur J Radiol 79:e90–e95
Suri JS, Setarehdan SK, Singh S (2002) Advanced algorithmic approaches to medical image segmentation: state-of-the-art applications in cardiology, neurology, mammography and pathology. Springer-Verlag London, ISBN 9781447110439, P 636
Suri JS, Wilson D, Laxminarayan S (2005) Handbook of biomedical image analysis: vol II: segmentation models. Springer-Verlag, London ISBN 978-0-306-48606-7, P 816
Svensson J (2003) Contrast-enhanced magnetic resonance angiography: development and optimization of techniques for paramagnetic and hyperpolarized contrast media. Acta Radiol 429:1–30
Tada S, Tarbell JM (2005) A computational study of flow in a compliant carotid bifurcation-stress phase angle correlation with shear stress. Ann Biomed Eng 33:1202–1212
Takahashi S, Yagi T, Umezu M (2011) Experimental insight into modeling of cerebral aneurysm hemodynamics: comparison between stereo PIV and CFD, in 11th Asian Symposium on visualization, Niigata, Japan
Takumi T, Lee S, Hamasaki S, Toyonaga K, Kanda D, Kusumoto K, Toda H, Takenaka T, Miyata M, Anan R, Otsuji Y, Tei C (2007) Limitation of angiography to identify the culprit plaque in acutemyocardial infarction with coronary total occlusion. J Am Coll Cardiol 50:2197–2203
Tan FP, Soloperto G, Bashford S, Wood NB, Thom S, Hughes A, Xu XY (2008) Analysis of flow disturbance in a stenosed carotid artery bifurcation using two-equation transitional and turbulence models. J Biomech Eng 130:061008

Tang D, Yang C, Zheng J, Woodard P, Saffitz J, Sicard G, Pilgram T, Yuan C (2005a) Quantifying effects of plaque structure and material properties on stress distributions in human atherosclerotic plaques using 3D FSI models. J Biomech Eng 127:1185–1194
Tang D, Yang C, Zheng J, Woodard PK, Saffitz JE, Petruccelli JD, Sicard GA, Yuan C (2005b) Local maximal stress hypothesis and computational plaque vulnerability index for atherosclerotic plaque assessment. Ann Biomed Eng 33:1789–1801
Tang D, Yang C, Mondal S, Liu F, Canton G, Hatsukami TS, Yuan C (2008) A negative correlation between human carotid atherosclerotic plaque progression and plaque wall stress: in vivo MRI-based 2D/3D FSI models. J Biomech 41:727–736
Tang DL, Teng ZZ, Canton G, Hatsukami TS, Dong L, Huang XY, Yuan C (2009) Local critical stress correlates better than global maximum stress with plaque morphological features linked to atherosclerotic plaque vulnerability: an in vivo multi-patient study. Biomed Eng Online 8:15
Taylor CA, Draney MT (2004) Experimental and computational methods in cardiovascular fluid mechanics. Annu Rev Fluid Mech 36:197–231
Taylor CA, Hughes TJR, Zarins CK (1998) Finite element modeling of three-dimensional pulsatile flow in the abdominal aorta: relevance to atherosclerosis. Ann Biomed Eng 26:975–987
Taylor CA, Hughes TJR, Zarins CK (1999) Effect of exercise on hemodynamic conditions in the abdominal aorta. J Vasc Surg 29:1077–1089
Thompson RB, McVeigh ER (2004) Flow-gated phase-contrast MRI using radial acquisitions. Magn Reson Med 52:598–604
Thompson JF, Warsi ZUA, Mastin CW (1985) Numerical grid generation—foundations and applications. Elsevier, New York
Thubrikar MJ, Robicsek F (1995) Pressure-induced arterial wall stress and atherosclerosis. Ann Thorac Surg 59:1594–1603
Tominaga R, Kambic HE, Emoto H, Harasaki H, Sutton C, Hollman J (1992) Effects of design geometry of intravascular endoprostheses on stenosis rate in normal rabbits. Am Heart J 123:21–28
Torii R, Wood NB, Hadjiloizou N, Dowsey AW, Wright AR, Hughes AD, Davies J, Francis DP, Mayet J, Yang GZ, Thom SAM, Xu XY (2009) Fluid-structure interaction analysis of a patient-specific right coronary artery with physiological velocity and pressure waveforms. Commun Numer Methods Eng 25:565–580
Tremeau A, Borel N (1997) A region growing and merging algorithm to color segmentation. Pattern Recognit 30:1191–1203
Tu JY, Fuchs L (1992) Overlapping grids and multigrid methods for three-dimensional unsteady flow calculations in IC engines. Int J Numer Methods Fluids 15:693–714
Tu JY, Wong KKL, Cheung SCP, R., Beare, R., Phan, T. (2011) Analysis of patient-specific carotid bifurcation models using computational fluid dynamics. J Med Imaging and Health Inform
Tu J, Yeoh GH, Liu C (2012) Computational fluid dynamics: a practical approach. Butterworth-Heinemann, ISBN: 978-0-7506-8563-4, P 480
Ufuk O, Vartan K, Poulikakos D (2008) Computational modeling of coupled blood-wall mass transport of LDL: effects of local wall shear stress. Am J Physiol Heart Circ Physiol 294:H909–H919
Umbaugh SE (2005) Computer imaging: digital image analysis and processing. CRC Press Taylor & Francis Group Florida, ISBN: 9780849329197, P 688
Utter B, Rossmann JS (2007) Numerical simulation of saccular aneurysm hemodynamics: influence of morphology on rupture risk. J Biomech 40:2716–2722
Vaidyanathan M, Clarke LP, Velthuizen RP, Phuphanich S, Bensaid AM, Hall LO, Bezdek JC, Greenberg H, Trotti A, Silbiger M (1995) Comparison of supervised MRI segmentation methods for tumor volume determination during therapy. Mag Res Imaging 13:719–728
Valencia A, Villanueva M (2006) Unsteady flow and mass transfer in models of stenotic arteries considering fluid-structure interaction. Int Commun Heat Mass Trans 33:966–975
Valencia A, Morales H, Rivera R, Bravo E, Galvez M (2008) Blood flow dynamics in patient-specific cerebral aneurysm models: the relationship between wall shear stress and aneurysm area index. Med Eng Phys 30:329–340

van Wyk S Prahl Wittberg L Do-Quang M Fuchs LF (2013) Rheology of red blood cell flow in large geometries, in 8th International Conference on Multiphase Flow (ICMF), Jeju, Korea

Varnava AM, Mills PG, Davies MJ (2002) Relationship between coronary artery remodeling and plaque vulnerability. Circulation 105:939–943

Vengrenyuk Y, Carlier S, Xanthos S, Cardoso L, Ganatos P, Virmani R, Einav S, Gilchrist L, Weinbaum S (2006) A hypothesis for vulnerable plaque rupture due to stress-induced debonding around cellular microcalcifications in thin fibrous caps. Proc Natl Acad Sci 103:14678–14683

Versteeg HK, Malalasekera W (2007) An introduction to computational fluid dynamics: the finite volume method. Pearson Education Limited

Vetel J, Garon A, Pelletier D (2009) Lagrangian coherent structures in the human carotid artery bifurcation. Exp Fluids 46:1067–1079

Vorp DA, Lee PC, Wang DHJ, Makaroun MS, Nemoto EM, Ogawa S (2001) Association of intraluminal thrombus in abdominal aortic aneurysm with local hypoxia and wall weakening. J Vasc Surg 34:291–299

Wankat PC (2002) Integrating the use of commercial simulators into lecture courses. J Eng Educ 91:19–23

Westerhof N, Stergiopulos N, Noble MIM (2010) Snapshots of hemodynamics: an aid for clinical research and graduate education. Springer, Florida, ISBN 978-0-387-23346-8, P 192

Wexler L, Bergel DH, Gabe IT, Makin GS, Mills CJ (1968) Velocity of blood flow in normal human venae cavae. Circ Res 23:349–359

Wilcox DC (1998) Turbulence modeling for CFD. DCW industries, California

Womersley JR (1955) Method for the calculation of velocity, rate of flow and viscous drag in arteries when the pressure gradient is known. J Physiol 127:553–563

Wong KKL, Mazumdar J, Pincombe B, Worthley SG, Sanders P, Abbott D (2006) Theoretical modeling of micro-scale biological phenomena in human coronary arteries. Med Biol Eng Comput 44:971–982

Wong KKL, Kelso RM, Worthley SG, Sanders P, Mazumdar J, Abbott D (2009a) Cardiac flow analysis applied to phase contrast magnetic resonance imaging of the heart. Ann Biomed Eng 37:1495–1515

Wong KKL, Kelso RM, Worthley SG, Sanders P, Mazumdar J, Abbott D (2009b) Noninvasive cardiac flow assessment using high speed magnetic resonance fluid motion tracking. PLoS ONE 4:e5688

Wong KKL, Cheung SCP, Yang W, Tu JY (2010a) Numerical simulation and experimental validation of swirling flow in spiral vortex ventricular assist device. Int J Artif Organs 33:856–867

Wong KKL, Tu JY, Kelso RM (2010b) Vortical flow analysis. J Mech Med Biol 10:191–212

Wong KKL, Tu JY, Kelso RM, Worthley SG, Sanders P, Mazumdar J, Abbott D (2010c) Cardiac flow component analysis. Med Eng Phys 32:174–188

Wong K, Thavornpattanapong P, Cheung S, Sun Z, Tu J (2012a) Effect of calcification on the mechanical stability of plaque based on a three-dimensional carotid bifurcation model. BMC Cardiovasc Disord 12:7

Wong KKL, Sun Z, Tu J, Worthley SG, Mazumdar J, Abbott D (2012b) Medical image diagnostics based on computer-aided flow analysis using magnetic resonance images. Comput Med Imaging Graph 36:527–541

Wong KKL, Thavornpattanapong P, Cheung SCP, Tu J (2013a) Numerical stability of partitioned approach in fluid-structure interaction for a deformable thin-walled vessel. Comput Math Methods Med 2013:10

Wong KKL, Thavornpattanapong P, Cheung SCP, Tu JY (2013b) Biomechanical investigation of pulsatile flow in a three-dimensional artherosclerotic carotid bifurcation model. J Mech Med Biol 13:1350001

Worthley S, Helft G, Fuster V, Zaman A, Fayad Z, Fallon J, Badimon J (2000a) Serial in vivo MRI documents arterial remodeling in experimental atherosclerosis. Circulation 101:586–589

Worthley SG, Helft G, Fuster V, Fayad ZA, Fallon JT, Osende JI, Roque M, Shinnar M, Zaman AG, Rodriguez OJ (2000b) High resolution ex vivo magnetic resonance imaging of in situ coronary and aortic atherosclerotic plaque in a porcine model. Atherosclerosis 150:321–329

Wu J, Aidun CK (2010) Simulating 3D deformable particle suspensions using lattice Boltzmann method with discrete external boundary force. Int J Numer Methods Fluids 62:765–783

Xia G, Lin C-L (2008) An unstructured finite volume approach for structural dynamics in response to fluid motions. Comput Struct 86:684–701

Xu M (2002) Local measurement of the pulse wave velocity using doppler ultrasound. In Dept. of Electrical Engineering and Computer Science, Massachusetts Institute of Technology, Massachusetts Institute of Technology, 79

Yang C, Canton G, Yuan C, Ferguson M, Hatsukami TS, Tang D (2010) Advanced human carotid plaque progression correlates positively with flow shear stress using follow-up scan data: an in vivo MRI multi-patient 3D FSI study. J Biomech 43:2530–2538

Yerry M, Shephard M (1984) A modified-quadtree approach to finite element mesh generation. IEEE Comput Graph Appl 3:39–46

Younis BA, Berger SA (2004) A turbulence model for pulsatile arterial flows. J Biomech Eng 126:578–584

Younis HF, Kaazempur-Mofrad MR, Chung C, Chan RC, Kamm RD (2003) Computational analysis of the effects of exercise on hemodynamics in the carotid bifurcation. Ann Biomed Eng 31:995–1006

Yu Q, Kong X, Liu D (2003) Differential diagnosis of arachnoid cyst from subarachnoid space enlargement by phase-contrast cine MRI. Chin Med J (Engl) 116:116–120

Yuan C, Murakami JW, Hayes CE (1995) Phased-array magnetic resonance imaging of the carotid artery bifurcation: preliminary results in healthy volunteers and a patient with atherosclerotic disease. J Magn Reson Imaging 5:561–565

Yuan C, Mitsumori L, Beach K, Maravilla K (2001) Carotid atherosclerotic plaque: noninvasive MR characterization and identification of vulnearable lesions. Radiology 221:285–299

Yun BM, Dasia LP, Aiduna CK, Yoganathan AP (2014) Computational modelling of flow through prosthetic heart valves using the entropic lattice-Boltzmann method. J Fluid Mech 743:170–201

Zhang JM, Chua LP, Ghista DN, Zhou TM, Tan YS (2008) Validation of numerical simulation with PIV measurements for two anastomosis models. Med Eng Phys 30:226–247

Zhao SZ, Xu XY, Collins MW, Stanton AV, Hughes AD, Thom SA (1999) Flow in carotid bifurcations: effect of the superior thyroid artery. Med Eng Phys 21:207–214

Zhao SZ, Ariff B, Long Q, Hughes AD, Thom SA, Stanton AV, Xu XY (2002) Inter-individual variations in wall shear stress and mechanical stress distributions at the carotid artery bifurcation of healthy humans. J Biomech 35:1367–1377

Zhao H, Shaqfeh ESG, Narsimhan V (2012) Shear-induced particle migration and margination in a cellular suspension. Phys Fluids 24(1):1–21

Zhu C, Patterson AJ, Thomas OM, Sadat U, Graves MJ, Gillard JH (2013) Carotid stenosis assessment with multi-detector CT angiography: comparison between manual and automatic segmentation methods. Int J Cardiovasc Imaging 29:899–905

Zienkiewicz OC, Taylor RL (2000) The finite element method vol 1: the basis, 5th edn. Butterworth-Heinemann, Oxford

Zohdi TI (2005) A simple model for shear stress mediated lumen reduction in blood vessels. Biomech Model Mechanobiol 4(1):57–61

Index

J. Tu et al., *Computational Hemodynamics—Theory, Modelling and Applications*,
Biological and Medical Physics, Biomedical Engineering,
DOI 10.1007/978-94-017-9594-4

CPSIA information can be obtained at www.ICGtesting.com
Printed in the USA
LVOW02*1749290315

432485LV00001B/41/P

9 789401 795937